THE DESK REFERENCE OF
STATISTICAL QUALITY METHODS

THE DESK REFERENCE OF STATISTICAL QUALITY METHODS

MARK L. CROSSLEY

ASQ Quality Press
Milwaukee, Wisconsin

Library of Congress Cataloging-in-Publication Data
Crossley, Mark L., 1941–
   The desk reference of statistical quality methods / Mark L. Crossley.
     p. cm.
   Includes bibliographical references and index.
   ISBN 0-87389-472-3
   1. Quality control—Statistical methods. 2. Process control—Statistical methods. I. Title.

   TS156 .C79 2000
   658.5'62—dc21   99-057262

© 2000 by ASQ

10 9 8 7 6 5 4 3 2 1

ISBN 0-87389-472-3

Acquisitions Editor: Ken Zielske
Project Editor: Annemieke Koudstaal
Production Administrator: Shawn Dohogne
Special Marketing Representative: Matt Meinholz

ASQ Mission: The American Society for Quality advances individual and organizational performance excellence worldwide by providing opportunities for learning, quality improvement, and knowledge exchange.

Attention: Bookstores, Wholesalers, Schools and Corporations: ASQ Quality Press books, videotapes, audiotapes, and software are available at quantity discounts with bulk purchases for business, educational, or instructional use. For information, please contact ASQ Quality Press at 800-248-1946, or write to ASQ Quality Press, P.O. Box 3005, Milwaukee, WI 53201-3005.

To place orders or to request a free copy of the ASQ Quality Press Publications Catalog, including ASQ membership information, call 800-248-1946. Visit our web site at www.asq.org. or qualitypress.asq.org.

Printed in the United States of America

 Printed on acid-free paper

American Society for Quality

Quality Press
611 East Wisconsin Avenue
Milwaukee, Wisconsin 53202
Call toll free 800-248-1946
http://www.asq.org
http://standardsgroup.asq.org

Contents

Usage Matrix Table

	Acceptance Sampling Plans	Comparative Data Analysis	Descriptive Methods	Process Improvement	Reliability	SPC (Basic)	SPC (Special Applications)
Acceptance Control Chart							●
Acceptance Sampling for Attributes	●						
Acceptance Sampling for Variables	●						
Average/Range Control Chart						●	
Average/Standard Deviation Control Chart						●	
Chart Resolution						●	
Chi-Square Control Chart							●
Chi-Square Contingency and Goodness-of-Fit		●					
Confidence Interval for the Average			●				
Confidence Interval for the Proportion			●				
Confidence Interval for the Standard Deviation			●				
Defects/Unit Control Chart						●	
Demerit/Unit Control Chart							●
Descriptive Statistics			●				
Designed Experiments				●			
Discrete Distributions			●				
Evolutionary Operations, EVOP				●			
Exponentially Weighted Moving Average Chart							●
F-Test		●					
Histograms			●				
Hypothesis Testing		●					
Individual-Median/Range Control Chart							●
Individual/Moving Range Control Chart						●	
Measurement Error Assessment			●				
Multivariate Control Charts							●
Nonnormal Distribution C_{pk}			●				
Nonparametric Statistics		●					
Normal Distribution			●				
Pareto Analysis			●				
Precontrol			●				
Process Capability Indices			●				
Proportion Defective Control Chart						●	
Regression and Correlation			●				
Reliability					●		
Sequential Simplex Optimization				●			
Short-Run Attribute Control Charts							●
Short-Run Average/Range Control Chart							●
Short-Run Individual/Moving Range Control Chart							●
SPC Chart Interpretation						●	
Taguchi Loss Function				●			
Test for Normal Distribution		●					
Weibull Analysis					●		
Zone Format Control Chart							●

PREFACE

This desk reference provides the quality practitioner with a single resource that will illustrate in a practical manner exactly how to execute specific statistical methods frequently used in the quality sciences. This reference is not intended to be a rigorously theoretical treatise on the subject; rather, it provides a brief presentation on how to use the tool or technique. A plethora of topics have been arranged in alphabetical order, ranging from *Acceptance Sampling Control Charts* to *Zone Format Control Charts*. At the end of each topic a bibliography has been provided to direct the reader to more sources for inquiry. All of the examples contained in this reference are based on data that are purely hypothetical and fictitious in nature. This desk reference does not assume that the reader has already been exposed to the details of a specific topic but perhaps has some familiarity with the topic and wants more information regarding its application. This reference is accessible for the average quality practitioner who will need a minimal prior understanding of the techniques discussed to benefit from them. Each topic is presented in a stand-alone fashion with, in most cases, several examples detailing computational steps and application comments.

HOW TO USE THIS REFERENCE

There are forty-one individual modules, or topics, in this desk reference, see Usage Matrix Table. These modules fall into one of the following seven major categories that are defined as a function of the topic's application.

DESCRIPTIVE METHODS

These tools and techniques are used to measure selected parameters such as the central tendency, variation, probabilities, data presentation methods, and other single-value responses such as measurement error and process capability indices. Topics in this category answer such questions as: What size sample should I take to determine the average of a process? How much error do I have using my current measurement system? How large of a sample do I take to be 90 percent confident that the true C_{pk} is not less than 1.20? If the measured Kurtosis of my data distribution is +0.75, is the distribution considered normal? What is the Hypergeometric distribution, and when do I use it? Will my error for the proportion be affected if my sample size represents a significant proportion of my population?

STATISTICAL PROCESS CONTROL (BASIC)

Traditional methods of statistic process control (SPC) and its general use are presented. SPC is used to assist in the determination of process changes. Topics in this category answer such questions as: What is the probability that I will detect a process shift of +0.53 using an average/range control chart with $n = 5$ within seven samples after the shift has occurred?

Why is the average/range control chart more robust with respect to a nonnormally distributed population than the individual/moving range control chart?

STATISTICAL PROCESS CONTROL (SPECIAL APPLICATIONS)

This is an extension of the traditional methods to include short-run applications, detection of changes in distributions, and non-Shewhart methods. Topics in this section answer such questions as: What control chart do I use if I only want to detect a change in the distribution of the data? How do I use a p chart when my defect rate is extremely low? What control chart can I use to make the charting process "user friendly"? Is there a control chart based on specification limits?

PROCESS IMPROVEMENT

Process improvement techniques can be used to improve processes by reduction of variation or by optimization of operating conditions. These techniques provide the means to improve processes by addressing such questions as: Can I use design of experiments on a continuous basis for process improvement?

How do I use design of experiment to reduce variation?

RELIABILITY

This section provides analysis of life data and applications related to the understanding of product reliability. These topics address such issues as: I have calculated the MTBF from 12 failure times. How do I determine the 90 percent confidence interval? What is the bathtub curve, and how does failure rate relate to the age of a product? How can I graphically estimate the parameters of the Weibull distribution?

COMPARATIVE DATA ANALYSIS

Comparative data analysis includes methods to compare observed data to expected values and to compare sets of data to each other. These methods are frequently used to determine if differences in populations exist. Typical questions answered using these techniques are: I have determined that my process average is 12.0 using 25 samples. Have I achieved my goal of 11.5 minimum based on this sample data? The old process was running at 7.6 percent defective (based on a sample of $n = 450$), and the new and improved process is 5.6 percent defective (based on 155 samples). Have I actually improved the process?

ACCEPTANCE SAMPLING PLANS

Acceptance sampling plans are schemes used to assist in the decision to accept or reject lots of materials based on sampling data for either variables or attributes. This section addresses such concerns as: Can I develop my own sampling plans, or must I depend on published plans? What affects the operating characteristic more—the sample size or the lot size?

The following list organizes the modules according to the aforementioned categories:

Descriptive Methods and Data Presentation:

Descriptive Statistics
Confidence Interval for the Average
Confidence Interval for the Proportion
Confidence Interval for the Standard Deviation
Measurement Error Assessment
Normal Distribution
Process Capability Indices
Nonnormal Distribution C_{pk}
Discrete Distributions
Histograms
Pareto Analysis
Pre-Control
Regression and Correlation

Statistical Process Control (Basic):

Average/Range Control Chart
Defects/Unit Control Chart
Individual/Moving Range Control Chart
Average/Standard Deviation Control Chart
Proportion Defective Control Chart
Chart Resolution
SPC Chart Interpretation

Statistical Process Control (Special Applications):

Chi-Square Control Chart
Demerit/Unit Control Chart
Exponential Weighted Moving Average, EWMA
Individual-Median/Range Control Chart
Multivariate Control Chart
Short-Run Average/Range Control Chart
Short-Run Individual/Moving Range Chart
Short-Run Attribute Control Charts
Zone Format Control Chart
Acceptance Control Chart

Process Improvement:

Designed Experiments
Evolutionary Operations, EVOP
Sequential Simplex Optimization
Taguchi Loss Function

Reliability:

> Reliability
> Weibull Analysis

Comparative Data Analysis:

> Chi-Square Contingency and Goodness-of-Fit
> Nonparametric Statistics
> F-Test
> Hypothesis Testing
> Tests for Normal Distribution

Acceptance Sampling Plans:

> Acceptance Sampling for Attributes
> Acceptance Sampling for Variables

While this reference does not exhaust the list of applicable techniques and tools to improve quality, it is hoped that those included will provide the reader with a more in-depth understanding on their use.

> Mark L. Crossley, M.S., C.Q.E., C.R.E.
> Quality Management Associates, Inc.
> Website: www.Qualman.com
> e-mail: QMA@interpath.com
> Salisbury, N.C. 28146

DEDICATION

Dedicated to Betty, my wife,
whose abundant support
has made this book possible.

ACCEPTANCE CONTROL CHART

The following sections describe control charts based on specification limits.

MODIFIED CONTROL LIMITS

In those situations where meeting the specification is considered economically sufficient, the use of *reject limits* in place of control limits may be used. These types of control charts are called *modified control limit charts or acceptance sampling control charts* (ACC). The use of this approach is practical where the spread of the process (6σ) is significantly less than the specification limits (upper spec − lower spec). That is, the percent Cr normal process variation (6s) as a percent of total tolerances is small, less than 75 percent. The concept behind the use of modified limits is based on the assumption that limited shifts in the process average will occur.

The only Statistical Process Control detection rule (SPC Rule) used with modified control limits is when a single point is greater than the upper reject limit (URL) or below the lower reject limit (LRL). Control charts based on modified control limits will give some assurance that the average proportional defective shipped does not exceed a predetermined level.

To specify the modified control limits, we assume that the process data is normally distributed. For the process nonconforming to be less than δ, the true process mean μ must be within the interval $\mu_L \leq \mu \leq \mu_U$. Therefore, we must have

$$\mu_L = \text{LSL} + Z_\delta \sigma$$

and

$$\mu_U = \text{USL} - Z_\delta \sigma$$

where: μ_L = the minimum process average allowed
μ_U = the maximum process average allowed
Z_δ = the upper $100(1 - \delta)$ percentage point of the standard normal distribution
δ = the maximum percentage nonconforming allowed for the process

If a specified level for a type I error or an alpha risk α is given, then the upper control limit (UCL) and the lower control limit (LCL) are:

$$UCL = \mu_U + \frac{Z_\alpha \sigma}{\sqrt{n}}$$

$$UCL = USL - Z_\delta \sigma + \frac{Z_\alpha \sigma}{\sqrt{n}}$$

$$UCL = USL - \left(Z_\delta - \frac{Z_\alpha}{\sqrt{n}} \right) \sigma$$

and

$$LCL = \mu_L - \frac{Z_\alpha \sigma}{\sqrt{n}}$$

$$LCL = LSL + Z_\delta \sigma + \frac{Z_\alpha \sigma}{\sqrt{n}}$$

$$LCL = LSL + \left(Z_\delta - \frac{Z_\alpha}{\sqrt{n}} \right) \sigma$$

where: LCL = lower control limit (modified)
UCL = upper control limit (modified)
LSL = lower specification limit
USL = upper specification limit
σ = process standard deviation
n = sample size
δ = maximum proportion nonconforming allowed
α = probability of a type I error, alpha risk

Example:
The product specification is 10.00 ± 4.0, $\alpha = .0025$
Maximum proportion nonconforming, $\delta = 0.0010$
Data:

	#1	#2	#3	#4	#5	#6	#7	#8	#9	#10
	9.5	8.0	10.4	8.8	10.5	10.1	8.9	9.9	10.4	9.8
	10.5	10.4	11.5	9.2	10.1	11.1	9.5	10.9	9.3	10.2
	11.2	9.4	11.2	9.8	10.4	9.8	10.0	12.3	8.9	10.6
	9.0	7.9	9.6	8.0	9.2	10.7	11.6	11.3	10.2	8.7
$\bar{X}$:	10.05	8.93	10.68	8.95	10.05	10.43	10.00	11.10	9.7	9.8
R:	2.2	2.5	1.9	1.8	1.3	1.3	2.7	2.4	1.5	1.9

Grand average, $\bar{\bar{X}}$: = 9.97
Average range, $\bar{R}$ = 1.95

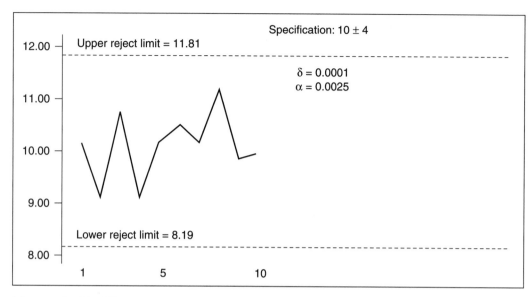

Figure 1. Modified Limits Control Chart.

Estimate of sigma, $\hat{\sigma} = \dfrac{\overline{R}}{d_2}$

d_2 values can be found in the "Table of constants for control charts"

$$\hat{\sigma} = 0.95$$

$$\mathrm{UCL} = \mathrm{USL} - \left(Z_\delta - \frac{Z_\alpha}{\sqrt{n}} \right)\sigma$$

$\delta = 0.00010$ $Z_\delta = 3.71$ from standard normal distribution
$\alpha = 0.0025$ $Z_\alpha = 2.81$ $n = 4$ $\sigma = 0.95$ $\mathrm{USL} = 14.0$

$$\mathrm{UCL} = \mathrm{USL} - \left(Z_\delta - \frac{Z_\alpha}{\sqrt{n}} \right)\sigma \quad \mathrm{UCL} = 14.0 - \left(3.71 - \frac{2.81}{2} \right)0.95 \quad \mathrm{UCL} = 11.81$$

$$\mathrm{LCL} = \mathrm{LSL} + \left(Z_\delta - \frac{Z_\alpha}{\sqrt{n}} \right)\sigma$$

$\delta = 0.00010$ $Z_\delta = 3.71$ from standard normal distribution
$\alpha = 0.0025$ $Z_\alpha = 2.81$ $n = 4$ $\sigma = 0.95$ $\mathrm{LSL} = 6.0$

$$\mathrm{LCL} = \mathrm{LSL} + \left(Z_\delta - \frac{Z_\alpha}{\sqrt{n}} \right)\sigma \quad \mathrm{LCL} = 6.0 + \left(3.71 - \frac{2.81}{2} \right)0.95 \quad \mathrm{LCL} = 8.19$$

The completed Modified Control Chart is illustrated in Fig. 1.

ACCEPTANCE CONTROL CHARTS

An extension to the modified control limits chart is the ACC. This technique allows the introduction of the four parameters associated with development of operating characteristic curves for describing acceptance sampling plans for attributes and variables. The four characteristics that completely define any sampling plan are

1. p_1 = the proportion nonconforming deemed acceptable (AQL)
2. α = the risk of rejection of an acceptable process or the risk that a point will fall outside the acceptance control limit (ACL) when the process is truly acceptable
3. p_2 = the proportion nonconforming deemed not acceptable (RQL)
4. β = the risk associated with the probability of acceptance of a non-acceptable process or the probability of a point not falling outside the ACL when the process is not acceptable

The risk factors, α and β should be the same for both the lower and upper specifications. In addition, the difference between the upper and lower specification must exceed 6σ.
Calculations:
The sample or subgroup size is found from

$$n = \left(\frac{Z_\alpha + Z_\beta}{Z_{p1} - Z_{p2}} \right)^2$$

The value for n chosen should be the next greatest integer.
The upper and lower control limits are designated as UACL and LACL, respectively, and are found from

$$\text{UACL} = \text{USL} - \left(Z_{p1} - \frac{Z_\alpha}{\sqrt{n}} \right)\sigma \quad \text{or} \quad \text{UACL} = \text{USL} - \left(Z_{p2} + \frac{Z_\beta}{\sqrt{n}} \right)\sigma$$

$$\text{LACL} = \text{LSL} + \left(Z_{p1} - \frac{Z_\alpha}{\sqrt{n}} \right)\sigma \quad \text{or} \quad \text{LACL} = \text{LSL} + \left(Z_{p2} + \frac{Z_\beta}{\sqrt{n}} \right)\sigma$$

The reason two formulas are required for each UACL and LACL is that the solution for n almost never will yield an integer value. The conservative action would be to select the lower value of the UACL and the higher value of the LACL.
As an example for the ACC, we will develop a chart using the data from the previous case (the modified control limit example), adding the additional parameters of α, β, p_1, and p_2.

USL $= 14.0$

LSL $= 6.0$

σ $= 0.95$

p_1 $= 0.0005$, $Z_{p1} = 3.29$

p_2 $= 0.0150$, $Z_{p2} = 2.17$

$\alpha = 0.01, Z_\alpha = 2.33$

$\beta = 0.05, Z_\beta = 1.65$

From this data:

$$n = \left(\frac{Z_\alpha + Z_\beta}{Z_{p1} - Z_{p2}} \right)^2 \quad n = \left(\frac{2.33 + 1.65}{3.29 - 2.17} \right)^2 \quad n = 12.6$$

Therefore, $n = 13$

$$\text{UACL} = \text{USL} - \left(Z_{p1} - \frac{Z_\alpha}{\sqrt{n}} \right)\sigma \quad \text{UACL} = 14.0 - \left(3.29 - \frac{2.33}{\sqrt{13}} \right)0.95 \quad \text{UACL} = 11.36$$

and

$$\text{UACL} = \text{USL} - \left(Z_{p2} + \frac{Z_\beta}{\sqrt{n}} \right)\sigma \quad \text{UACL} = 14.0 - \left(2.17 + \frac{1.65}{\sqrt{13}} \right)0.95 \quad \text{UACL} = 11.50$$

Use 11.36 for UACL (the lower of the two UACLs).

$$\text{LACL} = \text{LSL} + \left(Z_{p1} - \frac{Z_\alpha}{\sqrt{n}} \right)\sigma \quad \text{LACL} = 6.0 + \left(3.29 - \frac{2.33}{\sqrt{13}} \right)0.95 \quad \text{LACL} = 8.51$$

and

$$\text{LACL} = \text{LSL} + \left(Z_{p2} + \frac{Z_\beta}{\sqrt{n}} \right)\sigma \quad \text{LACL} = 6.0 + \left(2.17 + \frac{1.65}{\sqrt{13}} \right)0.95 \quad \text{LACL} = 8.49$$

For LACL use the greater of the two values, LACL = 8.51.

BIBLIOGRAPHY

Besterfield, D. H. 1994. *Quality Control.* 4th edition. Englewood Cliffs, NJ: Prentice Hall.

Freund, R. A. Acceptance Control Charts. *Industrial Quality Control* (October 1957).

Grant, E. L. and R. S. Leavenworth. 1996. *Statistical Quality Control.* 7th edition. New York: McGraw-Hill.

Montgomery, D. C. 1996. *Introduction to Statistical Quality Control.* 3rd edition. New York: John Wiley and Sons.

Schilling, E. G. 1982. *Acceptance Sampling in Quality Control.* New York: Marcel Dekker.

ACCEPTANCE SAMPLING FOR ATTRIBUTES

Lot acceptance sampling plans are used in the decisions regarding acceptance or rejection of a lot of material, either as an incoming lot or as one being released from manufacturing.

For example, you are responsible for the screening and acceptance of materials in receiving inspection and want to make a decision about acceptance of a shipment of parts. The incoming lot has 2500 units. Assume that there are no variables for the inspection and that the criteria for inspection is simply a functional test; either the parts perform or they do not. There are several options from which to choose:

1. Perform no inspection, and trust that the supplier has done a good job. The supplier is ISO-9000 certified.
2. Inspect 100 percent of the units.
3. Sample 10 percent of the lot, and accept it if no defective units are found.
4. Derive a statistically based sampling plan that defines all the risk factors for a given level of quality, and perform a sampling inspection.

In this module, the following topics will be discussed:

1. Sample plan parameters and operating characteristic curves
2. Development of attribute sampling plans using the Poisson Unity Value method
3. Using published sampling plans such as MIL-STD-105E (ANSI/ASQC Z1.4)
4. Continuous sampling plans such as CSP-1

Sampling plans for attributes are completely defined by their performance relative to seven characteristics or operational parameters:

Acceptable quality level (AQL):	That percent defective that, for purposes of acceptance sampling, will be considered acceptable; that is, "good". Lots at the AQL percent defective will be accepted by the plan the predominate amount of the time (such as 90 percent to 99 percent of the time). An AQL must have a defining alpha risk.
Manufacturer's risk, α:	The probability that a lot containing the AQL percent defective or less and that the lot will be rejected by the plan. The manufacturer's risk is also known as the alpha, or α, risk. The α risks are generally chosen in the range of 1 percent to 10 percent.

Rejectable quality level (RQL):

That percent defective that, for purposes of acceptance sampling, will be considered not acceptable; that is, "bad" or not acceptable. Lots that are at the RQL percent defective or greater will be rejected by the plan the predominate amount of the time. An RQL must have a defining beta risk.

Consumer's risk, β:

The probability that a lot containing the RQL percent defective or more will be accepted (by the consumer). The consumer's risk is also known as the beta, or β, risk. It is the probability of accepting a "bad" lot. Beta risks are generally chosen in the range of 1 percent to 10 percent.

Lot size, N:

The population from which the sample for examination will be chosen.

Sample size, n:

The number of items that will be examined for compliance to the specification.

Acceptance number, Ac or C:

The maximum number of nonconforming units that will be permitted in the sample. Exceeding this number will lead to rejection of the lot (for single sampling plans) or the selection of additional samples (for multiple sampling plans).

Note: The first four characteristics define the performance of the plan (the operating characteristic curve, or OC curve).

Operating Characteristic (OC) Curve

A graph of all possibilities of accepting a lot for each percent defective (usually unknown in practice) that can occur for a given sampling plan. Sample plans consist of the following two parts:

1. The sample size
2. The acceptance/rejection criteria

Example:
The following figure shows the OC curve for a sampling plan where the lot size is $N = 200$, the sample size is $n = 32$, and the acceptance number is $Ac = 3$. This plan provides a 4.39 percent AQL at a manufacturer's risk of 5.0 percent and an RQL of 19.7 percent at a consumer's risk of 10.0 percent.

$$\text{AQL} = 4.39\% \quad \text{RQL} = 19.7\%$$
$$\alpha = 5.0\% \quad \quad \beta = 10.0\%$$

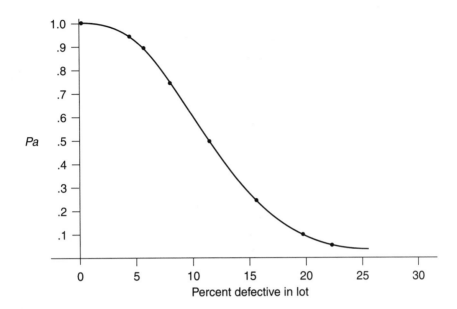

One useful method that may be utilized to create an attribute sampling plan is the poisson unity value method. This method uses an infinite lot size and is based on the four characteristics of AQL, RQL, α, and β.

Example:

Develop an attribute sampling plan that will satisfy the conditions of AQL = 4.0%, RQL = 16.0%, α = 5%, and β = 10%.

Step 1. Divide the RQL by the AQL.

$$\frac{\text{RQL}}{\text{AQL}} = \frac{16.0}{4.0} = 4.000 = \text{discrimination ratio}$$

Step 2. Look up the discrimination ratio by choosing the closest value available in the column provided for α = 5% and β = 10%. The closest value in the table for this case is 4.057.

Having located the closest discrimination ratio, go to the extreme left column to find the C = 4 value, and then go to the extreme right to locate the np_1 = 1.970 value.

C	$\alpha = 0.05$ $\beta = 0.10$	np_1
0	44.890	0.052
1	10.946	0.355
2	6.509	0.818
3	4.890	1.366
4	4.057	1.970
5	3.549	2.613
6	3.206	3.386

The maximum number of defective in the sample is $C = 4$. If the critical number of defective C is exceeded, the lot is rejected; otherwise, accept the lot.

Step 3: Calculate the sample size n.

The np_1 value is the product of the sample size n and the AQL (AQL $= p_1$).

In this example, $np_1 = 1.97$ and $n = \dfrac{np_1}{p_1}$; therefore, $n = \dfrac{1.97}{0.04} = 49.25 \approx 49$.

The sample plan is $n = 49$ and $C = 4$. The four plan criteria of AQL $= 4.0\%$, RQL $= 16.0\%$, $\alpha = 5\%$, and $\beta = 10\%$ will be satisfied by this plan.

Additional points to assist in the construction of the OC curve can be calculated using the supplemental table, "Poisson Unity Values, Pa."

Sample calculation:

In the previous example, $C = 4$ and $n = 49$. Looking in the table in the column headed by $Pa = 0.75$ and the row for $C = 4$, the value of 3.369 is found. This value represents the product of the lot percent defective and the sample size that will yield a 75 percent probability of acceptance Pa.

Dividing this value by the sample size $n = 49$ will give the percent defective that corresponds to a Pa of 0.75, or 75 percent.

$$np = 3.369 \text{ for } Pa = 0.75 \text{ where } n = 49 \text{ and } C = 4$$

$$p = \frac{3.369}{49} = 0.069$$

Other values for P at various Pas can be determined. The other associated values for the example sample plan of $C = 4$ and $n = 49$ and the graph of the OC curve follow.

Percent probability of acceptance Pa	Percent defective for lot
100	0
95	4.02
90	4.96
75	6.90
50	9.53
25	12.80
10	16.31
5	18.68

In the original calculation for the sample plan, the $\alpha = 0.05$ for an AQL of 4.00 percent was used to give a sample plan of $n = 49.25$ using an integer sample size of $n = 49$. This rounding off results in the Pa table for $Pa = 95\%$ to be 4.02 rather than the 4.0 percent AQL as specified. Also, the Poisson Unity Value of 4.000 could not be found exactly, and the default was 4.057. This same rounding off of the Poisson Unity Value also results in the $p = 16.31$ for a $Pa = 10\%$ where the original plan was established for a $p = 16.0\%$ for the RQL at $Pa = 10\%$.

OC curve for $n = 49$, $C = 4$ sampling plan:

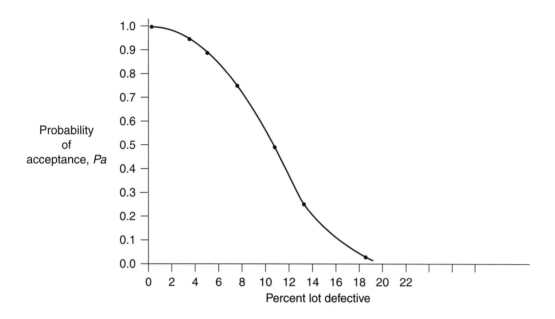

Percent probability of acceptance Pa	Percent defective for lot
100	0
95	4.02
90	4.96
75	6.90
50	9.53
25	12.80
10	16.31
5	18.68

Calculation of Poisson Unity Values and np_1 for given C, AQL, and RQL values: Given $\alpha = 0.05$, $\beta = 0.10$, and $C = 7$, calculate the Poisson Unity Value and np_1.

$$\text{Poisson Unity Value} = \frac{X^2_{\beta,2c+2}}{X^2_{1-\alpha,2c+2}} \text{ and } np_1 = \frac{1}{2}X^2_{1-\alpha,2c+2}$$

$$\text{Poisson Unity Value} = \frac{X^2_{0.10,16}}{X^2_{0.95,16}} = \frac{23.5}{7.96} = 2.95 \text{ and } np_1 = \frac{1}{2}X^2_{0.95,16} = 3.98$$

MIL-STD-105E (ANSI/ASQC Z1.4)

Sampling procedures for preset AQLs and various discrimination ratios for single, double, and multiple samples are available. These plans are AQL oriented and address the concerns related to RQLs and risk factors indirectly by allowing the user to select various *levels* of inspection.

Table 1. Sample Size Code Letters.

(See 9.2 and 9.3)

Lot or batch size	Special inspection levels				General inspection levels		
	S-1	S-2	S-3	S-4	I	II	III
2 to 8	A	A	A	A	A	A	B
9 to 15	A	A	A	A	A	B	C
16 to 25	A	A	B	B	B	C	D
26 to 50	A	B	B	C	C	D	E
51 to 90	B	B	C	C	C	D	E
91 to 150	B	B	C	D	D	E	F
151 to 280	B	C	D	E	E	G	H
281 to 500	B	C	D	E	F	H	J
501 to 1200	C	C	E	F	G	J	K
1201 to 3200	C	D	E	G	H	K	L
3201 to 10,000	C	D	F	G	J	L	M
10,001 to 35,000	C	D	F	H	K	M	N
35,001 to 150,000	D	E	G	J	L	N	P
150,001 to 500,000	D	E	G	J	M	P	Q
500,001 and over	D	E	H	K	N	Q	R

Adapted from: *ANSI/ASQC Z1.4-1993 Sampling Procedures and Tables for Inspection by Attributes.*
Milwaukee, WI: ASQC Quality Press, page 9. Used with permission.

The levels of inspection range from the nondiscriminating set of special inspection levels of S-1, S-2, S-3, and S-4 to the general inspection levels that are more discriminating. The default level of inspection is general level II.

Format for Use and Discussion

In Table 1, sample size code letters are arranged according to the lot size group and the level of inspection. For an illustration case, a single sample plant will be evaluated using general level II and an AQL = 4%. The lot size is 450.

For a lot size of $N = 450$ and general level II, the appropriate sample letter is H. The letter designates the appropriate sample size for the specified lot size and level of inspection. In addition to the levels that are reported as a letter, there are three sets of conditions that also affect the sampling and acceptance criteria. These conditions are

1. Normal inspection
2. Reduced inspection
3. Tightened inspection

During the beginning of the inspection scheme, normal inspection is carried out. Depending on the results of the history, the inspection may be switched to reduced or tightened based on performance. The directions regarding the choice of these degrees of inspection are outlined in Figure 1.

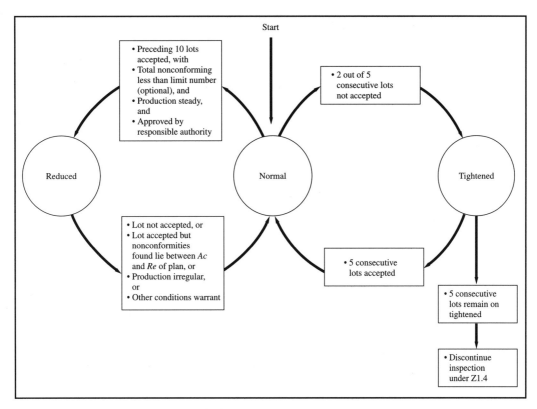

Figure 1. Switching Rules for ANSI Z1.4 System.

Source: *ANSI/ASQC Z1.4-1993. Sampling Procedures and Tables for Inspection by Attributes.* Milwaukee, WI: ASQC Quality Press, page 9. Used with permission.

The following example illustrates the use of the MIL-STD-105E sampling plan.

A shipment of 1000 bolts has been received. The AQL = 4.0%, and the level of inspection is general level II.

Step 1. Look up the appropriate sampling letter using Table 2.

The correct sample letter is J.

Step 2. Using Table 3, look up the sample size for letter J, then go across the top heading to AQL = 4.0 to locate the *Ac* and *Re* criteria.

The sample size for letter J is $n = 80$, and the acceptance criteria is that the lot is accepted if seven or less defective are found and is rejected if more than seven defective are found. Determine the sample size and acceptance criteria for the following:

A. Lot size $N = 250$, level = S-4, AQL = 6.5%
B. Lot size $N = 1800$, level II, AQL = 0.65%
C. Lot size $N = 1100$, level I, AQL = 0.65%

Table 2. Sample Size Code Letters.

(See 9.2 and 9.3)

Lot or batch size			Special inspection levels				General inspection levels		
			S-1	S-2	S-3	S-4	I	II	III
2	to	8	A	A	A	A	A	A	B
9	to	15	A	A	A	A	A	B	C
16	to	25	A	A	B	B	B	C	D
26	to	50	A	B	B	C	C	D	E
51	to	90	B	B	C	C	C	E	F
91	to	150	B	B	C	D	D	F	G
151	to	280	B	C	D	E	E	G	H
281	to	500	B	C	D	E	F	H	J
501	to	1200	C	C	E	F	G	J	K
1201	to	3200	C	D	E	G	H	K	L
3201	to	10,000	C	D	F	G	J	L	M
10,001	to	35,000	C	D	F	H	K	M	N
35,001	to	150,000	D	E	G	J	L	N	P
150,001	to	500,000	D	E	G	J	M	P	Q
500,001	and	over	D	E	H	K	N	Q	R

Adapted from: *ANSI/ASQC Z1.4-1993 Sampling Procedures and Tables for Inspection by Attributes.*
Milwaukee, WI: ASQC Quality Press, Page 9. Used with permission.

Table 3. Single Sampling Plans for Normal Inspection.

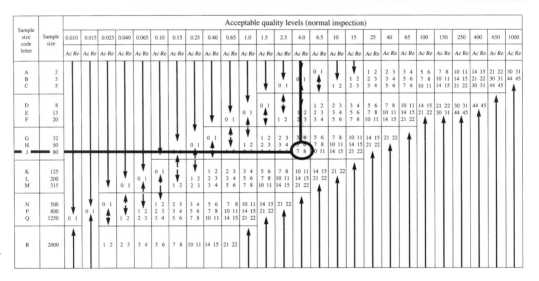

▼ = Use first sampling plan below arrow. If sample size equals, or exceeds, lot or batch size, do 100 percent inspection.
▲ = Use first sampling plan above arrow.
Ac = Acceptance number.
Re = Rejection number.

Adapted from: *ANSI/ASQC Z1.4-1993 Sampling Procedures and Tables for Inspection by Attributes.*
Milwaukee, WI: ASQC Quality Press, Page 12. Used with permission.

Table 4. Tables for Sample Size Code letter: J.

Percent of lots
expected to be
accepted (P_a)

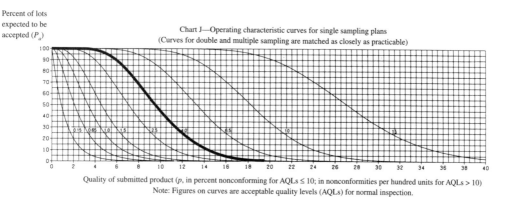

Chart J—Operating characteristic curves for single sampling plans
(Curves for double and multiple sampling are matched as closely as practicable)

Quality of submitted product (p, in percent nonconforming for AQLs ≤ 10; in nonconformities per hundred units for AQLs > 10)
Note: Figures on curves are acceptable quality levels (AQLs) for normal inspection.

Tabulated values for operating characteristic curves for single sampling plans

P_a	\multicolumn{22}{c}{Acceptable quality levels (normal inspection)}
	0.15	0.65	1.0	1.5	2.5	4.0	X	6.5	X	10	0.15	0.65	1.0	1.5	2.5	4.0	X	6.5	X	10	X	15
	\multicolumn{10}{c}{p (in percent nonconforming)}										\multicolumn{12}{c}{p (in nonconformities per hundred units)}											
99.0	0.0126	0.187	0.550	1.04	2.28	3.73	4.51	6.17	7.93	9.76	0.0126	0.186	0.545	1.03	2.23	3.63	4.38	5.96	7.62	9.35	12.9	15.7
95.0	0.0641	0.446	1.03	1.73	3.32	5.07	6.00	7.91	9.89	11.9	0.064	0.444	1.02	1.71	3.27	4.98	5.87	7.71	9.61	11.6	15.6	18.6
90.0	0.132	0.667	1.39	2.20	3.99	5.91	6.90	8.95	11.0	13.2	0.132	0.665	1.38	2.18	3.94	5.82	6.79	8.78	10.8	12.9	17.1	20.3
75.0	0.359	1.201	2.16	3.18	5.30	7.50	8.61	10.9	13.2	15.5	0.360	1.20	2.16	3.17	5.27	7.45	8.55	10.8	13.0	15.3	19.9	23.4
50.0	0.863	2.09	3.33	4.57	7.06	9.55	10.8	13.3	15.8	18.3	0.866	2.10	3.34	4.59	7.09	9.59	10.8	13.3	15.8	18.3	23.3	27.1
25.0	1.72	3.33	4.84	6.30	9.14	11.9	13.3	16.0	18.6	21.3	1.73	3.37	4.90	6.39	9.28	12.1	13.5	16.3	19.0	21.7	27.2	31.2
10.0	2.84	4.78	6.52	8.16	11.3	14.3	15.7	18.6	21.4	24.2	2.88	4.86	6.65	8.35	11.6	14.7	16.2	19.3	22.2	25.2	30.9	35.2
5.0	3.68	5.79	7.66	9.41	12.7	15.8	17.3	20.3	23.2	26.0	3.74	5.93	7.87	9.69	13.1	16.4	18.0	21.2	24.3	27.4	33.4	37.8
1.0	5.59	8.01	10.1	12.0	15.6	18.9	20.5	23.6	26.6	29.5	5.76	8.30	10.5	12.6	16.4	20.0	21.8	25.2	28.5	31.8	38.2	42.9
	0.25	1.0	1.5	2.5	4.0	X	6.5	X	10	X	0.25	1.0	1.5	2.5	4.0	X	6.5	X	10	X	15	X
	\multicolumn{22}{c}{Acceptable quality levels (tightened inspection)}																					

Note: Binomial distribution used for percent nonconforming computations; Poisson for nonconformities per hundred units.

Adapted from: *ANSI/ASQC Z1.4-1993 Sampling Procedures and Tables for Inspection by Attributes.*
Milwaukee, WI: ASQC Quality Press, page 12. Used with permission.

The actual operating characteristic can be found in the Table 4. The subject standards have OC curves for each code letter. The operating characteristic data is presented both graphically and tabularly.

Continuous Sampling Plans (CSPs)

Sampling plans defined by MIL-STD-105 and MIL-STD-414 are applicable when the lot of units to be examined is in one large collection, or *lot*, of material. Many manufacturing processes do not accumulate the items to be inspected in a lot format but rather are continuous. When production is continuous and discrete lots are formed, there are several disadvantages:

1. The rejection of the lot can lead to timely reinspection of the entire lot.
2. Additional space is required for the accumulated lot.

Using continuous sampling allows the lot to be judged on a continuous basis with no "surprises" upon conclusion of the lot.

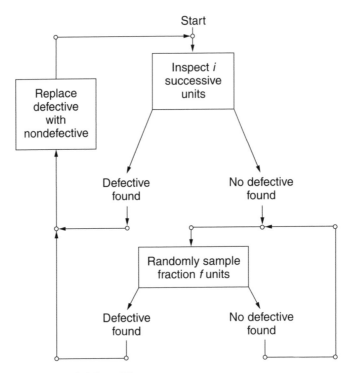

Figure 2. Procedure for CSP-1 Plan.

CSP-1 Plan

This plan was first introduced by Harold F. Dodge in 1943 (Figure 2). Initially, 100 percent inspection is required until i consecutive units have been found to be satisfactory. Upon reaching the clearing number i, the 100 percent inspection decreases to a fraction f of the units. Discovery of a defective unit during the frequency inspection results in 100 percent inspection until i consecutive units are found, in which case interval inspection of every fth unit resumes. Table values for various combinations of i and f for designated average outgoing quality limits (AOQL) can be found in Table 5.

Application example:

1. An electronic assembly line produces 6500 units per day. What CSP-1 plan will yield a 0.33 percent AOQL?

Referring to the CSP-1 table, there are several choices. One plan would be to inspect 100 percent until 335 consecutive nondefective units were found, and then inspect every tenth unit. If any defective units were found during the initial 100 percent inspection, the defective unit would be replaced with a good unit, and the accumulation count would be restarted until 335 consecutive good units were found. After that, any tenth sample found defective will result in the reinstatement of the 100 percent inspection. The AOQL for this plan is 0.33 percent nonconforming.

Table 5. Values of *i* and *f* for CSP-1 Plans.

										AOQL (%)									
f	.018	.033	.046	.074	.113	.143	.198	.33	.53	.79	1.22	1.90	2.90	4.94	7.12	11.46
1/2	1540	840	600	375	245	194	140	84	53	36	23	15	10	6	5	3
1/3	2550	1390	1000	620	405	321	232	140	87	59	38	25	16	10	7	5
1/4	3340	1820	1310	810	530	420	303	182	113	76	49	32	21	13	9	6
1/5	3960	2160	1550	965	630	498	360	217	135	91	58	38	25	15	11	7
1/7	4950	2700	1940	1205	790	623	450	270	168	113	73	47	31	18	13	8
1/10	6050	3300	2370	1470	965	762	550	335	207	138	89	57	38	22	16	10
1/15	7390	4030	2890	1800	1180	930	672	410	255	170	108	70	46	27	19	12
1/25	9110	4970	3570	2215	1450	1147	828	500	315	210	134	86	57	33	23	14
1/50	11730	6400	4590	2855	1870	1477	1067	640	400	270	175	110	72	42	29	18
1/100	14320	7810	5600	3485	2305	1820	1302	790	500	330	215	135	89	52	36	22
1/200	17420	9500	6810	4235	2760	2178	1583	950	590	400	255	165	106	62	43	26
	0.010	0.015	0.025	0.040	0.065	0.10	0.15	0.25	0.40	0.65	1.0	1.5	2.5	4.0	6.5	10.0
								AQL (%)								

The average number of units inspected in a 100 percent screening sequence following the occurrence of a defect is equal to

$$u = \frac{1 - Q^i}{PQ^i}$$

Where P = proportion defective for process or lot
$\quad\quad Q = 1 - P$
$\quad\quad i$ = initial clearing sample

If the process proportion defective for the example case had been 0.80 percent, then the average number of units inspected would be

$$u = \frac{1 - Q^i}{PQ^i} \quad u = \frac{1 - .992^{335}}{(.008)(.992^{335})} \quad u = 1718$$

The average number of units passed under CSP-1 plans before a defective unit v is found is given by

$$v = \frac{1}{fp} \quad f = 1/10 = 0.1 \quad \text{and } p = .008$$

$$v = \frac{1}{(0.1)(.008)} \quad v = 1250$$

The average fraction of the total manufactured units inspected (AFI) in the long run is given by

$$\text{AFI} = \frac{u + fv}{u + v} \quad u = 1718 \quad f = 0.10$$
$$\quad\quad\quad\quad\quad\quad v = 1250$$

$$\text{AFI} = \frac{1718 + (0.1)(1250)}{1718 + 1250} \quad \text{AFI} = 0.621$$

The average fraction of manufactured units passed under the CSP-1 plan is given by

$$Pa = \frac{v}{u + v} \quad Pa = 0.42 \text{ or } 42\%$$

When Pa is plotted as a function of p, the OC curve for the subject plan is derived. In a traditional lot acceptance sampling plan, Pa equals the probability of accepting lots that are p proportional defective.

The OC curve for CSP-1 plans gives the probability or proportion of units passed under the subject plan as a function of process proportion defective.

Example of OC Curve Development:

Draw the OC curve for a CSP-1 plan where $i = 38, f = 1/10$, and the AOQL = 2.90.

The parameters u and v are calculated for several values of p, the process proportion defective. The proportion of units passing the sampling plan Pa is determined using the u and v values for each proportion defective p. A plot of Pa as a function of p defines the OC curve for the plan. The values for the required factors are listed in a tabular form.

P	Q	$u = \dfrac{1-Q^i}{PQ^i}$	$v = \dfrac{1}{fp}$	$Pa = \dfrac{v}{u+v}$
0.01	0.99	47	1000	0.9551

Calculation for $p = 0.01$:

$$Q = 1 - p \quad Q = 0.99$$

$$u = \frac{1-Q^i}{PQ^i} \quad u = \frac{1-.99^{38}}{(.01)(.99^{38})} \quad u = 47$$

$$v = \frac{1}{fp} \quad v = \frac{1}{(.10)(.01)} \quad v = 1000$$

$$Pa = \frac{v}{u+v} \quad Pa = \frac{1000}{47+1000} \quad Pa = 0.9551$$

Continue the calculations of u, v, and Pa increasing P in increments of 0.01.

P	Q	$u = \dfrac{1-Q^i}{PQ^i}$	$v = \dfrac{1}{fp}$	$Pa = \dfrac{v}{u+v}$
0.01	0.99	47	1000	0.9551
0.02	0.98	58	500	0.8961
0.03	0.97	73	333	0.8202
0.04	0.96	93	250	0.7289
0.05	0.95	120	200	0.6250
0.06	0.94	158	167	0.5138
0.07	0.93	211	143	0.4040
0.08	0.92	285	125	0.3049
0.09	0.91	389	111	0.2220
0.10	0.90	538	100	0.1567
0.11	0.89	753	91	0.1078
0.12	0.88	1064	83	0.0724
0.13	0.87	1521	77	0.0481
0.14	0.86	2195	71	0.0313
0.15	0.85	3200	67	0.0205

Plot *Pa* as a function of *p:*

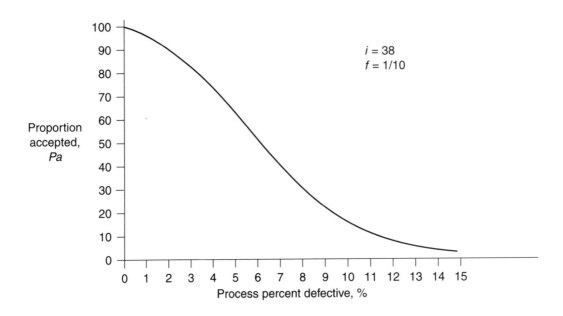

Calculation of the percent AOQL.

The percent AOQL is the percent defective that the sample plan overall will allow to pass through the system. It is determined by plotting the average outgoing quality (AOQ) as a function of the proportion defective of the process. The maximum AOQ for the subject plan defines the percent AOQL.

The AOQ is determined by

$$AOQ = \frac{P(1-f)Q^i}{f+(1-f)Q^i}$$

For the example given, where $i = 38$, $f = 1/10$, and $P = 0.03$ (process is 3% defective), the AOQ is

$$AOQ = \frac{0.03(1-0.10)Q^{38}}{0.10+(1-0.10)0.97^{38}} = 0.0222$$

$$\%\,AOQ = 2.22\%$$

Calculating all the AOQs from $P = 1.0\%$ to 15%, we have

% P	% AOQ
1	0.86
2	1.61
3	2.22
4	2.62
5	2.81
6	2.77
7	2.54
8	2.20
9	1.80
10	1.41
11	1.07
12	0.78
13	0.56
14	0.40
15	0.28

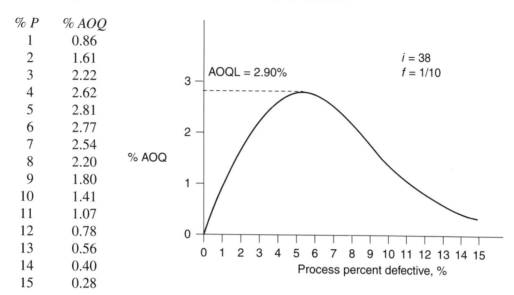

CSP-2 Plan

As a modification to the original CSP-1 plan, the CSP-2 was developed by Dodge and Torrey (1951). Under the CSP-2 mode, 100 percent inspection is performed until i units are found to be defect free. The inspection switches to a fraction f of the units. If a second defective is found during the sampling inspection of the i units, then 100 percent is resumed until ith consecutive units are found to be defect free. If no defectives are found during the sampling inspection, then resume the frequency inspection of a fraction f of the units.

The following table gives approximate values for i as a function of the AOQL.

	AOQL, %					
f	0.5	1.0	2.0	3.0	4.0	5.0
1/2	86	43	21	14	10	9
1/3	140	70	35	23	16	13
1/4	175	85	43	28	21	17
1/5	200	100	48	33	24	19
1/7	250	125	62	40	30	24
1/10	290	148	73	47	36	28
1/15	350	175	88	57	44	34
1/25	450	215	105	70	53	42
1/50	540	270	135	86	64	52

Example:

A 1.0 percent AOQL plan is desired with a clearing cycle of $n = 100$. Because 100 units are inspected and found to be defect free, every fifth unit is now inspected. During this frequency inspection, a defective is found. If during the next 100 consecutive units a second defective is found, 100 percent inspection is required until 100 consecutive units are found defect free, in which case the frequency inspection of every fifth is reinstated.

The values for initial clearing sample i and the fractional sample frequency f are given for several AOQLs in the following table.

Values of i for CSP-2 Plans

f	\multicolumn{8}{c}{AOQL, %}
	0.53	**0.79**	**1.22**	**1.90**	**2.90**	**4.94**	**7.12**	**11.46**
1/2	80	54	35	23	15	9	7	4
1/3	128	86	55	36	24	14	10	7
1/4	162	109	70	45	30	18	12	8
1/5	190	127	81	52	35	20	14	9
1/7	230	155	99	64	42	25	17	11
1/10	275	185	118	76	50	29	20	13
1/15	330	220	140	90	59	35	24	15
1/25	395	265	170	109	71	42	29	18
1/50	490	330	210	134	88	52	36	22

Effects of Lot Size N, Sample Size n, and Acceptance Criteria C on the OC Curve.

Traditional acceptance sampling plans such as ANSI Z1.4 and Z1.6 specify sampling plans as a function of the lot size. While lot size influences the overall OC curve, it is not the most contributing factor with respect to the shape of the OC curve. Other factors such as the sample size and the acceptance criteria C (the number of defective at which the lot will be accepted) are equally or more contributory.

Consider the following OC curves where the lot size, sample size, and acceptance criteria are varied.

Case I: Constant $C = 1$ and constant $n = 100$ with variable lot size $N = 200$ to 1000

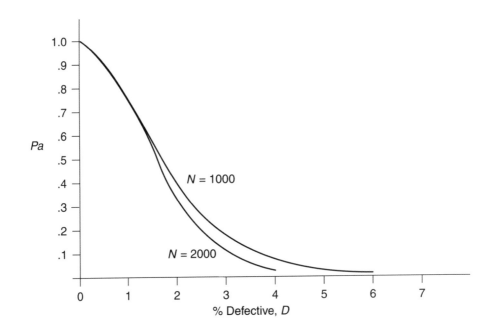

Case II: Constant $N = 1000$ and constant $n = 100$ with variable $C = 0$ to 4

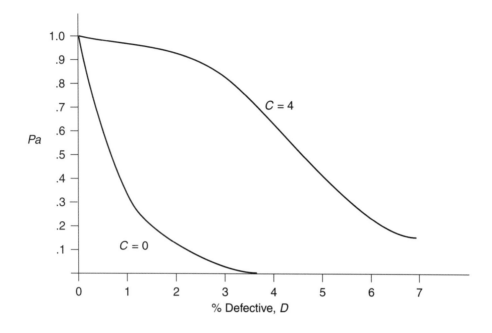

Case III: Constant $N = 1000$ and constant $c = 1$ with variable $n = 50$ to 200

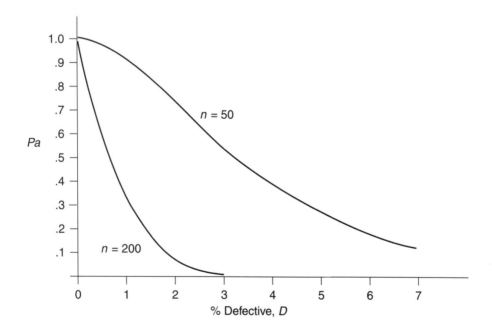

Comparatively, we can see that changes in the lot size have a relatively small effect on the overall shape of the OC curve. All of the above calculations are based on the hypergeometric distribution incorporating both the lot size N and the sample size n.

A Method for Selecting the AQL

If we consider the AQL as a *break-even* point associated with two intersecting relationships; The first relationship is the total, or lot, cost of 100 percent inspecting as a function of the percent defective, and the second relationship is the total, or lot, cost associated with the release of a lot that contains defective units. The cost of inspection will be set at k_1, and the cost of a defective unit by the consumer will be set at k_2. This break-even point can be suggested as an initial AQL as follows:

$$AQL = \frac{k_1}{k_2} \times 100$$

For example, if the cost of inspection is $0.50 and the cost of a defective unit received by the consumer is $60.00, the suggested AQL would be

$$AQL = \frac{0.50}{60} \times 100 = 0.83\%$$

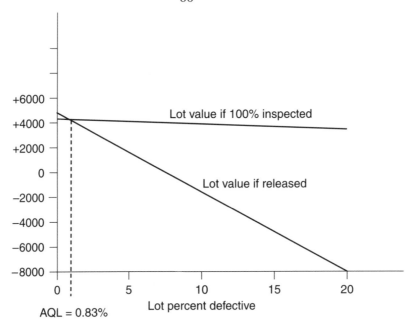

Lot size = 1000
Cost of inspection = $0.50 if unit is defect free and $5.80 if unit is defective
Value to consumer = $5.00 if defect free and $60.00 if defective

BIBLIOGRAPHY

ASQC Quality Press. 1993. *Sampling Procedures and Tables for Inspection by Attributes, ANSI/ASQC Z1.4-1993*. Milwaukee, WI: ASQC Quality Press.

Besterfield, D. H. 1994. *Quality Control*. 4th edition. Englewood Cliffs, NJ: Prentice Hall.

Dodge, H. F., and M. N. Torrey. 1951. Additional Continuous Sampling Inspection Plans. *Industrial Quality Control* 7, no. 5 (March): 7–12.

Grant, E. L. and R. S. Leavenworth, 1996. *Statistical Quality Control*. 7th Edition. New York: McGraw-Hill.

Schilling, E. G. 1982. *Acceptance Sampling in Quality Control*. New York: Marcel Dekker.

Acceptance Sampling for Variables

Lot acceptance plans for attribute data are based on acceptance criteria where a specified number of nonconforming units and a specified sample size are given. If this number of nonconforming units is exceeded in a sample, decisions are made to reject the lot from which the samples were taken. Such inspection plans are predefined under MIL-STD-105E (ANSI/ASQC Z1.4-1993). In acceptance sampling for attributes, the unit of inspection either conforms to a requirement or it does not. The decision criteria for each individual inspection are based on attributes. In many cases, the unit of inspection is measured as a direct variable such as weight, diameter, and angle or viscosity. If this measurement is compared to a specification requirement, then the observation is converted to an attribute. If the variable observation is not treated as an attribute, then the acceptance criteria can be based on the average of the observations and a variation statistic such as the standard deviation as related to the specification requirement. The following inspection plans are those covered by MIL-STD-414, sampling plans for variable data.

I. Single specification limit, form I, variability unknown, standard deviation method

Levels of inspection determine the sample size as a function of the submitted lot. Levels of inspection are directly related to the ratio of the acceptable quality level (AQL) and the repeatable quality level (RQL). The closer the RQL is to the AQL, the more discriminating the sample plan. Levels available are

$$S3 \rightarrow S4 \rightarrow I \rightarrow II \rightarrow III$$
Increasing discrimination $\rightarrow$

For a given lot size and level of inspection, the sample size is determined and designated as a letter B through P.

Lot size	S3	S4	I	II	III	Letter	Sample size
2 to 8	B	B	B	B	C	B	3
9 to 15	B	B	B	B	D	C	4
16 to 25	B	B	B	C	E	D	5
26 to 50	B	B	C	D	F	E	7
51 to 90	B	B	D	E	G	F	10
91 to 150	B	C	E	F	H	G	15
151 to 280	B	D	F	G	I	H	20
281 to 400	C	E	G	H	J	I	25
401 to 500	C	E	G	I	J	J	35
501 to 1200	D	F	H	J	K	K	50
1201 to 3200	E	G	I	K	L	L	75
3201 to 10,000	F	H	J	L	M	M	100
10,001 to 35,000	G	I	K	M	N	N	150
35,001 to 150,000	H	J	L	N	P	P	200
150,001 to 500,000	H	K	M	P	P	Note: there is no letter "O"	
≥500,001	H	K	N	P	P		

Step 1. Select a level of inspection and sample size.

A lot size of 80 rods has been received. For a general level II, what will the sample size be?

Lot size	S3	S4	I	II	III	Letter	Sample size
2 to 8	B	B	B	B	C	B	3
9 to 15	B	B	B	B	D	C	4
16 to 25	B	B	B	C	E	D	5
26 to 50	B	B	C	D	F	*E*	*7*
51 to 90	B	B	D	*E*	G	F	10
91 to 150	B	C	E	F	H	G	15
151 to 280	B	D	F	G	I	H	20
281 to 400	C	E	G	H	J	I	25
401 to 500	C	E	G	I	J	J	35
501 to 1200	D	F	H	J	K	K	50
1201 to 3200	E	G	I	K	L	L	75
3201 to 10,000	F	H	J	L	M	M	100
10,001 to 35,000	G	I	K	M	N	N	150
35,001 to 150,000	H	J	L	N	P	P	200
150,001 to 500,000	H	K	M	P	P	Note: there is no letter "O"	
≥500,001	H	K	N	P	P		

Sample size E = 7

Step 2. Select the sample, and calculate the average and sample standard deviation.

Seven samples are chosen and measured. The characteristic for this example is the diameter of the rods.

0.503
0.502
0.503 $\overline{X} = 0.503$
0.504 $S = 0.001$
0.505
0.501
0.503

Step 3. Select an AQL, and look up the appropriate k value from Table 1.

The specification for this example is a minimum diameter of 0.500 or lower specification limit of 0.500, and the selected AQL is 2.5 percent.

k for sample size letter E (sample size $n = 7$) and a 2.5 percent AQL is 1.33.

Step 4. Calculate critical decision factor (CDF).

For lower specifications: *For upper specifications:*
$CDF_L = \bar{X} - kS$ $CDF_U = \bar{X} + kS$

For this example, where there is a lower specification
$CDF_L = \bar{X} - kS$ $CDF_L = 0.503 - (1.33)(0.001)$ $CDF_L = 0.502$

Step 5: Confront CDF with the acceptance criteria.

For lower specifications:

If CDF_L is less than the lower specification, reject the lot.
If CDF_L is greater than the lower specification, accept the lot.

For upper specifications:

If CDF_U is greater than the upper specification, reject the lot.
If CDF_U is less than the upper specification, accept the lot.

For this example, CDF_L is greater than the lower specification; therefore, the lot is accepted.

Problem:

A lot of 200 units has been received. The specification is 25.0 maximum. Perform a level II, AQL = 1.00 inspection using the single specification, form 1 (k method).

Randomly select the sample from the 200 values listed below.

21	20	22	23	17	20	22	21	21	14	19	22	19	19	20	25	18	19	25	19
19	23	12	23	20	25	21	17	22	18	23	19	21	22	20	18	23	17	19	20
16	22	20	22	23	21	17	14	18	20	24	21	18	20	24	14	17	19	20	22
18	19	23	22	21	21	20	16	24	20	18	15	20	21	22	18	18	19	21	18
20	22	21	20	17	18	21	15	19	19	17	22	19	19	22	21	16	18	18	24
16	18	20	20	23	21	27	15	19	20	22	23	22	21	26	23	19	21	19	23
20	24	16	21	22	23	19	22	15	14	21	19	18	20	20	23	23	17	17	20
18	20	20	22	19	18	19	21	20	17	24	17	19	23	20	20	17	15	19	16
21	15	17	22	22	19	21	21	16	23	24	20	22	14	18	20	19	21	20	21
20	23	20	22	20	24	14	17	18	19	21	23	19	22	19	20	20	20	22	19

Sampling Plans Based on Specified AQL, RQL, α Risk, and β Risk

The sample size n and the k factor can be estimated using the following relationships:

$$k = \frac{Z_{P_2} Z_\alpha + Z_{P_1} Z_\beta}{Z_\alpha + Z_\beta} \quad \text{and} \quad n = \left(\frac{Z_\alpha + Z_\beta}{Z_{P_1} + Z_{P_2}} \right)^2 + \left(\frac{k^2}{2} \right)$$

Table 1. Single Specification, Variability Unknown, Form I, Standard Deviation Method.

Sample letter	Sample size	AQL (Normal)											
		T	.10	.15	.25	.40	.65	1.00	1.50	2.50	4.00	6.50	10.0
B	3									1.12	0.96	0.77	0.57
C	4							1.45	1.34	1.17	1.01	0.81	0.62
D	5						1.65	1.53	1.40	1.24	1.07	0.87	0.68
E	7				2.00	1.88	1.75	1.62	1.50	1.33	1.15	0.96	0.76
F	10			2.24	2.11	1.98	1.84	1.72	1.58	1.41	1.23	1.03	0.83
G	15	2.53	2.42	2.32	2.20	2.06	1.91	1.79	1.65	1.47	1.30	1.09	0.89
H	20	2.58	2.47	2.36	2.24	2.11	1.96	1.82	1.69	1.51	1.33	1.12	0.92
I	25	2.61	2.50	2.40	2.26	2.14	1.98	1.85	1.72	1.53	1.35	1.14	0.94
J	35	2.65	2.54	2.45	2.31	2.18	2.03	1.89	1.76	1.57	1.39	1.18	0.97
K	50	2.71	2.60	2.50	2.35	2.22	2.08	1.93	1.80	1.61	1.42	1.21	1.00
L	75	2.77	2.66	2.55	2.41	2.27	2.12	1.98	1.84	1.65	1.46	1.24	1.03
M	100	2.80	2.69	2.58	2.43	2.29	2.14	2.00	1.86	1.67	1.48	1.26	1.05
N	150	2.84	2.73	2.61	2.47	2.33	2.18	2.03	1.89	1.70	1.51	1.29	1.07
P	200	2.85	2.73	2.26	2.47	2.33	2.18	2.04	1.89	1.70	1.51	1.29	1.07
		.10	.15	.25	.40	.65	1.00	1.50	2.50	4.00	6.50	10.0	—
		AQL (Tightened)											

All AQL values are in percent nonconforming. T denotes plan used on tightened inspection.

where: $P_1 = \text{AQL}$ $\alpha = \text{alpha, or manufacturer's, risk}$
$P_2 = \text{RQL}$ $\beta = \text{beta, or consumer's, risk}$

The equation for n assumes that σ is unknown. The calculations for n and k assume a normal distribution and are independent of the lot size.

The criteria for acceptance single-sided specifications for either a lower specification limit or an upper specification limit are as follows:

Calculate the sample average and standard deviation from a sample of size n. If $\bar{X} - kS >$ lower specification or $\bar{X} + kS <$ upper specification, accept the lot.

Example:
Decide whether or not to accept a lot given the following conditions:

AQL = 4.0%   Manufacturer's risk = 5%
RQL = 12.0%   Consumer's risk = 10%

Lookup the following Z scores corresponding to the AQL, RQL, α, and β values:

$$Z_{P_1} = Z_{.04} = 1.75$$
$$Z_{P_2} = Z_{.12} = 1.18$$

$$Z_{\alpha} = Z_{.05} = 1.65$$
$$Z_{\beta} = Z_{.10} = 1.28$$

$$k = \frac{Z_{P_2} Z_{\alpha} + Z_{P_1} Z_{\beta}}{Z_{\alpha} + Z_{\beta}} \qquad k = \frac{(1.18)(1.65) + (1.75)(1.28)}{1.65 + 1.28} = 1.43$$

$$n = \left(\frac{Z_{\alpha} + Z_{\beta}}{Z_{P_1} - Z_{P2}}\right)^2 + \left(\frac{k^2}{2}\right) \quad \left(\frac{1.65 + 1.28}{1.75 - 1.18}\right)^2 + \left(\frac{(1.83)^2}{2}\right) = 28.09 = 28$$

Sampling Plans Based on Range *R*

Sometimes it is desirable to use the range rather than the standard deviation. The range has several desirable properties over the standard deviation such as the following:

1. It is easy to compute.
2. It is quick to determine.
3. It is easy to explain.

The calculations n and k are carried out as normally done. The acceptance criteria are modified to use multiples of average range in lieu of the standard deviation.

The relationship of standard deviation and range are defined from SPC applications

$$\sigma = \frac{\bar{R}}{d_2}$$

where d_2 depends upon the subgroup sample size.

The traditional d_2 assumes that the number of ranges used in the determination of $\bar{R}$ is large. If the number of subgroups is small, then the d_2 value should be adjusted. This adjustment results in a d_2* value.

where m = number of subgroups, and n_r = number of subgroups.

$$d_2* = d_2\left(1+\frac{0.2778}{m(n_r-1)}\right)$$

The number of subgroups m:

where $n = 28$ and $n_r = 3$ (arbitrarily chosen), then the number of subgroups would be

$$m = \frac{n-1}{0.9(n_r-1)}$$

$$m = \frac{n-1}{0.9(n_r-1)} = 15$$

For this example, 15 subgroups of 3 each would have been chosen, and the d_2* value is given by

$$d_2^* = d_2\left(1+\frac{0.2778}{m(n_r-1)}\right)$$

$$d_2^* = 1.693\left(1+\frac{0.2778}{15(3-1)}\right) \quad d_2^* = 1.708$$

Substitute $\bar{R}$ and d_2* into the acceptance criteria:

If $\dfrac{\overline{X}-\overline{R}}{d_2{}^*}$ < lower specification or if $\dfrac{\overline{X}+\overline{R}}{d_2{}^*}$ > upper specification, reject the lot;

otherwise, accept the lot.

Application Example

A lot of small electric motors is being tested for acceptance. The specification for the resistance is 1245 maximum. A lot of 150 items is submitted for inspection. Assume that the standard deviation is unknown and that the following characteristics for the plan are required:

P_1 = AQL = 2.5%
α = manufacturer's risk = 5%
P_2 = RQL = 15.0%
β = consumer's risk = 10%

$$k = \frac{Z_{P_2}Z_\alpha+Z_{P_1}Z_\beta}{Z_\alpha+Z_\beta} \quad k = \frac{(1.04)(1.65)+(1.96)(1.28)}{(1.65)+(1.28)} \quad k = 1.44$$

Sample size n:

$$n = \left(\frac{Z_\alpha + Z_\beta}{Z_{P_1} - Z_{P_2}} \right)^2 + \left(\frac{k^2}{2} \right) \quad n = \left(\frac{1.65 + 1.28}{1.96 - 1.04} \right)^2 + \frac{(1.44)^2}{2} \quad n = 11.18 = 12$$

The select sample is a subgroup of $n_r = 5$ (arbitrarily chosen) number of subgroups,

$$m = \frac{n-1}{0.9(n_r - 1)} \quad m = \frac{11}{0.9(5-1)} \quad m = 3$$

Three subgroups of subgroup sample size 5 are chosen from the lot.
The appropriate value for d_2^* is given by

$$d_2^* = d_2 \left(1 + \frac{0.2778}{m(n_r - 1)} \right)$$

$$d_2^* = 2.326 \left(1 + \frac{0.2778}{3(5-1)} \right)$$

$$d_2^* = 2.380$$

Acceptance criteria:
Three samples of five each are chosen, and the average range and grand average are calculated.

If $\dfrac{\overline{X} + \overline{R}}{2.380} > 1245$, reject the lot; otherwise, accept the lot.

BIBLIOGRAPHY

ASQC Quality Press. 1993. *Acceptance Sampling for Variables Data, ANSI/ASQC Z1.4-1993*. Milwaukee, WI: ASQC Quality Press.

Besterfeld, D. H. 1994. *Quality Control*. 4th edition. Englewood Cliffs, NJ: Prentice Hall.

Grant, E. L., and R. S. Leavenworth. 1996. *Statistical Quality Control*. 7th edition. New York: McGraw-Hill.

Schilling, E. G. 1982. *Acceptance Sampling in Quality Control*. New York: Marcel Dekker.

AVERAGE/RANGE CONTROL CHART

Shewhart control charts are used to detect changes in processes. Changes can occur in both the average of a process and the variation of the response variable of a process. The simplest control chart for variables data is the individual/moving range control chart. This chart is relatively insensitive to small changes and subject to distorted results when the process is significantly nonnormal with respect to the distribution of the data. While all control charts are robust with respect to nonnormal distributions, the individual/moving range chart is the one that is most affected by nonnormality of the data. This is due to the lack of the effect of the central limit theorem, which states that the distribution of averages will be more normally distributed than the individuals from which they came.

Control charts can be made more sensitive to small process changes by increasing the sample size (sometimes referred to as subgroup sample size). For sample sizes of 2 to 10, we may use the average/range control chart.

The central tendency or location of the process will be monitored by tracking the average of individual subgroup averages and their behavior relative to a set of control limits based upon the overall process average plus or minus three standard deviations (for the variation of averages, not for the variation of individuals).

The variation of the process will be monitored by tracking the range of individual subgroups and their behavior relative to a set of control limits.

The control limits for both the averages and ranges will be based upon three standard deviations and will, therefore, reflect the 99.7 percent probability limits. Changes in these parameters will be noted when certain *rules* of statistical conduct are violated.

If we elect to have even greater sensitivity for detecting process changes, the choice for control charts will be the average/standard deviation control chart. For this chart, the subgroup sample size may be from 2 to 25. See the module entitled *Average/Standard Deviation Chart* for a more detailed discussion of these control charts.

The steps for the construction and interpretation of the average/range control chart are the same as for all variables control charts. The following example will illustrate the steps for establishing an average/range control chart.

Step 1. Collect historical data.

A minimum of 25 subgroup samples of $n = 4$ to $n = 5$ are taken over the period of time that will serve as a baseline period. For our example, measurements will be made approximately every 3 hours using a subgroup sample size of $n = 5$ with a total of 16 samples ($k = 16$). The total number of actual individual observations will be 5×16 ($n \times k$), or 80.

For each sample of data collected, record the date, time of sample, the individual observations, the average of the subgroup sample, and the range of the subgroup sample.

The range of a subgroup is calculated by taking the difference between the smallest and largest values within a subgroup. Ranges are always positive.

	Sample #1 Date: 10/2/94 Time: 7:00 AM	Sample #2 Date: 10/2/94 Time: 11:00 AM	Sample #3 Date: 10/2/94 Time: 2:00 PM	Sample #4 Date: 10/2/94 Time: 5:00 PM
	22.4	19.5	26.1	24.3
	31.8	19.9	31.2	17.3
	20.9	25.0	25.2	26.9
	23.6	23.1	31.4	26.6
	31.4	27.8	24.5	21.1
Average:	26.0	23.1	27.7	23.3
Range:	10.9	8.3	6.9	9.6
	Sample #5 Date: 10/3/94 Time: 7:15 AM	Sample #6 Date: 10/3/94 Time: 10:30 AM	Sample #7 Date: 10/3/94 Time: 1:10 PM	Sample #8 Date: 10/3/94 Time: 4:20 PM
	25.2	19.7	20.1	25.5
	22.4	25.4	24.9	23.6
	20.9	23.0	27.9	25.2
	26.6	26.5	28.8	25.7
	27.6	21.9	26.4	23.6
Average:	24.5	23.3	25.6	24.7
Range:	6.7	6.8	8.7	2.1
	Sample #9 Date: 10/4/94 Time: 8:10 AM	Sample #10 Date: 10/4/94 Time: 11:20 AM	Sample #11 Date: 10/4/94 Time: 2:09 PM	Sample #12 Date: 10/4/94 Time: 5:15 PM
	27.5	23.7	25.4	21.1
	28.5	19.6	25.6	27.4
	25.2	32.6	27.6	33.1
	28.7	20.9	18.6	27.2
	28.0	23.4	31.3	26.3
Average:	27.6	24.0	25.7	27.0
Range:	3.5	13.0	12.7	12.0
	Sample #13 Date: 10/5/94 Time: 7:10 AM	Sample #14 Date: 10/5/94 Time: 10:00 AM	Sample #15 Date: 10/5/94 Time: 1:00 PM	Sample #16 Date: 10/5/94 Time: 4:25 PM
	26.3	25.9	27.4	22.4
	22.4	27.2	24.8	19.5
	19.5	22.7	25.8	26.1
	23.5	18.5	25.5	24.3
	21.1	23.5	27.1	25.2
Average:	22.6	23.6	26.1	23.5
Range:	6.8	8.7	2.6	5.7

Step 2. Calculate a location and variation statistic.

The location statistic will be the average of the 16 subgroup averages:

$$\bar{\bar{X}} = \frac{\bar{X}_1 + \bar{X}_2 + \bar{X}_3 + \ldots + \bar{X}_k}{k}$$

where: k = total number of subgroups

For our example:

$$\bar{\bar{X}} = \frac{26.0 + 23.1 + 27.7 + \ldots + 23.5}{16} = \frac{398.3}{16} = 24.89$$

The variation statistic will be the average range:

$$\bar{R} = \frac{R_1 + R_2 + R_3 + \ldots + R_k}{k}$$

where: k = total number of subgroups

For our example:

$$\bar{R} = \frac{10.9 + 8.3 + 6.9 + \ldots + 5.7}{16} = \frac{125.0}{16} = 7.81$$

Step 3. Determine the control limits for the location and variation statistics.

Sometimes referred to as the grand average, the average of the averages $\bar{\bar{X}}$ represents the best overall location statistic for the process. This statistic will be monitored by charting the individual averages X and their relationship to the grand average $\bar{\bar{X}}$. Control limits are defined as $\bar{\bar{X}} \pm 3S_{\bar{x}}$, and the probability of finding individual averages of subgroups inside these limits is 99.7 percent. The 0.3 percent probability of getting an average outside is so small that when averages outside these limits are obtained, it is assumed that the process average has changed.

In a similar manner, we will calculate the three standard deviation control limits for the distribution of the ranges R. Any individual range falling outside the upper control limits (UCL) or lower control limits (LCL) will serve as evidence that the variation of the process has changed.

The changes in the process will be monitored in the following two areas:

1. Location
2. Variation

Control limits for the location statistic $\bar{X}$:

$$\mathrm{UCL}_{\bar{X}} = \bar{\bar{X}} + 3S_{\bar{X}}$$

$$3S_{\bar{X}} = A_2 \bar{R}$$

where A_2 = a factor whose value depends upon the subgroup sample size. Control chart factors can be found in table I of the appendix. For our example, the subgroup sample size is 5; therefore, $A_2 = 0.577$.

$$3S_{\bar{x}} = A_2 \bar{R} = 0.577 \times 7.81 = 4.51$$

$$\mathrm{UCL} = 24.89 + 4.51 = 29.40$$

The LCL is calculated in a similar manner, except that rather than add the three standard deviations (4.51) to the grand average (24.89), we will subtract.

$$\mathrm{LCL}_{\bar{x}} = \bar{\bar{X}} - A_2 \bar{R} = 24.89 - 4.51 = 20.39$$

There is a 99.7 percent probability that a sample of five observations chosen from this process will yield an average value between 20.39 and 29.40, provided that the process average remains 24.89 and that the standard deviation of averages for subgroup size $n = 5$ remains 1.50 (one-third of the three standard deviations, which is estimated to be 4.51).

Since we are using the average of the subgroup ranges $\bar{R}$ to estimate the three standard deviations, we need a statistical check to determine if each of the individual ranges used to calculate the average range is appropriate or statistically acceptable. We check the individual ranges by comparing them to a UCL and LCL for ranges.

$$\mathrm{UCL}_R = D_4 \bar{R} = 2.114 \times 7.81 = 16.51$$

where the value D_4 depends on the subgroup sample size. For our case, $D_4 = 2.114$.

$$\mathrm{LCL}_R = D_3 \bar{R}$$

For a subgroup size $n = 5$, there is no defined value for D_3; therefore, there is no LCL for the range. There is no LCL for the range until we have a subgroup sample size of $n \geq 7$.

With our example, we expect 99.7 percent of the time to find subgroup ranges between 0 and 16.51.

Step 4. Construct the control chart, and plot the data.

Appropriate plotting scales are determined. Control limits for both the averages and ranges are drawn using a broken line, and the grand average or center line for the averages chart is drawn using a solid line. The completed chart using the historical data follows. The vertical wavy line separates the *History* from the future data. The control limits are derived using only the historical data. If there is supporting evidence in the subsequent future that there has been a real and continuing change in the process, the control limits may be recalculated using the recent historical data.

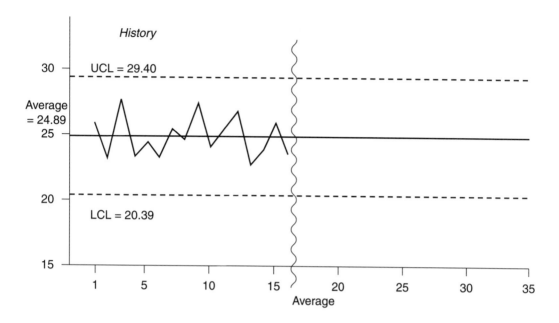

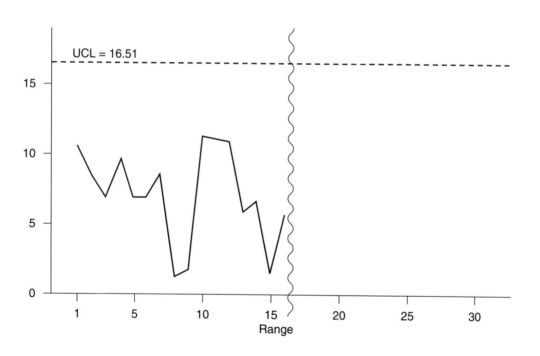

Step 5. Continue plotting data, looking for changes in the future.

There are several *rules* or patterns of data as they appear on the control chart that indicate a lack of control or give us evidence that the process has changed. The rules serve as an aid to interpreting the control charts. Some of the rules are as follows:

Rule 1. A single point outside the UCL or LCL.

This rule or test was originally devised by W. A. Shewhart in 1931 and was defined as *Criterion I.* In a normal distribution where no change in the average had occurred, this rule would be invoked 0.3 percent of the time, or 1 in 333 observations. This is the rate at which a false alarm would be seen. If the range chart indicates no change, then we assume that a point outside the average control limit indicates a change in the process average.

Rule 2. Seven points in a row on the same side of the average or centerline with no points outside the control limits.

The probability of the pattern of no process change occurring is approximately the same as getting seven consecutive heads in the toss of a fair coin. This probability is 1 in 128, or .78 percent. This likelihood is so remote that if it occurs, we assume that there has been a process change.

Rule 3. Seven consecutive points steadily increasing or decreasing.

This is a test for a trend or drift in the process. A small trend in a process will be indicated by this rule violation before rule 1 can react. Causes for a trend can be improvement in skills over time or tool wear.

Rule 4. Two out of three points greater than the two standard deviation limits.

Any two points between the two and three standard deviation limit lines with the third point being anywhere is an indication of process change. The probability of getting a false alarm is the same as rule 1.

Rule 5. Four out of five points greater than the one standard deviation limit. Similar to rule 4 above, the fifth point can be located anywhere.

Rule 1 through rule 3 utilize only the control limits based on ±3 standard deviations and are traditionally the major ones used in SPC. However if additional limits based on ±1 and ±2 standard deviations are chosen, the sensitivity of the control chart can be increased by invoking the additional rules. For a more detailed discussion, see the module entitled *SPC Chart Interpretation.*

Assume that the following additional data points are collected in the future after establishing the control limits and operating characteristics based on the historical data. Use them to determine if you evidence that the process has changed.

Sample #17 Date: 10/6/94 Time: 8:00 AM	Sample #18 Date: 10/6/94 Time: 11:15 AM	Sample #19 Date: 10/6/94 Time: 2:08 PM	Sample #20 Date: 10/6/94 Time: 4:48 PM
21.8	19.7	22.9	18.7
25.0	22.1	24.1	23.3
17.0	21.7	19.7	23.5
24.4	24.9	28.2	25.2
21.1	26.2	23.7	28.8

Sample #21 Date: 10/7/94 Time: 6:48 AM	Sample #22 Date: 10/7/94 Time: 10:25 AM	Sample #23 Date: 10/7/94 Time: 2:00 PM	Sample #24 Date: 10/7/94 Time: 5:12 PM
23.3	21.8	21.2	27.6
19.4	21.7	18.9	27.7
24.6	28.3	27.8	30.3
30.2	23.0	23.9	21.7
30.4	27.5	24.6	25.0

Sample #25 Date: 10/8/94 Time: 7:45 AM	Sample #26 Date: 10/8/94 Time: 11:15 AM	Sample #27 Date: 10/8/94 Time: 1:55 PM	Sample #28 Date: 10/8/94 Time: 4:48 PM
18.6	27.2	22.8	15.2
23.6	24.8	24.4	25.9
29.1	24.0	21.1	24.1
30.7	28.2	28.7	20.2
22.2	23.2	26.7	24.7

Sample #29 Date: 10/9/94 Time: 8:10 AM	Sample #30 Date: 10/9/94 Time: 11:09 AM	Sample #31 Date: 10/9/94 Time: 2:20 PM	
31.5	22.7	25.7	
28.0	17.8	21.5	
24.7	29.5	28.0	
27.2	25.4	25.5	
25.9	28.3	26.2	

Based on the new information, do you feel that the process average and/or standard deviation have changed?

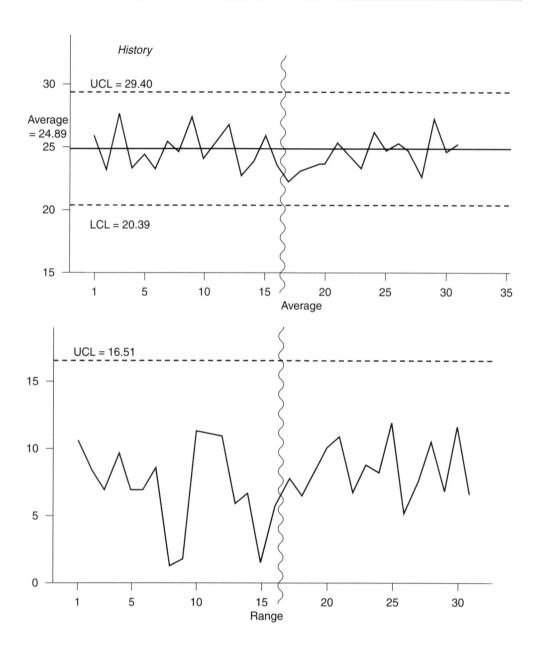

There is no evidence of a process change.

BIBLIOGRAPHY

Besterfield, D. H. 1994. *Quality Control.* 4th edition. Englewood Cliffs, NJ: Prentice Hall.

Grant, E. L. and R. S. Leavenworth. 1996. *Statistical Quality Control.* 7th edition. New York: Mc Graw-Hill.

Montgomery, D. C. 1996. *Introduction to Statistical Quality Control.* 3rd edition. New York: John Wiley and Sons.

Wheeler, D. J., and D. S. Chambers. 1992. *Understanding Statistical Process Control.* 2nd edition. Knoxville, TN: SPC Press.

AVERAGE/STANDARD DEVIATION CONTROL CHART

The traditional Shewhart control charts used to monitor variables are the individual/moving range, average/range, and the average/standard deviation. The individual/moving range chart is used when the observations are treated as individuals. In order to increase the sensitivity of the chart to detect a process change, the average/range control chart is used. The process variation is measured using the range. The average range $\bar{R}$ is used to monitor the process variation and is also used to estimate the variation of the averages using the relationship

$$3S_{\bar{x}} = A_2 \bar{R}$$

where A_2 is a constant whose value depends upon the subgroup sample size.

This relationship is effective as an estimate of the standard deviation for the distribution of averages provided that the subgroup sample size is 10 or less. Sample subgroup sizes greater than 10 render this relationship less efficient and the following alternative estimate must be used:

$$3S_{\bar{x}} = A_3 \bar{S}$$

Increasing the sample size also increases the sensitivity of the control chart. The average/standard deviation chart may be used for any subgroup sample size of $n \geq 2$.

The five steps for establishing the average/standard deviation control chart are as follows:

1. Collect historical data.
2. Select a location and variation statistic.
3. Determine control limits for both the location and variation statistic.
4. Construct the control chart.
5. Continue data collection and plotting, looking for signs of a process change.

The following example will illustrate the average/standard deviation control chart.

Background:

An injection molding process utilizes a nine-cavity mold. The characteristic being measured is the part weight for each of the nine parts. All nine cavities are measured for part weight every three hours. Weights are recorded in grams.

Step 1. Collect historical data. This data will serve as a historical characterization.

Typically 25 subgroups are taken. For this demonstration, only 12 samples will be taken.

For each subgroup sample of $n = 9$ taken, the average and standard deviation will be determined.

41

	1	2	3	4	5	6	7	8	9	10	11	12
	49.7	53.0	43.4	50.4	52.3	54.0	47.3	49.6	51.8	51.9	48.2	49.4
	46.7	48.2	51.3	55.1	49.4	56.5	50.0	48.6	52.6	48.2	46.5	46.3
	46.5	51.4	47.0	43.8	48.3	52.6	51.7	49.3	45.4	50.4	45.5	52.9
	46.2	45.9	50.5	49.8	48.2	50.9	56.3	51.5	51.1	51.2	47.7	53.5
	52.0	52.3	49.3	47.9	56.5	46.8	53.0	50.8	48.0	51.8	50.9	55.3
	46.3	50.0	48.9	50.9	47.3	43.9	48.3	49.4	51.3	50.5	49.9	50.7
	47.0	52.9	48.6	48.7	52.4	45.4	47.8	51.1	52.1	50.9	53.5	46.8
	51.8	48.3	53.4	53.2	52.9	44.1	49.7	51.5	47.2	47.4	45.5	51.7
	47.0	45.7	53.1	57.3	51.4	48.8	52.6	50.4	51.3	47.7	51.7	50.5
$\bar{X}$:	48.1	49.7	49.5	50.8	51.0	49.2	50.7	50.2	50.1	50.0	48.8	50.8
S:	2.4	2.9	3.1	4.0	2.9	4.5	2.9	1.1	2.5	1.8	2.8	2.9

Step 2. Select a location and variation statistic.

The selection of the location and variation statistic from which the process will be monitored is derived from the name of the chart. In this case, the chart is the average/standard deviation. The location statistic will be the average of the averages, and the variation statistic will be the average of the standard deviations.

$$\bar{\bar{X}} = \frac{48.1 + 49.7 + 49.5 + \ldots + 50.8}{12} = 49.9$$

$$\bar{S} = \frac{2.9 + 2.9 + 3.1 + \ldots + 2.4}{12} = 2.8$$

Step 3. Determine control limits for both the location and variation statistic.

A. Control limits for the location statistic:

The location or central tendency of this process will be monitored by the movement of the averages about the grand average of the averages. Each of the individual averages do vary. Upper control limits (UCLs) and lower control limits (LCLs) for these individual averages will be determined based on the grand average ±3 standard deviations (standard deviation of the averages, *not* of the individual observations). The three standard deviations will be determined from

$$3S_{\bar{x}} = A_3 \bar{S}$$

The value of A_3 is dependent on the subgroup sample size *n*. For this case, $n = 9$, and $A_3 = 1.032$ (see Control Chart Factors, Table F).

$$3S_{\bar{x}} = (1.032)(2.8) = 2.89 = 2.9$$

The UCLs and LCLs for the distribution of averages are, respectively

$$UCL = 49.9 + 2.9 = 52.8$$

$$LCL = 49.9 - 2.9 = 47.0$$

Since these limits are based on ±3 standard deviations, 99.7 percent of the *averages* are expected to fall between 47.0 and 52.8.

B. Control limits for the variation statistic:

The process variation will be monitored by the movement of the individual standard deviations about the average standard deviation S.

The UCLs and LCLs for the average standard deviation are determined by

$$UCL_s = B_4 \overline{S} \quad UCL_s = (1.761)(2.8) = 4.9$$

$$LCL_s = B_3 \overline{S} \quad LCL_s = (0.239)(2.8) = 1.1$$

Approximately 99.7 percent of the sample standard deviations will fall within the limits of 1.1 to 4.9.

Step 4. Construct the control chart.

The averages, control limits, and data values are plotted on the chart. Control limits are normally drawn using a broken line, and averages are drawn using a solid line.

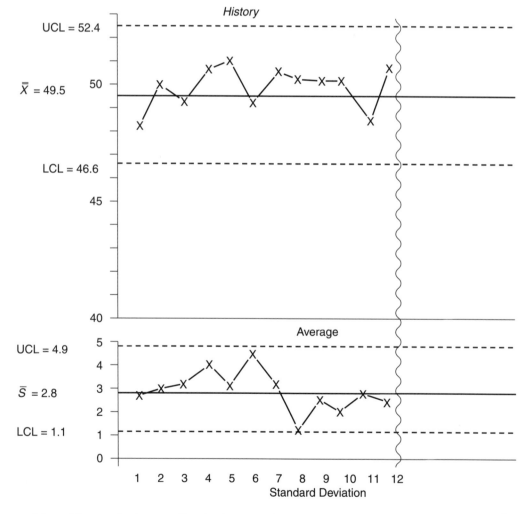

Note: The vertical wavy line is drawn only to show the period from which the chart parameters were determined (averages and control limits).

Step 5. Continue data collection and plotting, looking for signs of a process change.

	13	14	15	16	17	18
	46.9	52.4	52.3	51.9	50.4	50.2
	53.3	54.0	49.4	50.7	49.7	46.7
	47.5	51.5	50.3	51.2	51.5	46.5
	52.5	51.8	51.3	53.5	51.1	52.6
	50.7	50.6	56.5	53.0	48.0	55.9
	47.9	52.7	50.3	51.2	51.3	52.0
	50.1	50.7	52.4	53.2	52.1	53.4
	54.9	55.9	52.9	53.8	47.2	48.9
	54.5	55.5	51.4	52.4	51.3	49.2
$\bar{X}$:	50.9	52.5	51.9	52.3	50.3	50.6
S:	3.1	2.3	2.1	1.1	1.7	3.1

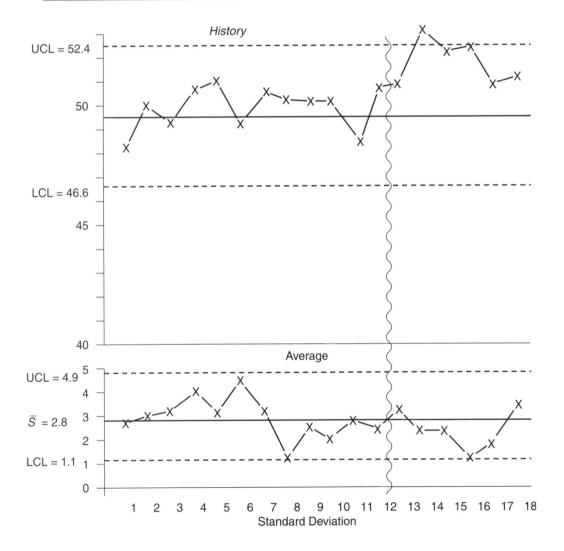

Sample #14 has the average greater than the UCL for averages, indicating that the process, relative to the historical characterization, has experienced a change. It is suspected that the process average has increased. This is an example of the violation of one of the *SPC detection* rules (a single point outside the UCL or LCL. The process variation remains stable with no indication of change.

BIBLIOGRAPHY

Besterfield, D. H. 1994. *Quality Control.* 4th edition. Englewood Cliffs, NJ: Prentice Hall.

Grant, E. L., and R. S. Leavenworth. 1996. *Statistical Quality Control.* 7th edition. New York: McGraw-Hill.

Montgomery, D. C. 1996. *Introduction to Statistical Quality Control.* 3rd edition. New York: John Wiley and Sons.

Wheeler, D. J., and D. S. Chambers. 1992. *Understanding Statistical Process Control.* 2nd edition. Knoxville, TN: SPC Press.

CHART RESOLUTION

The ability of a control chart to resolve the difference between its historical characterization (the statistics used to establish the chart) and the future process data is a measure of the power or resolution of the chart. Several interrelated factors affect this resolution:

1. The magnitude of the change being detected
2. The subgroup sample size, n
3. The quickness in the response time desired to detect the change
4. The number of detection criteria used to indicate a change

Two families of control charts will be reviewed for their resolving power as related to these factors:

1. Control charts for variables
   A. Individuals
   B. Averages
2. Control charts for attributes
   A. Proportion (defective)
   B. Nonconformities (defects)

VARIABLES CONTROL CHARTS: INDIVIDUALS

Consider an individuals control chart developed from a process that has an average $\bar{X} = 35.00$ and a moving range $\overline{MR} = 9.0$.

The control limits are calculated upon $\bar{X} \pm 3S$, where $3S$ is estimated by $2.66\,\overline{MR}$.

$$3S = (2.66)(9.9) = 23.94 \Rightarrow 24.00$$

$$S = 8.00$$

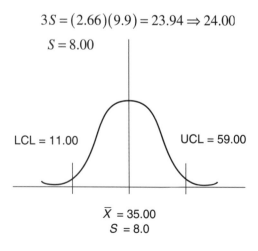

LCL = 11.00   UCL = 59.00

$\bar{X} = 35.00$
$S = 8.0$

One of the signals indicating an out-of-control condition is the occurrence of a single point falling outside the control limits. The probability of this occurrence when the process change is zero is:

$$Z = \frac{59.0 - 35.0}{8.00} = 3.00$$

Looking up this value in a standard normal distribution, the probability of getting a value greater than 59.00 is 0.00135, or a percent probability of 0.135 percent. The probability of getting a value less than 11.00 is also 0.00135, for a combined probability of 0.0027.

$$\frac{1}{0.0027} = 370$$

There is a probability that 1 sample in 370 will exceed the control limit when there has, in fact, been no process change. The value of 370 is also called the average run length (ARL).

Once points of interest such as the upper control limits (UCLs) and lower control limits (LCLs) have been determined, then the probability of exceeding either or both can be determined assuming any future process average.

For example, if the UCLs and LCLs for an individual chart are 59.00 and 11.00, respectively, and the standard deviation of individuals is 8.00, what is the probability of getting a single point above the UCL if the process average shifts from 35.00 to 43.00 (an increase of $+ 1.5\sigma$)?

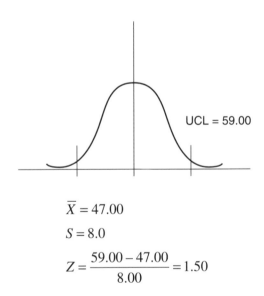

UCL = 59.00

$$\overline{X} = 47.00$$
$$S = 8.0$$
$$Z = \frac{59.00 - 47.00}{8.00} = 1.50$$

The probability of $Z = 1.50$ is 0.0668, or 6.68 percent, and the ARL = 15. This means that if a control chart is established with UCLs and LCLs of 59.00 and 11.00, respectively, and the process average shifts from 35.00 to 47.00, there will be a probability of .0668 that a single point will exceed the UCL. One sample in 15 will exceed the UCL.

Calculations can be performed in a similar manner to determine the probability of detection with various degrees of process shift (in units of σ). Table 1 lists several degrees of process change and the corresponding probability of exceeding a control limit and the ARL. A plot of Table 1 is seen in Figure 1.

Table 1. Probability of Exceeding Control Limit as a Function of Shift.

Amount of shift, σ	Probability of detection	ARL
0.00	.00135	741
0.20	.00256	391
0.40	.00466	215
0.60	.00820	122
0.80	.01390	72
1.00	.02275	44
1.20	.03593	28
1.40	.05480	18
1.60	.08076	12
1.80	.11507	9
2.00	.15866	6
2.20	.21186	5
2.40	.27425	4

Figure 1. Probability of Shift Detection $n = 1$.

To this point, all the probabilities have been based on exceeding a limit with a single sample. We can determine the probability of exceeding a control limit within any number of samples k using the relationship

$$P_{k=i} = 1 - (1 - P_{k=1})^k$$

where: $P_{k=1}$ = probability of exceeding a limit in one sample
k = number of samples after the shift
$P_{k=i}$ = probability of exceeding limit within k samples

Example:
A process has been defined by an average of 26.5 and a standard deviation of 3.8. What is the probability of detecting a shift in the average to 25.5 within 10 samples after the shift has occurred?

$$UCL = \bar{X} + 3S = 26.5 + 3(3.8) = 37.9$$

$$LCL = \bar{X} - 3S = 26.5 - 3(3.8) = 22.7$$

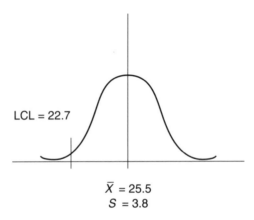

LCL = 22.7

$\bar{X}$ = 25.5
S = 3.8

The probability of getting a point below the LCL is

$$Z = \frac{25.5 - 22.7}{3.8} = 0.74$$

The probability of getting a 0.74 is 0.2297, or 22.97 percent.
The probability of getting a single point below the LCL within 10 samples after the shift occurred is

$$P_{k=i} = 1 - (1 - P_{k-1})^k$$
$$P_{10} = 1 - (1 - 0.2297)^{10} \quad P_{k=10} = 0.9264$$

Table 2. Probability of Detection within k Samples Following a Shift σ.

shift σ	$k =$ 1	2	3	4	5	10	15	20
0.2	.0026	.0051	.0077	.0102	.0129	.0253	.0377	.0500
0.4	.0047	.0093	.0139	.0185	.0231	.0456	.0677	.0892
0.6	.0082	.0163	.0244	.0324	.0403	.0790	.1162	.1518
0.8	.0139	.0276	.0411	.0544	.0676	.1306	.1894	.2442
1.0	.0228	.0450	.0667	.0879	.1087	.2056	.2919	.3689
1.2	.0359	.0706	.1040	.1362	.1672	.3064	.4224	.5190
1.4	.0548	.1066	.1556	.2018	.2456	.4308	.5706	.6761
1.6	.0808	.1550	.2232	.2860	.3436	.5692	.7172	.8144
1.8	.1151	.2169	.3070	.3868	.4573	.7055	.8402	.9133
2.0	.1587	.2922	.4045	.4989	.5784	.8223	.9251	.9684
2.2	.2119	.3788	.5104	.6142	.6959	.9075	.9719	.9915
2.4	.2743	.4733	.6177	.7226	.7986	.9595	.9918	.9984

Table 2 gives the probability of a single point being outside a control limit within k samples of a process shift of σ units.

VARIABLES CONTROL CHARTS: AVERAGES

Just as we can determine the probability of detecting process shifts using individuals, we can do the same using averages. The only difference is in the calculation of the standard deviation for the distribution of averages $S_{\bar{x}}$. Two ways to determine $S_{\bar{x}}$ are:

1. From the SPC relationship $3S_{\bar{x}} = A_2\bar{R}$, where A_2 = a constant dependent on the subgroup sample size n and $\bar{R}$ = the average range.

2. From the relationship defined by the central limit theorem $S_{\bar{x}} = \dfrac{\sigma}{\sqrt{n}}$, where σ is the true standard deviation (estimated by S), and n = the subgroup sample size.

Example of application 1:
An average/range control chart is constructed using a subgroup sample size of $n = 5$. The process average $\bar{X} = 15.00$, and the average range $\bar{R} = 4.8$. What is the standard deviation of the averages?

$$3S_{\bar{x}} = A_2\bar{R} \quad 3S_{\bar{x}} = (0.577)(4.8) = 2.77$$

$$S_{\bar{x}} = \frac{2.77}{3} = 0.92$$

Example of application 2:
The standard deviation for the individuals of a process is $S = 8.55$. What is the standard deviation for the distribution of averages of $n = 7$ from this process?

$$S_{\bar{x}} = \frac{\sigma}{\sqrt{n}}$$

$$S \approx \sigma = 8.55 \quad n = 7$$

$$S_{\bar{x}} = \frac{8.55}{\sqrt{7}} = 3.23$$

Having the standard deviation for the distribution of averages, we can calculate the probability of a single point exceeding a control limit in one sample or the probability of a single point exceeding a limit within k samples of a process change.

Example:
What is the probability of an average of $n = 6$ exceeding the UCL within eight samples after the process shifts from the historical average of $\bar{X} = 45.00$ to 47.50? The UCL is 50.0, and the LCL is 40.00.

$$6S_{\bar{x}} = \text{UCL} - \text{LCL} = 10.0 \quad \therefore \quad S_{\bar{x}} = 1.67$$

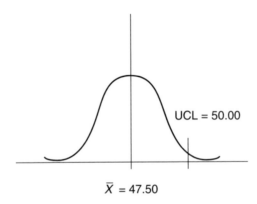

UCL = 50.00

$\bar{X} = 47.50$

The Z score is

$$Z = \frac{50.00 - 47.50}{1.67} = 1.50$$

The probability of getting a $Z = 1.50$ is 0.0668. The probability of exceeding the UCL on a single sample average is 0.0668.

If the probability of exceeding the UCL on a single sample is 0.0668, then the probability of exceeding the UCL within eight samples after the process shift is

$$P_{k=i} = 1-(1-P_{k=1})^k$$

where: $\quad k = 8$
$$P_{k=1} = 0.0668$$
$$P_8 = 1 - (1 - 0.0668)^8$$
$$P_8 = 0.4248$$

Table 3. Probability of Exceeding a Limit for Averages of Subgroup Size n and Shift σ.

Shift σ	Subgroup n								
	2	**3**	**4**	**5**	**6**	**7**	**8**	**9**	**10**
0.2	.0033	.0040	.0047	.0054	.0060	.0069	.0076	.0081	.0089
0.4	.0075	.0105	.0139	.0176	.0216	.0267	.0311	.0358	.0412
0.6	.0157	.0249	.0359	.0487	.0628	.0804	.0969	.1148	.1347
0.8	.0308	.0531	.0808	.1131	.1488	.1913	.2313	.2741	.3186
1.0	.0564	.1023	.1587	.2228	.2907	.3655	.4327	.5000	.5641
1.2	.0963	.1782	.2743	.3762	.4756	.4265	.6538	.7259	.7865
1.4	.1539	.2825	.4207	.5525	.6660	.7625	.8316	.8852	.9232
1.6	.2306	.4095	.5793	.7188	.8210	.8931	.9364	.9642	.9803
1.8	.3249	.5468	.7257	.8478	.9206	.9618	.9817	.9919	.9965
2.0	.4321	.6788	.8413	.9298	.9713	.9893	.9960	.9987	.9996
2.2	.5445	.7913	.9192	.9727	.9916	.9977	.9994	.9998	.9999
2.4	.6534	.8765	.9641	.9911	.9980	.9996	.9999	.9999	.9999

Table 3 gives the probability of exceeding a control limit for various subgroup sample sizes and various degrees of process shift in units of σ in a single sample.

ATTRIBUTE CHART: *P* CHARTS

The resolution for a p chart can be determined just as the case for variables. The normal distribution, however, is not used to define the variation for attribute data. The distribution used for attribute data is the binomial and Poisson.

The following example will be used to illustrate the calculation of an operating characteristic curve (OCC) for a p chart. The OCC will define the resolution or power of the p chart to detect a process shift in units of percent nonconforming for a p chart. For this example, the historical process percent nonconforming is $\bar{P} = 0.18$, and the sample size is $n = 100$.

The control chart limits are calculated as follows:

$$\bar{P} \pm 3\sqrt{\frac{\bar{P}(1-P)}{n}} \rightarrow 0.18 \pm 3\sqrt{\frac{0.18(1-0.18)}{100}} \rightarrow 0.18 \pm 0.115$$

$$\text{UCL} = 0.295$$
$$\text{LCL} = 0.065$$

If the process average proportion $\bar{P}$ shifts, we may calculate the probability of exceeding a control limit. For proportions less than or equal to .10, we will use the Poisson distribution to estimate probabilities, and for proportions greater than 0.10, we will use the normal distribution as an approximation to the binomial.

FOR $P \leq 0.10$, POISSON DISTRIBUTION

Sample Calculations:

Assume that the process average has shifted from the historical average of 0.10 to 0.08. What is the probability of getting a single point below the LCL, or 0.065?

Using the Poisson model, we are determining the probability of getting less than (0.065)(100) defective (or to the nearest integer, seven, defective) in a sample chosen from a process that has an average number defective $np = (100)(.08) = 8.0$.

Using the Poisson relationship, we can determine the probability of getting seven defective or less in a sample of 100 chosen from a process that is .08 proportional defective. This is done by obtaining the probability of getting exactly zero defective plus exactly one defective plus exactly two defective, etc., until the cumulative probability of getting seven defective has been reached. This cumulative probability will represent the probability of getting seven or less defective in a sample of $n = 100$ and will represent the probability of **not** exceeding the UCL. 1.00 minus this probability will be the probability of exceeding the UCL.

Poisson Formula:

$$P_x = \frac{e^{-np}(np)^x}{x!}$$

where: n = sample size
x = exact number defective in sample
p = proportion defective of process

The probability of getting **exactly** three defective in a sample of 100 chosen from a process that is 0.05 proportion defective is given by

$$P_x = \frac{e^{-np}(np)^x}{x!} \qquad P_{x=3} = \frac{e^{-[(100)(0.05)]}[(100)(0.05)]^3}{3!} \qquad P_{x=3} = \frac{e^{-5}(5)^3}{3!} \qquad P_{x=3} = 0.1404$$

The probability of getting three or less defective from a sample of $n = 100$ chosen from a process that is 0.05 proportional defective is the sum of all the individual probabilities from zero defective through three defective:

$$P_{x \leq 3} = \frac{e^{-5}(5)^0}{0} + \frac{e^{-5}(5)^1}{1!} + \frac{e^{-5}(5)^2}{2!} + \frac{e^{-5}(5)^3}{3!}$$

$$P_{x=3} = 0.00674 + 0.03369 + 0.08422 + 0.14037 = 0.265$$

This result can also be obtained more readily by using a cumulative Poisson distribution table and looking up $np = 5.0$ and $x = 3$. The result is 0.265.

FOR $P > 0.10$, NORMAL DISTRIBUTION

What is the probability of getting a single proportion defective greater than the UCL of 0.295 from a sample of $n = 100$ chosen from a process that has proportional defective of 0.250? The control limit was based on a historical process proportional defective of 0.180 and a sample size of $n = 100$.

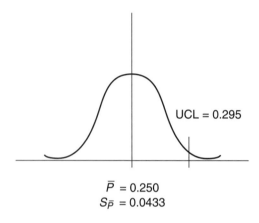

$$\bar{P} = 0.250$$
$$S_{\bar{p}} = 0.0433$$

$$S_{\bar{p}} = \sqrt{\frac{\bar{P}\left(1-\bar{P}\right)}{n}} \quad S_{\bar{P}} = \sqrt{\frac{0.25\left(1-0.25\right)}{100}}$$

Calculation of the standard deviation was based upon a binomial distribution for attribute data. The probability that a given sample proportion will exceed the UCL will be determined by using the normal distribution.

$$Z = \frac{0.295 - 0.250}{0.0433} = 1.04$$

Looking up $Z = 1.04$, the probability of occurrence is 0.149.

Additional values for various process shifts have been made for the control chart based on a sample size of $n = 100$ and a historical process proportional defective of $P = 0.18$. Table 4 lists the probabilities of exceeding a limit for those cases. All probabilities for shifts to $p \leq 0.10$ are based on the Poisson, and $P > 0.10$ are based on the normal distribution.

Table 4. Resolution of a $n = 100$, $P = 0.18$ P Chart.

P	Probability < LCL	Probability > UCL	Total probability	P	Probability < LCL	Probability > UCL	Total probability
0.01	1.000	0.000	1.000	0.20	0.001	0.009	0.010
0.02	0.995	0.000	0.995	0.21	0.000	0.184	0.184
0.03	0.988	0.000	0.988	0.22	0.000	0.035	0.035
0.04	0.949	0.000	0.949	0.23	0.000	0.061	0.061
0.05	0.867	0.000	0.867	0.24	0.000	0.099	0.099
0.06	0.744	0.000	0.744	0.25	0.000	0.149	0.149
0.07	0.599	0.000	0.599	0.26	0.000	0.213	0.213
0.08	0.453	0.000	0.453	0.27	0.000	0.287	0.287
0.09	0.324	0.000	0.324	0.28	0.000	0.369	0.369
0.10	0.220	0.000	0.220	0.29	0.000	0.456	0.456
0.11	0.075	0.000	0.075	0.30	0.000	0.543	0.543
0.12	0.045	0.000	0.045	0.31	0.000	0.627	0.627
0.13	0.027	0.000	0.027	0.32	0.000	0.704	0.704
0.14	0.018	0.000	0.018	0.33	0.000	0.772	0.772
0.15	0.009	0.000	0.009	0.34	0.000	0.829	0.829
0.16	0.005	0.000	0.005	0.35	0.000	0.876	0.876
0.17	0.003	0.000	0.003	0.36	0.000	0.912	0.912
0.18	0.001	0.001	0.002	0.37	0.000	0.940	0.940
0.19	0.001	0.004	0.005	0.38	0.000	0.960	0.960

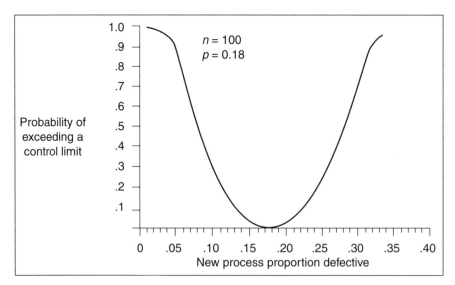

Figure 2. Operating Characteristic Curve for P chart, $n = 100$, $P = 0.18$.

A plot of total probability as a function of P from the data in Table 4 can be see in Figure 2. This plot is called an OCC and completely defines the resolution or power for the control chart.

BIBLIOGRAPHY

Grant, E. L, and R. S. Leavenworth. 1996. *Statistical Quality Control.* 7th edition. New York: McGraw-Hill.

Montgomery, D. C. 1996. *Introduction to Statistical Quality Control.* 3rd edition. New York: John Wiley and Sons.

Wheeler, D. J. 1995. *Advanced Topics in Statistical Process Control.* Knoxville, TN: SPC Press.

Wheeler, D. J, and D. S. Chambers. 1992. *Understanding Statistical Process Control.* 2nd edition. Knoxville, TN: SPC Press.

CHI-SQUARE CONTINGENCY AND GOODNESS-OF-FIT

CHI-SQUARE CONTINGENCY TABLES

The chi-square distribution can be used in two very useful manners. The first is to test for the independence of two factors. Data are arranged in a tabular fashion with rows and columns. In the following example, we want to test for the independence of the two factors of dealership and type of vehicles sold.

Example 1:
Four dealerships are evaluated by the number of vehicles of three different types sold in one month. The objective is to determine if there is a dependency of the type of vehicles sold as a function of the dealership.

We will construct a contingency table consisting of four columns (dealerships) and three rows (type of vehicles). The response will be the number of vehicles sold in one month.

| Type of Vehicle | Dealership | | | | |
	J.R.'s Auto	Crazy Harry's	Abe's	Reliable Rosco's	Row totals
Truck	26	38	75	53	192
Compact	18	24	32	28	102
Midsize	33	45	68	49	195
Column totals	77	107	175	130	489 (Grand total)

The null hypothesis assumes that the distribution of the number of cars sold of a particular type is independent of the dealership. That is

H_0: assumes no real differences between the dealerships with respect to distribution of sales

The expected frequency of sales within each class is given by:

$$\text{expected value} = \frac{\text{row totals} \times \text{column totals}}{\text{grand total}}$$

Thus for "Compact/Reliable Rosco's" the expected frequency is

$$\frac{102 \times 130}{489} = 27.1$$

The expected value for all dealerships and vehicle type are calculated. The values in parentheses are the expected frequencies.

	Dealership				
Type of vehicle	J.R.'s Auto	Crazy Harry's	Abe's	Reliable Rosco's	Row totals
Truck	26(30.2)	38(42.0)	75(68.7)	53(51.0)	192
Compact	18(16.1)	24(22.3)	32(36.5)	28(27.1)	102
Midsize	33(30.7)	45(42.7)	68(69.8)	49(51.8)	195
Column totals	77	107	175	130	489 (Grand total)

The chi-square value is now determined for all the data.

Chi-square is the sum of the differences between the observed value and the expected value divided by the expected values.

There will be 12 terms in the chi-square calculation, one for each class combination.

O	E	$O - E$	$(O - E)^2$	$(O - E)^2/E$
26	30.2	−4.2	17.64	0.58
38	42.0	−4.0	16.00	0.38
75	68.7	6.3	6.30	0.58
53	51.0	2.0	4.00	0.08
18	16.1	1.9	3.61	0.22
24	22.3	1.7	2.89	0.13
32	36.5	−4.5	20.25	0.55
28	27.1	0.9	0.81	0.03
33	30.7	2.3	5.29	0.17
45	42.7	2.3	5.29	0.12
68	69.8	−1.8	3.24	0.05
49	51.8	−2.8	7.84	0.15

$$\chi^2 = 3.04 \text{ (sum of all } (O - E)^2/E$$

We will reject the hypothesis of independency if the calculated chi-square value (3.04) is greater than the table value of critical chi-square value for a given level of risk (α).

The table value is a function of the total degrees of freedom (υ) and the risk (α). The degrees of freedom are determined by $(r - 1)(c - 1)$ where r = the number of rows and c = the number of columns in the contingency table. For our example, the degrees of freedom are

$$(3 - 1)(4 - 1) = 6$$

If we chose a risk of 5 percent then the critical chi-square is $\chi^2_{6,0.05} = 12.6$. Since the calculated chi-square is not greater than the critical chi-square, we cannot reject the hypothesis that the type of vehicles sold is independent of the dealership. The two factors are independent of each other at the level of significance or risk of 5 percent.

Example 2:

An experimenter wishes to determine whether there is a relationship between the type of UV inhibitor and the shade of paint with respect to the length of time required to cause a test specimen to lose 15 percent of its original gloss. One hundred and fifty samples were tested using UV inhibitor types A and B with both a light and dark color of paint. A summary of the test data in each of the four categories is shown below. Do the data provide sufficient evidence to indicate a relationship between UV inhibitor type and shade of color at a significance of $\alpha = 0.10$?

Shade of paint	Type of UV inhibitor		
	Type 101	**Type 320**	**Row totals**
Light	47	32	79
Dark	21	50	71
Column totals	68	82	150 (Grand total)

Calculation of expected values:

$$\text{Expected value for Light} / 101 = \frac{79 \times 68}{150} = 35.8$$

$$\text{Expected value for Dark} / 101 = \frac{79 \times 68}{150} = 32.2$$

$$\text{Expected value for Light} / 320 = \frac{79 \times 82}{150} = 43.2$$

$$\text{Expected value for Dark} / 320 = \frac{71 \times 82}{150} = 38.8$$

This is the completed contingency table. Expected values are in parentheses.

Shade of paint	Type of UV inhibitor		
	Type 101	**Type 320**	**Row totals**
Light	47(35.8)	32(43.2)	79
Dark	21(32.2)	50(38.8)	71
Column totals	68	82	150 (Grand total)

Calculation of χ^2:

$$\chi^2 = \frac{(47.0-35.8)^2}{35.8} + \frac{(21.0-32.2)^2}{32.2} + \frac{(32.0-43.2)^2}{43.2} + \frac{(50.0-38.8)^2}{38.8} = 13.53$$

Calculation of $\chi^2_{Critical}$:

$$\text{Degrees of freedom} = (r-1)(c-1) = (2-1)(2-1) = 1$$

$$\text{Significance or risk } = \alpha = 0.10$$

$$\chi^2_{1,0.10} = 2.71$$

Since the chi-square calculated (13.53) is greater than the chi-square critical (2.71), we reject the hypothesis that the type of UV inhibitor and shade are independent with respect to the time required to loose 15 percent of the original gloss.

Performance of the type of inhibitor is related to the shade of paint.

Note: There is an assumption that the sample size is sufficiently large such that each category will have at least five observations. This may require that some categories be combined to give a minimum count of five.

GOODNESS-OF-FIT

A goodness-of-fit test is a statistical test to determine the likelihood that sample data have been generated from a population that conforms to a particular distribution such as a normal, binomial, Poisson, uniform, etc.

The objective is to compare the sample distribution to the expected distribution. The null hypothesis would be that the sample data come from a specified distribution with a defined mean and standard deviation. The alternative hypothesis would be that the sample data come from some other type of distribution.

GOODNESS-OF-FIT TO POISSON

The following data represent the number of defects found on printed circuit boards.

Number of defects found:　　0　1　2　3　4　5

Number of boards affected:　32　44　41　22　4　3

As with the case of the chi-square contingency table, we must determine the expected frequency assuming that we have a Poisson distribution. In order to accomplish this, we must determine the average defect rate, λ.

$$\lambda = \frac{\text{total defects}}{\text{total units}} = \frac{(0)(32)+(1)(44)+(2)(41)+(3)(22)+(4)(4)+(5)(3)}{32+44+41+22+4+3} = \frac{223}{146} = 1.53$$

The probability of getting exactly 0 defects based upon the Poisson distribution where $\lambda = 1.53$ is

$$P_{x=0} = \frac{e^{-\lambda}\lambda^x}{x!} = 0.217$$

In a sample of $n = 146$, we would expect to find $(146)(0.217) = 31.6$ boards with 0 defects.

For this data the observed number of boards with 0 defects was 32, and the expected value is 31.6.

The probability of finding exactly one defect is given by:

$$P_{x=0} = \frac{e^{-\lambda}\lambda^x}{x!} = P_{x=1} = \frac{e^{-1.53}(1.53)^1}{1} = 0.331$$

In a sample of $n = 146$, we would expect to find $(146)(0.331)$ or 48.3 boards with one defect. For this data the observed number of boards with one defect was 44, and the expected value is 48.3.

The probability of finding exactly two defects is given by:

$$P_{x=0} = \frac{e^{-\lambda}\lambda^x}{x!} = P_{x=1} = \frac{e^{-1.53}(1.53)^2}{2} = 0.253$$

In a sample of $n = 146$, we would expect to find $(0.253)(146)$ or 36.9 boards with two defects.

For this data the observed number of boards with one defect was 41, and the expected value is 36.9.

The probability of finding exactly three defects is given by:

$$P_{x=0} = \frac{e^{-\lambda}\lambda^x}{x!} = P_{x=1} = \frac{e^{-1.53}(1.53)^3}{6} = 0.129$$

In a sample of $n = 146$, we would expect to find $(0.129)(146)$ or 18.8 boards with three defects.

For this data the observed number of boards with one defect was 22, and the expected value is 18.8.

The probability of finding exactly four defects is given by:

$$P_{x=0} = \frac{e^{-\lambda}\lambda^x}{x!} = P_{x=1} = \frac{e^{-1.53}(1.53)^4}{24} = 0.049$$

In a sample of $n = 146$, we would expect to find $(0.049)(146)$ or 7.2 boards with four defects.

For this data the observed number of boards with one defect was 4, and the expected value is 7.2.

The probability of finding exactly five defects is given by:

$$P_{x=0} = \frac{e^{-\lambda}\lambda^x}{x!} = P_{x=1} = \frac{e^{-1.53}(1.53)^5}{120} = 0.015$$

In a sample of $n = 146$, we would expect to find (0.015)(146) or 2.2 boards with five defects.

For this data the observed number of boards with one defect was 3, and the expected value is 2.2.

The following is a summary of the observed frequency (from the data) and the expected frequency (for a Poisson distribution where $\lambda = 1.52$). The chi-square value is calculated the same as in the example for the contingency table.

Number of defects	Frequency observed, O	Frequency expected, E	$O - E$	$(O - E)^2$	$(O - E)^2/E$
0	32	31.6	0.4	0.16	0.01
1	44	48.3	−4.3	18.49	0.38
2	41	36.9	4.1	16.81	0.46
3	22	18.8	3.2	10.24	0.54
4	4	7.2	−3.2	10.24	1.42
5	3	2.2	0.8	0.64	0.29

The last row has fewer than five observations in the expected column; therefore, we will combine this row with the previous row to increase the count to > 5 as required.

Number of defects	Frequency observed, O	Frequency expected, E	$O - E$	$(O - E)^2$	$(O - E)^2/E$
0	32	31.6	0.4	0.16	0.01
1	44	48.3	−4.3	18.49	0.38
2	41	36.9	4.1	16.81	0.46
3	22	18.8	3.2	10.24	0.54
4 and 5	7	9.4	−2.4	5.76	0.61
					$\chi^2 = 2.00$

Chi-square calculated = 2.00.

We will now look up the critical chi-square. The risk (r) level of significance is chosen to be 10 percent or $\alpha = 0.10$. The degrees of freedom are determined by

$$df = \text{number of classes (adjusted) - 1 - number of estimated parameters}$$

$$\text{classes adjusted} = 5 \qquad \text{number of parameters estimated} = 1$$

$$df = 5 - 1 - 1 = 3$$

The critical chi-square value from the table is $\chi^2_{3,0,0.10} = 6.25$. Since 2.00 is less than the critical value of 6.25, we cannot reject the null hypothesis that the data comes from a Poisson distribution.

GOODNESS-OF-FIT FOR THE UNIFORM DISTRIBUTION

The following serial numbers were obtained from 20 one-dollar bills:

12049847	72772262	32549642	25178095
78711872	26509623	58136745	87717396
11054247	17503581	34286999	92733162
04301842	56704709	84544014	17194985
63460081	24052919	13103479	46878096

The objective is to confirm a level of significance of 5 percent that the distribution of integers is uniform.

Integer	Observed frequency, O	Expected frequency, E	$O-E$	$(O-E)^2$	$(O-E)^2/E$
0	15	16	−1	1	0.0625
1	19	16	3	9	0.5625
2	19	16	3	9	0.5625
3	11	16	−5	25	1.5625
4	20	16	−4	16	1.0000
5	13	16	−3	9	0.5625
6	13	16	−3	9	0.5625
7	20	16	4	16	1.0000
8	14	16	−2	4	0.2500
9	16	16	0	0	0.0000
Total:	160				$\chi^2 = 6.125$

If the frequency were uniform, there would be the same frequency for all integer values, 16.

The critical chi-square value is determined with a significance of $\alpha = 0.05$ and the degrees of freedom of:

df = number of classes (adjusted) - 1 - number of estimated parameters

$df = 10 - 1 - 0 = 9$

Note that no parameters are estimated from the sample data.

Critical chi-square, $\chi^2_{9,0.05} = 16.9$. Since the calculated chi-square is less than the critical chi-square, we cannot reject the hypothesis that the distribution is uniform.

GOODNESS-OF-FIT FOR ANY SPECIFIED DISTRIBUTION

Goodness-of-fit tests can be applied to any specified distribution, not just those that are well known. Consider the following case. A collector of old books has been approached by a seller of what has been claimed to be a rare, but otherwise original, work of William Shakespeare. In order to substantiate the claims, the owner of the bookstore determined the distribution of certain key words used by Shakespeare. Fifteen works of Shakespeare were examined for the frequency of the use of the words *thou, withal, wouldst, thine,* and *hadst.*

The frequency of the words per 1,000 words of text were determined and compared to the suspect manuscript.

These data provide an opportunity to examine the true distribution to the suspect distribution. In this example the observed rate will be that of the suspect, and the expected will be that of the true Shakespeare works.

The following computation reveals the truth about the suspect manuscript.

Word	Observed rate (suspect)	Expected rate (true Shakespeare)	$O - E$	$(O - E)^2$	$(O-E)^2/E$
Thou	40	58	−18	324	5.58
Withal	18	24	−6	36	1.50
Wouldst	58	56	2	4	0.07
Thine	13	19	−6	36	1.89
Hadst	44	31	13	169	5.45

Calculated chi-square: 14.49

Setting the level of significance at 5 percent and the degrees of freedom at 4, the critical chi-square value is:

$$\chi^2_{4, 0.05} = 9.49$$

The calculated chi-square is greater than the critical chi-square; therefore, we conclude that the suspect manuscript is not a work of William Shakespeare with a risk of 5 percent.

BIBLIOGRAPHY

Walpole, R. E. *Probability and Statistics for Engineers and Scientists,* 5th edition. Englewood Clifts, NJ: Prentice-Hall, Inc.

CHI-SQUARE CONTROL CHART

In his pioneering book *Economic Control of Quality of Manufactured Product,* Dr. W. A. Shewhart (1931) stated that "perhaps the single statistic most sensitive to change is the Chi-square function." Shewhart went on to say, "One difficulty is that the Chi-square control chart can only be used for comparatively large samples." This control chart, however, has the advantage of combining data that otherwise would have to be monitored using several individual control charts into one. If the objective is to monitor the ***change in a distribution*** arising from either a change in the location statistic or the variation statistic and there is sufficient data to construct a frequency table such as a histogram, then the chi-square control chart is applicable. Examples for use of the chi-square control chart would be monitoring the weight of capsules or the distribution of particle sizes using sieve analysis data. One real advantage in the chi-square control chart is that it looks at the entire process and answers the question "Has the process changed?" It looks at the entire distribution rather than just the average and standard deviation as with conventional control charts.

The following illustration will demonstrate the concept of the chi-square control chart.

Background:

A powdered, clay material is used in the production of a resin filler product. The particle distribution affects the viscosity of the finished product for a given amount of clay added. Changes in the distribution of particle sizes will cause an adjustment in the concentration of clay to be added or might require adding a liquid to reduce the viscosity of the final product. These adjustments also affect the bonding characteristics of the resin.

Analysis of the clay is accomplished by sifting the clay through a series of sieves and measuring the percent retention on each sieve. There are nine sieves of various sizes used. Data for 15 lot analyses are collected. For each lot, the percent retained on each of the nine sieves is determined. The average percent retention for each sieve size is determined. This information will be used to determine the chi-square value for each lot analysis. Chi-square values are calculated using the following relationship:

$$X^2 = \left(\frac{(\text{Obs} - \text{Exp})^2}{\text{Exp}} \right)_1 + \left(\frac{(\text{Obs} - \text{Exp})^2}{\text{Exp}} \right)_2 + \ldots \left(\frac{(\text{Obs} - \text{Exp})^2}{\text{Exp}} \right)_k$$

where: Obs (observed) = the actual observed frequency of occurrence for strata or class k, and

Exp (expected) = the expected frequency for strata or class k (can be a count or a percent).

Data for 15 lot analyses:

Date	20	28	35	48	65	100	150	180	200
				Sieve size					
11/1	2	9	17	17	19	13	11	9	3
11/2	1	11	13	22	20	11	9	10	3
11/3	4	10	15	21	16	10	12	7	5
11/4	3	10	18	20	17	12	9	9	2
11/5	2	8	9	18	19	16	10	15	3
11/6	3	9	15	22	18	13	11	8	1
11/7	4	12	19	19	15	10	10	7	4
11/8	2	9	17	22	18	11	9	9	3
11/9	4	12	15	19	16	12	10	8	4
11/10	2	10	18	21	17	14	11	5	2
11/11	3	7	15	20	18	12	13	9	3
11/12	5	8	17	22	16	13	9	5	5
11/13	1	11	16	20	17	11	8	7	3
11/14	3	14	14	21	18	12	7	9	2
11/15	3	10	17	19	17	11	9	10	4
$\bar{X}$:	2.80	10.00	15.67	20.20	17.40	12.07	9.87	8.87	3.13

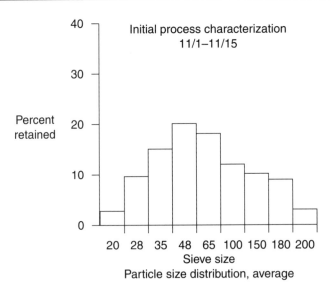

Initial process characterization
11/1–11/15

Percent retained

Sieve size
Particle size distribution, average

A chi-square value is determined for each lot analysis.

$$X^2{}_1 = \frac{(2.00-2.80)^2}{2.80} + \frac{(9.00-10.00)^2}{10.00} + \frac{(17.00-15.76)^2}{15.75} + \frac{(17.00-20.20)^2}{20.20}$$

$$+ \frac{(19.00-17.40)^2}{17.40} + \frac{(13.00-12.07)^2}{12.07} + \frac{(11.00-9.87)^2}{9.87}$$

$$+ \frac{(9.00-8.87)^2}{8.87} + \frac{(3.00-3.13)^2}{3.13}$$

$$X^2{}_1 = 1.2990$$

The remaining lot analysis results have the chi-square values calculated as in the previous example. A summary of all 15 chi-square values are as follows:

Date	Chi-square
11/1	1.299
11/2	2.582
11/3	2.990
11/4	0.859
11/5	9.378
11/6	2.060
11/7	3.017
11/8	0.821
11/9	1.456
11/10	3.150
11/11	1.966
11/12	5.469
11/13	2.124
11/14	3.089
11/15	0.765

The control limits for the chi-square control chart are based on the following relationships:

k = number of classes or strata in which data are reported. For this example, $k = 9$.

Center line = $k - 1 = 8$

Upper control limit (UCL) = $CL + 3\sqrt{2(k-1)} = 8 + 12 = 20$

Lower control limit (LCL) = $CL - 3\sqrt{2(k-1)} = 8 - 12 = -4 \Rightarrow 0$ (All chi-squares are positive; therefore, if a negative limit is derived, default to zero.)

We now establish the control chart and plot the calculated chi-square values.

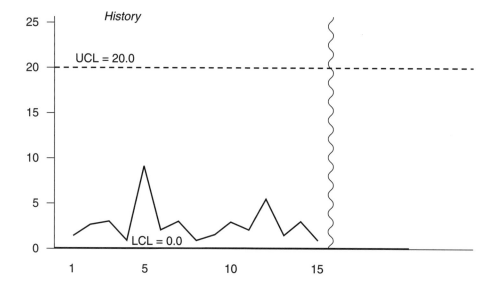

The control limits are based on $\pm 3S$, which means there is a 99.7 percent probability that samples chosen from the population with the historical characterization will have a 99.7 percent probability of having a calculated chi-square less than 20.0 (for this example).

The first 15 samples are compared to a parent distribution defined by an expected percent retention on sieve #20 of 2.8 percent, sieve #28 of 10.0 percent, sieve #35 of 15.56 percent, and so on. The greater the deviation from the expected distribution, the greater the chi-square value.

Additional data are taken, and signs of a process (distribution) change will be indicated if the chi-square calculated exceeds the calculated UCL of 20.0.

Calculate the chi-square values for the next five lots of material, and see if there is evidence of a process (distribution) change.

	Sieve size								
Date	**20**	**28**	**35**	**48**	**65**	**100**	**150**	**180**	**200**
11/16	3	9	17	11	12	21	13	6	8
11/17	2	7	14	14	9	23	16	9	6
11/18	1	11	15	10	11	19	15	11	7
11/19	4	9	12	15	12	17	17	9	5
11/20	5	11	16	14	11	22	14	6	1
Avg. (16–20)	3.0	9.4	14.8	13.6	11.0	20.4	15.0	8.2	5.4
Avg. (1–15)	2.8	10.0	15.67	20.2	17.4	12.07	9.87	8.87	3.13

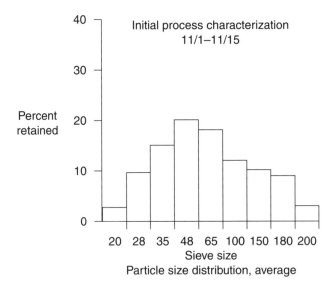

Particle size distribution, average

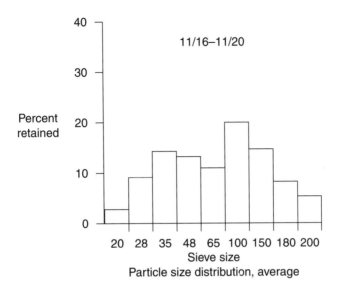

Particle size distribution, average

Note the difference in the two distributions.
Chi-square values for samples 11/16 through 11/20 are as follows:

Date	Chi-square
11/16	22.20
11/17	22.75
11/18	20.73
11/19	12.77
11/20	16.76

The new additional chi-square values are plotted on the original chart. The results can be seen in the following figure.

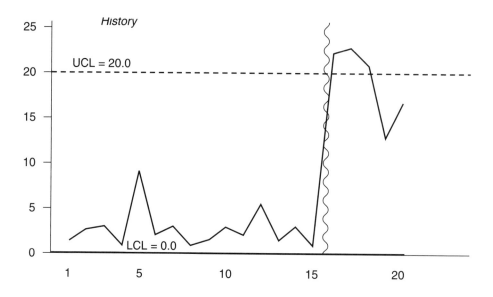

The chart has detected this change in the base or historical distribution.

An additional four lots are analyzed with the following results. Continue to calculate the chi-square values and plot the data.

Date	Sieve size								
	20	**28**	**35**	**48**	**65**	**100**	**150**	**180**	**200**
11/21	2	11	12	17	15	15	13	10	3
11/22	3	9	11	16	18	13	11	14	5
11/23	1	12	10	19	14	17	14	10	3
11/24	3	12	13	17	15	14	15	9	2
Avg. (21–24)	2.3	11.0	11.5	17.3	15.5	14.8	13.3	10.8	3.3
Avg. (16–20)	3.0	9.4	14.8	13.6	11.0	20.4	15.0	8.2	5.4
Avg. (1–15)	2.8	10.0	15.67	20.2	17.4	12.07	9.87	8.87	3.13

A significant change in the distribution of the data in samples 16–20 has caused the chart to be out of control. The distribution of data in samples 21–24 is not sufficient to cause a change in the control chart pattern relative to the historical data contained in samples 1–15.

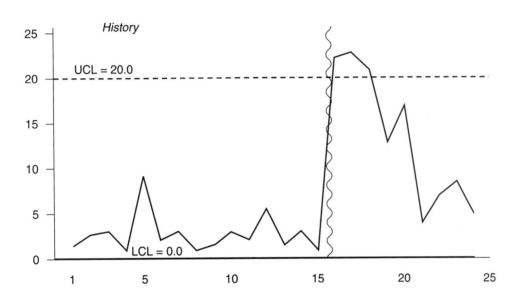

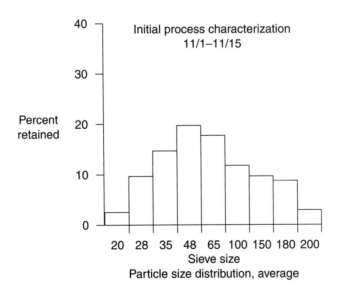

Particle size distribution, average

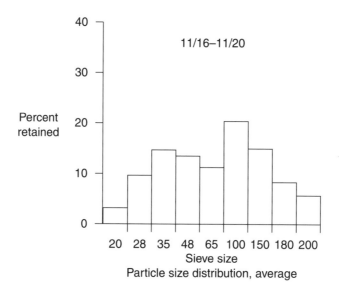

Particle size distribution, average

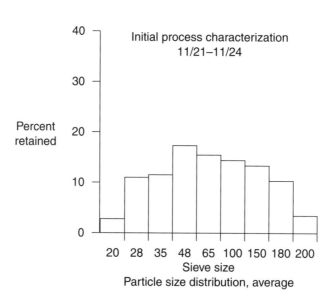

Particle size distribution, average

BIBLIOGRAPHY

Montgomery, D. C. 1996. *Introduction to Statistical Quality Control.* 3rd edition. New York: John Wiley and Sons.

Shewhart, W. A. 1980. *Economic Control of Quality of Manufactured Product.* Milwaukee, WI: ASQC Quality Press.

CONFIDENCE INTERVAL FOR THE AVERAGE

Two of the most used descriptive statistics are the average $\bar{X}$ and the standard deviation S. As with all statistics, these statistics are derived from samples taken from a larger population or universe and, as such, are only estimates of the true parameter. The true average is actually the mean and is abbreviated μ. All estimates are subject to error with respect to the degree to which they represent the truth. This discussion will focus on the confidence of the average as it represents the truth or parameter of the mean. Confidence is related to

1. Sample size n
2. Error of the estimate E

Confidence in the average can be stated in two specific ways.

1. Two-sided confidence intervals
2. Single-sided confidence limits

An example of the first case, the two-sided confidence interval, can be expressed in statements such as, "I don't know the true average, but I am 90 percent confident it is between 12.4 and 15.5."

An example of a single-sided confidence limit would be, "I do not know the true average, but I am 95 percent confident that it is less than 34.6," or "I am 90 percent confident that the true average is greater than 123.0."

Confidence can be thought of as a degree of certainty, and the opposite of confidence is risk or degree of uncertainty. The sum total of confidence and risk is 100 percent:

$$\text{Confidence} + \text{Risk} = 100\%$$

The risk as associated with confidence is abbreviated α.

Calculations for confidence are based upon probability distributions. For sample sizes less than 30, we use the t-distribution, and for sample sizes equal to and greater than 30, we use the normal distribution.

The following example will illustrate confidence calculations.

CONFIDENCE INTERVAL, $n \leq 30$

Example:

What is the 90 percent confidence interval for the average, where $n = 15$, $S = 1.2$, and the sample average $\overline{X} = 25.0$?

Since this is a confidence *interval,* we will be calculating an upper and lower limit. Confidence intervals for the average when the sample size is less than or equal to 30 are determined using the following relationship:

$$\overline{X} \pm t_{\alpha/2,n-1}\left(\frac{S}{\sqrt{n}}\right)$$

where: n = sample size

S = sample standard deviation

$\overline{X}$ = sample average

$t_{\alpha/2,\,n-1}$ = a factor based on sample size and a confidence of $1 - a$

For informational purposes, the quantity $\dfrac{S}{\sqrt{n}}$ is the standard error, and the quantity $t_{\alpha/2,n-1}$ represents the number of standard error to add and subtract to the estimate average to define the confidence interval. In selecting levels of confidence, we typically chose values of 99.9, 99.0, 95.0, and 90.0.

Step 1. Determine the t value.

The chosen level of confidence is 90 percent or 0.90; therefore, the risk $\alpha = 0.10$. Since this is a two-sided interval, we use half the risk ($\alpha/2 = 0.05$). The sample size is 15. Looking up the t value using a standard t-distribution, we have $t_{\alpha/2,n-1} = 1.761$.

Step 2. Determine the standard error.

The standard error is

$$\frac{S}{\sqrt{n}} = \frac{1.2}{3.87} = 0.31$$

Step 3. Calculate the error of the estimate E.

$$E = t_{\alpha/2,n-1}\left(\frac{S}{\sqrt{n}}\right) \quad E = 1.761 \times 0.31 \quad E = 0.55$$

Step 4. Determine the confidence interval.

Confidence interval = Point estimate $\pm$ error

Confidence interval = 25.0 ± 0.55

Confidence interval = 24.45 to 25.55

An appropriate statement for this case would be: "I do not know the true average (mean), but I am 90 percent confident that it is between 24.45 and 25.55."

Supplemental problem:

Determine the 95 percent confidence interval when $n = 22$, $S = 4.5$, and $\overline{X} = 45.00$.

CONFIDENCE INTERVAL, $n > 30$

The steps that evolve when the sample size is greater than 30 are exactly the same as with sample sizes less than or equal to 30, except we use the normal distribution rather than the t-distribution.

Unlike the t-distribution, the normal distribution is not dependent on sample size, as it assumes the sample size to be infinite. We use the standard normal distribution in reverse by looking up the Z-score that will produce the appropriate $\alpha/2$ value for a given level of confidence. This value is substituted for the quantity $t_{\alpha/2,n-1}$ used when n is less than or equal to 30.

Confidence C	Risk α	$\alpha/2$	$Z_{\alpha/2}$
0.999	0.001	0.0005	3.29
0.995	0.005	0.0025	2.80
0.990	0.010	0.0050	2.58
0.950	0.050	0.0250	1.96
0.900	0.100	0.0500	1.65

Example:
A sample $n = 150$ yields an average $\overline{X} = 80.00$ and standard deviation $S = 4.50$. What is the 99 percent confidence interval for this point estimate?

Step 1. Determine the $Z_{\alpha/2}$ value.

The chosen level of confidence is 99.0 percent or 0.990; the risk α is 0.010; and $\alpha/2$ is 0.0050. Looking in the body of a standard normal distribution, we find that the nearest Z value is 2.58.

Step 2. Calculate the standard error.
The standard error is given by

$$\frac{\sigma}{\sqrt{n}}$$

where: σ = true standard deviation
n = sample size

Our best estimate of σ is S, and in this case $S = 4.50$. Therefore, our estimate of the standard error is

$$\frac{4.50}{\sqrt{150}} = 0.37$$

Step 3. Calculate the error of the estimate E.

$$E = Z_{\alpha/2} \frac{S}{\sqrt{n}} \quad E = Z_{\alpha/2} \frac{S}{\sqrt{n}}$$

$$E = (2.58)\left(\frac{4.50}{\sqrt{150}}\right) \quad E = 0.95$$

Step 4. Determine the confidence interval.

Confidence interval = Point estimate ± error

Confidence interval = 80.00 ± 0.95

Confidence interval = 79.05 to 80.95

CONFIDENCE LIMIT, $n > 30$, SINGLE-SIDED LIMIT

The confidence *limit* differs from the confidence interval in that only one value is given—either an upper confidence limit or a lower confidence limit.

Example:
What is the 95 percent upper confidence limit given a point estimate average $\bar{X}$ of 65.0, a sample size of $n = 75$, and a standard deviation of $S = 3.00$?

Step 1. Determine the Z_{α} value.

Since the sample size is greater than 30, we will use the standard normal distribution. For large sample sizes, we assume that σ is known and the normal distribution is appropriate to use. However, σ is never really known, and we will assume that the sample standard deviation S will serve as a good estimate of σ.

Notice that we are determining a Z_{α} rather than a $Z_{\alpha/2}$ as with the case of a two-sided confidence interval. This is because we are only developing an upper 95 percent confidence limit.

We will use the standard normal distribution in reverse; that is, we will determine what Z-score will give us a value equal to the risk α. Since we have specified a confidence of 0.95, the risk α will be 0.05. The corresponding Z-score that will give a value of 0.0500 in the main body of the table is 1.65.

Other Z values for selected levels of confidence for a single-sided limit are as follow:

Confidence C	Risk α	Z_α
0.999	0.001	3.09
0.995	0.005	2.58
0.990	0.010	2.33
0.950	0.050	1.65
0.900	0.100	1.28

Step 2. Calculate the standard error.

$$\frac{S}{\sqrt{n}} = \frac{3}{\sqrt{75.0}} = 0.35$$

Step 3. Determine the error of the estimate E.

$$E = Z_\alpha \left(\frac{S}{\sqrt{n}} \right) \quad E = 1.65 \left(\frac{3.0}{\sqrt{75.0}} \right) \quad E = 0.57$$

Step 4. Calculate the upper 95 percent confidence limit.

Upper 95% confidence limit = Point estimate + Error

Upper 95% confidence limit = 65.00 + 0.57

Upper 95% confidence limit = 65.57

An appropriate statement for this example would be: "I do not know the true average (mean), but I am 95 percent confident that it is not greater than 65.57."

Supplemental problem:

What is the 90 percent lower confidence limit given a standard deviation of 12.00, an average of 150, and a sample size of 200?

CONFIDENCE LIMIT, $n \leq 30$, SINGLE-SIDED LIMIT

The steps are identical as with the sample $n > 30$, except use the t-distribution.

Example:
What is the lower 90 percent confidence limit where $n = 25$, $S = 10.0$, and $\bar{X} = 230.0$?

Step 1. Determine the t value.

Risk $\alpha = 0.10$, and $n = 25$ (we look up $n - 1$ or 24 degrees of freedom df). The t value from the t-table distribution is 1.318.

Step 2. Calculate the standard error.

$$\frac{S}{\sqrt{n}} = \frac{10}{\sqrt{25}} = 2.00$$

Step 3. Calculate the error of the estimate E.

$$E = t_{\alpha,n-1}\left(\frac{S}{\sqrt{n}}\right) \quad E = 1.318\left(\frac{10}{\sqrt{25}}\right) \quad E = 2.64$$

Step 4. Determine the 90 percent lower confidence limit.

Lower 90% confidence limit = Point estimate − Error

Lower 90% confidence limit = 230.0 − 2.64

Lower 90% confidence limit = 227.36

An appropriate statement would be: "We do not know the true mean, but we are 90 percent confident that it is not less than 227.36."

Problems:

1. If the error for an average is excessive, what two things could be done to reduce it?
2. Calculate the 85 percent upper confidence limit for the following data:

    22.5

    23.6

    20.6

    24.0

    25.8

    22.9

    23.1

3. Calculate the 90 percent confidence interval given the following:

    Average = 68.0

    Standard deviation = 3.7

    Sample size = 190

4. Perform the same calculation as Problem 3, except use a sample size of 25.

BIBLIOGRAPHY

Dovich, R. A. 1992. *Quality Engineering Statistics.* Milwaukee, WI: ASQC Quality Press.

Juran, J. M, and F. M. Gryna. 1980. *Quality Planning and Analysis.* 3rd edition. New York, NY: McGraw-Hill.

Petruccelli, J. D, B. Nandram, and M. Chen. 1999. *Applied Statistics for Engineers and Scientists.* Upper Saddle River, NJ: Prentice Hall.

CONFIDENCE INTERVAL FOR PROPORTIONS

Whenever we calculate a descriptive statistic such as an average, proportion, or standard deviation from a random sample selected from a larger set of observations, we assume that the result represents to some degree the true nature (parameter) of the population from which the sample was chosen. The degree to which the subject statistic represents the population is expressed in the confidence of the estimate (statistic).

For example, assume that we interview 150 individuals chosen at random to find out if they are satisfied with the service provided by a local hospital and find that 27 are dissatisfied. From this data, we calculate a proportion of 0.18, or 18 percent, are dissatisfied.

While this answer is absolutely correct for the sample, it may or may not reflect the true feelings for the entire community.

As with any statistic, there is always an associated risk of error. We will focus the following discussion on error as it relates to proportions. The following four distinct cases will be considered:

1. Two-sided confidence intervals where the sample size is relatively small compared to the population, $n \leq 0.05N$
2. Two-sided confidence intervals where the sample size is relatively large compared to the population, $n > 0.05N$
3. Single-sided confidence limit where the sample size is small relative to the population, $n \leq 0.05N$
4. Single-sided confidence limit where the sample size is large relative to the population, $n > 0.05N$

Case 1: Two-sided confidence interval where the sample size is small relative to the population, $n \leq 0.05N$

In the preceding example, the population of the community that the hospital served was 100,000. The sample size was 150, and the number of individuals that responded negatively was 27. The calculated proportion nonconforming to our requirement of satisfied customers is 0.18. This 0.18 is an estimate of the true condition and, as with any estimate, there is a risk of error. The only way to totally avoid error is to poll the entire population of 100,000.

The degree to which the sample estimate reflects the truth is defined by our level of confidence. All statistics are subject to error.

The error E (or sometimes referred to as standard error SE) for the proportion P is calculated as follows:

$$E = \sqrt{\frac{p(1-p)}{n}}$$

where: p = proportion
n = sample size

In order to express a level of confidence in the estimate of the proportion, we add and subtract a specific number of standard errors to the point estimate. The greater the level of confidence, the more standard errors we add and subtract. The number of standard errors is $Z_{\alpha/2}$, where α is the risk.

When the error is added and subtracted to the estimate, we are determining a two-sided limit or confidence interval. If all of the error is applied to one side, of the estimate we refer to a single-sided confidence limit and use Z_{α}.

Confidence and risk must total 100 percent. If we are 95 percent confident, we are assuming a risk of 5 percent. If we are 90 percent confident, we are assuming a risk of 10 percent.

$$\% \text{ Confidence} + \% \text{ Risk} = 100\%$$

$$\% \text{ Risk} = 100\% - \% \text{ Confidence}$$

Confidence and risk are expressed as a proportion rather than as a percentage. If we are 90 percent confident, we are assuming a 10 percent risk that the true proportion will not be inside the limits defined by the point estimate of $\pm Z_{\alpha/2}E$:

$$p \pm Z_{\alpha/2} \sqrt{\frac{p(1-p)}{n}}$$

Note: This equation is to be used when $n \leq 0.05N$

where: p = proportion
n = sample size
$Z_{\alpha/2}$ = factor dependent on risk assumed

The multiple of errors E to be added and subtracted can be found in Table 1. The quantities $Z_{\alpha/2}$ are to be used for upper and lower confidence intervals. The use of the Z_{α} values are used in single-sided confidence limits and will be discussed later.

Table 1. Factors for Confidence Intervals and Limits, Two-Sided Limits.

Confidence C	Risk α	α/2	$Z_{\alpha/2}$
.999	0.001	0.0005	3.291
.990	0.010	0.0050	2.576
.950	0.050	0.0250	1.960
.900	0.100	0.0500	1.645

Example 1:
Calculate the 90 percent confidence interval for the previous example, where $n = 150$ and $p = 0.18$. The total population is 100,000; therefore, $n < 0.05N$.

From Table 1 for a level of confidence of 90 percent, the appropriate $Z_{\alpha/2}$ is 1.645 and using the relationship

$$p \pm Z_{\alpha/2}\sqrt{\frac{p(1-p)}{n}} \rightarrow 0.18 \pm 1.645\sqrt{\frac{0.18(1-0.18)}{150}} \Rightarrow 0.18 \pm 0.05$$

the 90 percent confidence interval is 0.13 to 0.23.

An appropriate statement would be: "I do not know the true percentage of the population that is dissatisfied with the service, but I am 90 percent confident it is between 13.0 percent and 23.0 percent."

If the error is considered excessive at $\pm 5\%$ we can reduce it by

1. Increasing the sample size n
2. Reducing the level of confidence

Example 2:
Eighty-five records are sampled out of a total of 6000. From these 85 records, 15 are found to be incorrect. What is the 95 percent confidence interval for the proportion?

$$n \leq 0.05N \text{ and } p = \frac{15}{85} = 0.176$$

The standard error of the estimate E is given by

$$E = \sqrt{\frac{p(1-p)}{n}} = \sqrt{\frac{0.176(1-0.176)}{85}}$$

$$E = 0.041$$

The confidence interval is given by

$$p \pm Z_{\alpha/2}E$$

for a 95 percent confidence, $Z_{\alpha/2} = 1.96$.

99% confidence interval $= 0.176 \pm 1.96(0.041)$

$$= 0.176 \pm 0.080$$

The upper 95 percent confidence limit is $0.176 + 0.080 = 0.256$.

The lower 95 percent confidence limit is $0.172 - 0.080 = 0.096$.

An appropriate statement would be: "We do not know the true proportion of records that are incorrect, but we are 95 percent confident it is between 9.6 percent and 25.65 percent."

Example 3:
One hundred fifty items are randomly sampled from a population of 4500. A total of six items are found to be nonconforming to the specification. What is the 99 percent confidence interval?

$$p = \frac{6}{150} = 0.040$$

$Z_{\alpha/2}$ for 99 percent confidence is 2.576.

$$E = \sqrt{\frac{p(1-p)}{n}} = \sqrt{\frac{0.040(1-0.040)}{150}}$$

$$E = 0.016$$

99 percent confidence interval $= 0.040 \pm 2.576(0.014)$

$$= 0.040 \pm 0.041$$

The upper 99 percent confidence limit is $0.040 + 0.041 = 0.081$.

The lower 99 percent confidence limit is $0.040 - 0.041 = -0.001 \Rightarrow 0.000$.

When a negative lower control limit results, default to zero.

Case 2: Two-sided confidence interval where the sample size is large relative to the population, $n > 0.05N$

When the sample size n is greater than $0.05N$, a correction factor must be applied to the standard error. This correction factor is called the finite population correction factor. The finite population factor is given by

$$\sqrt{\frac{N-n}{N-1}} = \sqrt{1 - \frac{n}{N}}$$

The standard error must be multiplied by this correction factor anytime the sample size n is greater than $0.05N$.

The standard error for those cases now becomes

$$E = \sqrt{\frac{p(1-p)}{n}} \sqrt{1 - \frac{n}{N}}$$

The confidence interval is now determined by

$$p \pm Z_{\alpha/2} \sqrt{\frac{p(1-p)}{n}} \sqrt{1 - \frac{n}{N}}$$

Example 1:
What is the 90 percent confidence interval for the proportional defective where the sample size $n = 50$, the population $N = 350$, and the number defective $np = 12$.

$$p = \frac{12}{50} = 0.24$$

$$Z_{\alpha/2} = 1.645$$

$$E = \sqrt{\frac{p(1-p)}{n}} \sqrt{1 - \frac{n}{N}} \quad E = \sqrt{\frac{0.24(1-0.24)}{50}} \sqrt{1 - \frac{50}{350}}$$

$$E = 0.056$$

90% confidence interval = $0.24 \pm 1.645(0.056)$

$$= 0.24 \pm 0.092$$

The 90% upper confidence limit is $0.24 + 0.092 = 0.332$.

The 90% lower confidence limit is $0.24 - 0.092 = 0.148$.

Example 2:
A small engine-repair company serviced 6800 customers last quarter. A survey was sent to the entire customer base, and of the 1500 who responded, 60 indicated that they felt the length of time for engine repairs was excessively long. What is the 95 percent confidence interval for the resulting percent of dissatisfied customers?

$$N = 6800$$

$$n = 1500$$

$$np = 60$$

$$p = 0.04$$

$$C = 95\%$$

The sample size is 22 percent of the population, which exceeds the criteria of $n \geq 0.05\ N$; therefore, we must apply the finite population correction factor to the error calculation.

$$Z_{\alpha/2} = 1.960$$

$$E = \sqrt{\frac{p(1-p)}{n}} \sqrt{1 - \frac{n}{N}} \quad E = \sqrt{\frac{0.04(1-0.04)}{1500}} \sqrt{1 - \frac{1500}{6800}} \quad E = 0.004$$

The confidence interval is

$$p = Z_{\alpha/2}E = 0.040 \pm (1.96)(0.004) = 0.040 \pm 0.008$$

The upper 95 percent confidence limit is $0.040 + 0.008 = 0.048$.

The lower 95 percent confidence limit is $0.040 - 0.008 = 0.032$.

Case 3: Single-sided confidence interval where the sample size is small relative to the population, $n \leq 0.05\,N$

and

Case 4: Single-sided confidence interval where the sample size is large relative to the population, $n > 0.05\,N$

In all of the previous cases, we have been dividing the error of the estimate and applying it to the upper and lower limits to establish a confidence interval. If all of the error is associated with either the upper or lower end of a confidence interval, we are calculating a confidence limit rather than a confidence interval. The only parameter that changes is the term $Z_{\alpha/2}$, which is replaced with $Z\alpha$.

With confidence intervals, an appropriate statement would be: "I don't know the true percent nonconforming, but I am 90 percent confident it is between 12.7 percent and 18.5 percent."

With a single-sided confidence limit, or *simple confidence limit,* a similar statement would be: "I do not know the true percent nonconforming, but I am 95 percent confident it is not less than 23.5 percent."

confidence limit = $p + Z_{\alpha/2}E$ or confidence limit = $p - Z_{\alpha/2}E$

A list of selected Z_α can be found in Table 2.

Table 2. Factors for Single-Sided Confidence Limits.

Confidence *C*	Risk α	Z_α
0.999	0.001	3.090
0.990	0.010	2.326
0.950	0.050	1.645
0.900	0.100	1.282

Example 1:

A sample of 90 individuals are interviewed during the week as they are leaving the "Le Cash, Le Penty" restaurant. During the same week, 3500 customers exited the restaurant. Of those interviewed, 28 percent feel that the service was a little too slow. What is the upper 90 percent confidence limit for the estimated percent of dissatisfied customers?

Since $n < 0.05N$, the finite population correction factor is not required.

$$p + Z_\alpha \sqrt{\frac{p(1-p)}{n}} \quad 0.28 + 1.282 \sqrt{\frac{0.28(1-0.28)}{90}}$$

The upper 90 percent confidence limit is $0.28 + 0.06 = 0.34$.

An appropriate statement would be: "I do not know the true percentage of customers that feel that the service is too slow, but I am 90 percent confident that it is not greater than 34 percent."

Example 2:

What is the lower 95 percent confidence limit for a proportion if the total population $N = 1500$, the sample size $n = 400$, and the number of nonconforming is 36?

Since the sample size exceeds $0.05N$, the finite population correction factor must be used.

$$p = \frac{36}{400} = 0.09 \quad Z_\alpha = 1.645$$

$$p - Z_\alpha \sqrt{\frac{p(1-p)}{n}} \sqrt{1 - \frac{n}{N}} \Rightarrow 0.09 - 1.645 \sqrt{\frac{0.090(1-0.090)}{400}} \sqrt{1 - \frac{400}{1500}}$$

The lower 95 percent confidence limit is $0.090 - 0.020 = 0.070$.

An appropriate statement would be: "I do not know the true proportion, but I am 95 percent confident that it is not less than 0.070."

Additional problems:

Calculate the confidence intervals for each of the following:

$N = 12,000$	$N = 1800$	$N = 50,500$
$n = 120$	$n = 200$	$n = 160$
$np = 7$	$np = 12$	$np = 42$
$C = 90\%$	$C = 95\%$	$C = 99\%$

Sample size required given a level of confidence C and acceptable error *SE*

Up to this point, we have been determining the amount of error associated with a given level of confidence with the sample size specified.

From a more practical matter, we would like to know what sample size would be required to assure a level of confidence with a predetermined amount of acceptable error. This can be accomplished by simply rearranging the equations for the error and solving for the sample size n.

There are two cases to consider as follows:

1. Sample size n, where $n \leq 0.05\ N$:

$$n = \frac{(Z_{\alpha/2})^2\ p(1-p)}{(E)^2}$$

where: p = sample proportion
$Z_{\alpha/2}$ = factor for two-sided confidence interval with a level of confidence of $1 - \alpha$
E = acceptable error

2. Sample size $n > 0.05N$:

$$n = \frac{p(1-p)}{\left(\dfrac{E}{Z_{\alpha/2}}\right)^2 + \left(\dfrac{p(1-p)}{N}\right)}$$

where: p = sample proportion
$Z_{\alpha/2}$ = factor for two-sided confidence interval with a level of confidence of $1 - \alpha$
E = acceptable error

If we are calculating a confidence limit (single sided), then use Z_{α} rather than $Z_{\alpha/2}$.

The only obstacle in using these two relationships to solve for the correct sample size n is that we need to use the quantity p in the calculation, and it is the determination of p for which we want to know n.

We may use the following options:

1. Let $p = 0.50$. This will give us the absolute maximum sample size that we ever require. This sample size may be more than required.

   Since we do not know what percentage the sample will be of the population of $N = 3200$, we will assume that it will be less than 5 percent of the population. If the resulting sample is greater than or equal to 5 percent, the sample size will be recalculated.

2. Do a preliminary sampling to provide an initial estimate of the proportion p. This initial sampling should be sufficiently large as to provide at least two or three nonconforming units.

3. Use a previous sampling or survey data.

Example 1:
You have been asked to develop a customer-satisfaction survey to determine the percentage of customers dissatisfied with a certain aspect of a service. The total number of customers that receive service during the subject period of time was $N = 3200$. The level of confidence you desire is 95 percent, and the amount of error you are willing to tolerate is $E = \pm0.03$. A similar survey was conducted last quarter, and the results of that survey showed that 15 percent of the customers were dissatisfied. What is the correct sample size for the new survey?

$$n = \frac{(Z_{\alpha/2})^2\, p(1-p)}{(E)^2} \qquad n = \frac{(1.960)^2\, 0.15(1-0.15)}{(0.03)^2} \qquad n = 544$$

To test if the finite population factor is required, we determine the percentage of the population the sample represents:

$$\frac{544}{3200} = 0.17$$

Since $0.17 \geq 0.05$, we should recalculate using the correction factor version:

$$n = \frac{p(1-p)}{\left(\dfrac{E}{Z_{\alpha/2}}\right)^2 + \left(\dfrac{p(1-p)}{N}\right)} \qquad n = \frac{0.15(1-0.15)}{\left(\dfrac{0.03}{1.960}\right)^2 + \left(\dfrac{0.15(1-0.15)}{3200}\right)} \qquad n = 46$$

A sample size of 465 should be taken. We can be 95 percent confident that the resulting proportion will be within ± 0.03 of the true proportion.

Example 2:
What is the required sample size n for the two-sided confidence interval given

Confidence $C = .99$ Error $E = 0.05$ Population $N = 4500$ Estimate of $p = 0.05$

Should the finite population correction factor be applied for this case?

BIBLIOGRAPHY

Dovich, R. A. 1992. *Quality Engineering Statistics.* Milwaukee, WI: ASQC Quality Press.
Juran, J. M., and F. M. Gryna. 1980. *Quality Planning and Analysis.* 3rd edition. New York, NY: McGraw-Hill.
Petruccelli, J. D., B. Nandram, and M. Chen. 1999. *Applied Statistics for Engineers and Scientists.* Upper Saddle River, NJ: Prentice Hall.

CONFIDENCE INTERVAL FOR STANDARD DEVIATION

The standard deviation is one of several variation statistics (those descriptive statistics that measure variation). Others include the range and maximum absolute deviation. The square of the standard deviation is the variance. For each continuous and discrete distribution, there is an associated variance and standard deviation. For the normal continuous distribution, we can determine the sample standard deviation from the relationship

$$S = \sqrt{\frac{\Sigma(X - \bar{X})^2}{n-1}}$$

where: X = individual value
$\bar{X}$ = average of individuals
n = sample size
Σ = sum

The standard deviation can be described as a standard way of measuring the deviation (variation) of the individuals from their average.

As with all statistics, the standard deviation is subject to error. The error for the sample standard deviation is calculated using the chi-square distribution,

where: n = sample size
s = sample standard deviation
σ = population standard deviation (unknown)

Cumulative probabilities for the chi-square distribution can be determined as a function of the area associated with various tail areas designated as $\alpha/2$ for the upper portion of the distribution and $1 - \alpha/2$ for the lower portion of the distribution. α is the probability of risk.

Derive the confidence interval for the standard deviation by

$$X^2_{1-\alpha/2} < \frac{(n-1)s^2}{\sigma^2} < X^2_{\alpha/2}$$

We rewrite as follows so that σ^2 is isolated in the middle term:

$$\frac{X^2_{1-\alpha/2}}{(n-1)s^2} < \frac{1}{\sigma^2} < \frac{X^2_{\alpha/2}}{(n-1)s^2}$$

Taking the reciprocal of the three terms

$$\frac{(n-1)s^2}{X^2{}_{1-\alpha/2}} > \sigma^2 > \frac{(n-1)s^2}{X^2{}_{\alpha/2}}$$

Taking the reciprocals reverses the direction of the inequalities. If we reverse the order of the terms, the result is now

$$\frac{(n-1)s^2}{\chi^2{}_{\alpha/2}} < \sigma^2 < \frac{(n-1)s^2}{X^2{}_{1-\alpha/2}}$$

This expression represents the $100(1 - \alpha)$ percent confidence interval for the variance. Taking the square root of the three terms gives the confidence interval for the standard deviation.

$$\sqrt{\frac{(n-1)s^2}{\chi^2{}_{\alpha/2}}} < \sigma < \sqrt{\frac{(n-1)s^2}{X^2{}_{1-\alpha/2}}}$$

Example of confidence interval calculation:

As part of a study for variation reduction in an injection molding operation, the quality engineer has taken a sample of 30 parts (all of which have the same target value) and determined the standard deviation s to be 0.0358. What is the 90 percent confidence interval for the standard deviation?

$n = 30$

$s = 0.0358$

$C = .90$

$\alpha = .10$

$$\sqrt{\frac{(n-1)s^2}{x^2{}_{a/2}}} < \sigma < \sqrt{\frac{(n-1)s^2}{X^2{}_{1-a/2}}}$$

$$\sqrt{\frac{(29)(0.0358)^2}{x^2{}_{0.05,29}}} < \sigma < \sqrt{\frac{(29)(0.0358)^2}{X^2{}_{0.95,29}}}$$

Looking up the chi-square value in a chi-square table using $n - 1$ degrees of freedom, we find

$$X^2{}_{0.05,29} = 42.6 \text{ and } X^2{}_{0.95,29} = 17.7$$

$$\sqrt{\frac{(29)(0.0358)^2}{42.6}} < \sigma < \sqrt{\frac{(29)(0.0358)^2}{17.7}}$$

$$0.0295 < \sigma < 0.0458$$

We are 90 percent confident that the true standard deviation σ is between 0.0295 and 0.0458.

Notice that the interval is nonsymmetrical with respect to the point estimate of sigma $s = 0.0358$.

The difference between the lower 90 percent limit and the point estimate is 0.0043, where the difference between the point estimate and the upper 90 percent confidence limit is 0.0058.

Testing Homogeneity of Variances

On occasion we might want to know if the standard deviations are equal for several populations. One such case might be the use of the deviation from nominal as a response for a short-run SPC control chart. It would be appropriate to use this response provided that the ability to produce at a *target* were the same for all products being made. If the variation from target were different for one product than another, then the population standard deviations (and hence the variances) would also be different. A method for determining if the population variances are different for several populations is the method developed by M. S. Bartlett, which is referred to as the "Bartlett's test."

The derived statistic $K^2{}_{k-1}$ is approximately distributed as X^2 with $k - 1$ degrees of freedom, where k = the number of populations compared.

The Bartlett's test in this discussion requires that the sample sizes all be equal. The K statistic is calculated as follows:

$$K^2{}_{k-1} = \frac{\left[k(n-1)\ln\left(\sum \frac{S^2}{k} \right) - (n-1)\sum \ln s^2 \right]}{C}$$

$$C = 1 + \frac{k+1}{3k(n-1)}$$

where: k = number of populations or samples
n = sample size (same for all k samples)

Example:
The objective is to establish a short-run SPC control chart based on the deviation from nominal for several products. The specific characteristic will be the melt index.

The following table gives five measurements (sample size, $n = 5$) for each of nine products ($k = 9$).

Product	\multicolumn{5}{c	}{Sample number}			
	1	2	3	4	5	S	S^2	$\ln S^2$
A	123	127	132	125	126	3.362	11.3	2.425
B	48	50	55	46	50	3.347	11.2	2.416
C	228	230	225	229	222	3.271	10.7	2.370
D	245	247	249	239	241	4.147	17.2	2.845
E	67	70	73	65	71	3.194	10.2	2.322
F	155	150	143	160	162	7.714	59.5	4.086
G	88	89	81	76	92	6.535	42.7	3.754
H	101	113	96	115	106	7.981	63.7	4.154
I	132	132	137	138	130	3.493	12.2	2.501

$$\sum \ln S^2 = 26.873 \quad \sum \frac{S^2}{k} = 26.522$$

$$C = 1 + \frac{k+1}{3k(n-1)} \quad C = 1 + \frac{9+1}{27(5-1)} = 1.0926$$

where: $k = 9 \quad n = 5$

$$K^2{}_{k-1} = \frac{\left[k(n-1)\ln\left(\sum \frac{S^2}{k}\right) - (n-1)\sum \ln s^2 \right]}{C}$$

$$K^2{}_{k-1} = \frac{\left[9(4)\ln(26.522) - (4)(26.873) \right]}{1.0926} = 9.624$$

$$K^2{}_8 = 9.624 \text{ vs } x^2{}_{8,0.05} = 15.51$$

Since the calculated K^2 does not exceed $\chi^2_{8,\,0.05}$ homogeneity (constant or equal variance) is a tenable hypothesis. A deviation from nominal control chart would be plausible.

BIBLIOGRAPHY

Burr, E. W. 1974. *Applied Statistical Methods.* New York: Academic Press.

Dovich, R. A. 1992. *Quality Engineering Statistics.* Milwaukee, WI: ASQC Quality Press.

Petruccelli, J. D., B. Nandram, and M. Chen 1999. *Applied Statistics for Engineers and Scientists.* Upper Saddle River, NJ: Prentice Hall.

DEFECTS/UNIT CONTROL CHARTS

The selection of control charts depends on the following two factors:

1. Sample size: The sample size can vary in size or remain fixed.
2. Type of data collection: The data collected from the sample can be either a total number of defective units np in the sample or the total number of defects c.

If we want the opportunity to vary the sample size and we want to tally the number of defects, the resulting control chart will be the defects/unit control chart, or u chart.

The central tendency for the u chart is the average number of defects per unit for all of the samples used to provide the initial process characterization.

$$\text{Average defects per unit, } \bar{u} = \frac{\text{Total defects in } k \text{ samples}}{\text{Total number of units in } k \text{ samples}}$$

Examples:
Thirty-five CDs were inspected, and three defects were observed. The average number of defects per unit is

$$\bar{u} = \frac{3}{35} = 0.086$$

Fifty-eight customer-satisfaction questionnaires were returned. Each questionnaire had 12 response areas. Each question had a choice of five responses: poor, below average, average, above average, and excellent. A response of less than average was deemed *nonconformance* or a *defect*. A total of four defects were noted in the 58 returned questionnaires. What is the average defects per unit?

$$\bar{u} = \frac{4}{58} = 0.069$$

Normal variation for the average number of defects per unit u is defined as ± 3 standard deviations about the average defects per unit. The standard deviation for defects is based on the Poisson distribution and is dependent on the average defects per unit and the sample size.

The limits of this normal variation define the upper control limits (UCLs) and lower control limits (LCLs) for the u chart. Processes that are well behaved and statistically stable will have 99.7 percent of their data within these limits. That is to say that the UCLs and LCLs for the u chart are determined by

$$\text{UCL} = \bar{u} + 3\sqrt{\frac{\bar{u}}{n}} \ \text{ and } \ \text{LCL} = \bar{u} - 3\sqrt{\frac{\bar{u}}{n}}$$

Note: The control limits are different for different sample sizes.

Example:
The average number of nonconformities per unit $\bar{u}$ for customers checking out of a hotel has been determined to be 0.055. This value was determined from exit surveys during a three-month period. If the daily sample size is 25, what are the UCLs and LCLs for this sample size?

$$\text{UCL} = \bar{u} + 3\sqrt{\frac{\bar{u}}{n}} \quad \text{UCL} = 0.055 + 3\sqrt{\frac{0.055}{25}} = 0.196$$

$$\text{LCL} = \bar{u} - 3\sqrt{\frac{\bar{u}}{n}} \quad \text{LCL} = 0.055 - 3\sqrt{\frac{0.055}{25}} = -0.086 \equiv 0.000$$

Since the LCL is negative, we default to a zero control limit. The following steps to producing a u chart are the same as with any control chart for attributes:

Step 1. Collect historical data to provide a basis of process characterization.

Control charts are used to detect changes in processes. Before we can detect a change, we need to know how the process was performing in the past. Typically, 25 samples k of size n will be collected. The sample size, or subgroup sample size as it is frequently called, should be sufficiently large as to contain at least two or three defects. For each sample, the number of defects per unit will be determined.

Step 2. Determine a location statistic using all of the historical data.

For the u chart, this location statistic will be the average defects per unit of inspection or sampling. This statistic is determined by dividing the sum of all defects found by the sum of all samples taken to characterize the process.

$$\bar{u} = \frac{c_1 + c_2 + c_3 + \dots c_k}{n_1 + n_2 + n_3 + \dots n_k}$$

where: c = number of defects
n = sample size

Step 3. Calculate control limits based on the process average ±3 standard deviations.

Step 4. Construct the chart, and plot historical data.

Traditionally, average lines are drawn as solid lines and control limits are drawn using broken lines. The vertical plotting scale should be selected such that adequate room remains to plot future data points that might fall outside the control limits. Review the plotted points, looking for statistical stability and evidence that the process was under a state of statistical control (few or no points outside the control limits and no statistically rare patterns, or *rule violations*).

Step 5. Continue to collect and plot data into the future, looking for evidence of a process change.

Change is evident if any of the SPC *rule violations* have occurred including points outside the UCL or LCL. These rule violations will be discussed later.

CASE STUDY

As manager of a hotel, you want to establish a *u* chart to monitor the room-preparation activities. After a guest checks out, the rooms are cleaned and prepared for the next guest. An audit check sheet has been prepared with several items that relate to this preparation. The following nonconformities list will be used:

1. Ground fault failed test
2. Too few towels, face cloths in bathroom
3. Covers on bed not smooth or even
4. Dust on furniture
5. Floors dirty

6. No toiletries in bathroom
7. No house cleaning I.D./welcome card left
8. Bathroom not adequately cleaned

Each day for three weeks, 100 percent of the rooms have been audited for the eight items listed. Develop a *u* chart based on the data provided. In this example, only 21 samples will be used to characterize the process.

Step 1. Collect historical data.

Sample number	Date	Sample size n	Number of defects c	Defects/unit
1	11/1	75	6	0.080
2	11/2	103	8	0.075
3	11/3	53	2	0.038
4	11/4	68	6	0.088
5	11/5	88	5	0.057
6	11/6	115	11	0.096
7	11/7	90	5	0.056
8	11/8	48	2	0.042
9	11/9	65	5	0.077
10	11/10	100	3	0.030
11	11/11	85	6	0.071
12	11/12	75	4	0.053
13	11/13	55	2	0.036
14	11/14	90	5	0.056
15	11/15	60	2	0.033
16	11/16	95	7	0.074
17	11/17	70	6	0.086
18	11/18	105	4	0.038
19	11/19	87	4	0.046
20	11/20	92	6	0.065
21	11/21	70	3	0.043

Step 2. Determine the location statistic $\bar{u}$.

$$\bar{u} = \frac{c_1 + c_2 + c_3 + \ldots c_k}{n_1 + n_2 + n_3 + \ldots n_k} = \frac{6 + 8 + 2 + 6 + \ldots 3}{75 + 103 + 53 + 68 + \ldots 70} = 0.060$$

Step 3. Calculate control limits.

The statistical control limits are based on the average defects per unit for the historical data ±3 standard deviations. The three standard deviations are calculated using the relationship $3\sqrt{\dfrac{\bar{u}}{n}}$, where n = sample size. Technically speaking, every time the sample size changes, the standard deviation changes. From a practical perspective, the average sample size may be used in place of individual samples when the individual sample size is within the limits of $0.75\bar{n}$ to $1.25\bar{n}$. The average sample size for this example is $\bar{n} = 80.4$. For any samples falling within the limits of 60 to 101, we may use the average sample size of 80.4 for calculation of the three standard deviations. With the exception of samples 2, 3, 6, 8, 13, and 18, we may use a sample size of 80.4 for the standard deviation calculation without significantly affecting the chart.

A. Control limits for sample sizes 60 to 101 inclusive:

Let $\bar{n} = 80.4$

$$\text{UCL} = 0.060 + 3\sqrt{\frac{0.060}{80.4}} = 0.060 + 0.082 = 0.142$$

$$\text{LCL} = 0.060 - 3\sqrt{\frac{0.060}{80.4}} = 0.060 - 0.082 = -0.022 = 0.000$$

B. Control limits for remaining samples:

Sample #2, n = 103

Note: The LCLs will not be calculated. A default LCL value of zero will be used, since all LCLs using a process average defects/unit of 0.060 will yield a negative control limit when the sample size is less than 146. As a matter of fact, the relationship of average defects per unit and sample size give the following relationship.

For a given average defects/unit u, the minimum sample size to give a zero LCL is defined by the relationship

$$\frac{9\bar{u}}{(\bar{u})^2}$$

For example, a process with an average number of defects/unit of 0.15 would require a sample size of

$$n = \frac{9\bar{u}}{(\bar{u})^2} = \frac{(9)(0.15)}{(0.15)^2} = 60$$

The actual sample size should be a little larger than $n = 60$ such that the LCL will be slightly larger than zero. It is suggested that the sample size be increased approximately 10 percent to, for example, $n = 66$.

$$\text{UCL} = 0.060 + 3\sqrt{\frac{0.060}{103}} = 0.060 + 0.072 = 0.132$$

Sample #3, n = 53

$$\text{UCL} = 0.060 + 3\sqrt{\frac{0.060}{53}} = 0.060 + 0.101 = 0.161$$

Sample #6, n = 115

$$\text{UCL} = 0.060 + 3\sqrt{\frac{0.060}{115}} = 0.060 + 0.069 = 0.129$$

The remaining samples may be calculated in a similar manner. A summary of the historical data and UCLs follows. (All LCLs defaulted to 0.000.)

Sample number	Date	Sample size n	Number of defects c	Defects/unit	UCL
1	11/1	75	6	0.080	0.142
2	11/2	103	8	0.075	0.132
3	11/3	53	2	0.038	0.161
4	11/4	68	6	0.088	0.142
5	11/5	88	5	0.057	0.142
6	11/6	115	11	0.096	0.129
7	11/7	90	5	0.056	0.142
8	11/8	48	2	0.042	0.167
9	11/9	65	5	0.077	0.142
10	11/10	100	3	0.030	0.142
11	11/11	85	6	0.071	0.142
12	11/12	75	4	0.053	0.142
13	11/13	55	2	0.036	0.159
14	11/14	90	5	0.056	0.142
15	11/15	60	2	0.033	0.142
16	11/16	95	7	0.074	0.142
17	11/17	70	6	0.086	0.142
18	11/18	105	4	0.038	0.132
19	11/19	87	4	0.046	0.142
20	11/20	92	6	0.065	0.142
21	11/21	70	3	0.043	0.142

Notice that as the sample size increases, the control limits become closer to the average line. This increases the probability of detecting a process change as indicated by a point outside the control limit.

Step 4. Construct the chart, and plot historical data.

The control chart with the initial 21 data points plotted and the control limits with the process average defects per unit follows. Note the variable control limits for those data points resulting when the sample size exceeds the interval of $= 0.75\overline{n}$ to $1.25\overline{n}$. The process average defects line is drawn as a solid line and extends into the future beyond the last historical sample. This is done because it is this historical average to which we want to compare all future sample results.

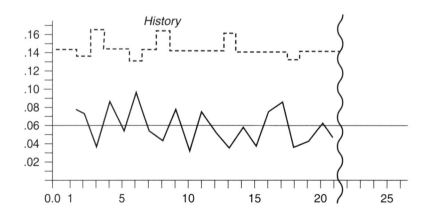

Step 5. Continue to collect and plot data.

A review of the historical data indicates a stable, in-control process. This conclusion is supported by no violations of the SPC rule violations. A detailed discussion of these *rules* can be found in the module entitled *SPC Chart Interpretation.*

On November 22, a quality-improvement team decided to address the issue of room preparation in an effort to improve the process. Each member of the cleaning and preparation staff was given a report of the more frequently occurring nonconformance and were also given more training. Each member of the staff was given a checklist that provided a sign-off to assist in making sure that all requirements were addressed. The following data were collected. Continue to collect data for the process, and plot the data. Do you feel that the improvement effort was beneficial?

Sample number	Date	Sample size *n*	Number of defects *c*	Defect/unit *u*
22	11/22	105	11	0.105
23	11/23	90	12	0.130
24	11/24	58	2	0.034
25	11/25	88	4	0.045
26	11/26	95	1	0.011
27	11/27	75	2	0.027
28	11/28	55	2	0.036
29	11/29	93	4	0.043

Upon completion of the control chart, it appears that the process average has decreased relative to the historical value of .060 defects/unit. This conclusion is supported by the violation of rule #2: seven consecutive points below the historical average.

Completed *u* chart:

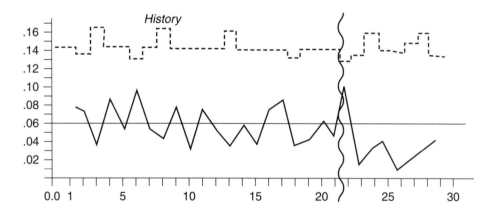

Optional problem:

The Stockton Plastics Corporation (SPC) has been collecting data for approximately one week on the frequency and nature of defects that occur in the injection molding department. Based on this data, do you feel that the process is in control, and where would you focus your attention to improve the process?

Defect list:

Characteristic	Defect code	Cost to repair, $
Cracked parts	23	8.00
Splay	44	1.25
Undersize	100	11.50
Chips	58	0.80
Flash	168	3.25
Color	135	0.75
Under Weight	75	2.00
Plagimitized	80	3.60

Inspection data from 100 percent inspection:

Date	Lot size n	Total defects	Type and frequency of defects							
			23	44	100	58	168	135	75	80
11/1	300	5	1	0	2	1	0	1	0	0
11/1	400	12	3	2	3	0	1	0	0	0
11/2	400	8	0	2	3	5	0	0	0	0
11/2	580	15	2	4	3	0	4	0	1	1
11/2	350	14	0	0	2	9	1	2	0	0
11/3	400	22	4	5	5	7	0	1	0	0
11/4	500	10	1	4	3	2	0	0	0	0
11/5	280	14	2	2	2	6	1	1	0	0
11/5	500	25	5	4	3	5	3	2	2	1
11/6	350	7	3	3	0	0	0	0	0	1
11/7	400	24	0	0	4	9	5	0	2	4
11/7	500	18	1	4	4	6	3	2	1	0

What is the most frequently occurring defect, and where would you focus your attention to improve quality? Why?

BIBLIOGRAPHY

Besterfield, D. H. 1994. *Quality Control.* 4th edition. Englewood Cliffs, NJ: Prentice Hall.

Grant, E. L, and R. S. Leavenworth. 1996. *Statistical Quality Control.* 7th edition. New York: McGraw-Hill.

Montgomery, D. C. 1996. *Introduction to Statistical Quality Control.* 3rd edition. New York: John Wiley and Sons.

Wheeler, D. J., and D. S. Chambers. 1992. *Understanding Statistical Process Control.* 2nd edition. Knoxville, TN: SPC Press.

DEMERIT/UNIT CHART

Control charts for attributes are broadly classified into the following four types:

1. Defective with constant sample size np
2. Defective with variable sample size p
3. Defects with constant sample size c
4. Defects with variable sample size u

For both the c chart and u chart there is a presumption that all defects are equal, or at least the severity of the nature of the defect will not influence behavior of the control chart. As a point of practical concern, we know that in some cases, defects are not always equal in their impact on the quality of the product or performance of the service to which they are associated. For example, in the audit of a sample of hotel rooms, if one finds that no hot water is available versus having two towels where three are required, the resulting level of customer satisfaction is significantly different. Another example would be the results of the final audit of the readiness of a New York to Seattle commercial flight where three defects were found: (1) inadequate pillows relative to the specification for the flight, (2) a shortage of 25 meals for the flight, and (3) in-flight movie equipment not working. With some imagination, defects in most operations and services can be divided into two or more classification levels with respect to the severity of the defects. By categorizing defects into classifications and giving a relative weighting scale to specific defects, we can develop a control chart that will not be overly sensitive to minor defects or under sensitive to critical defects. Such a treatment of defects will lead to the use of what is commonly referred to as a demerit/unit control chart. The following example will illustrate the construction of the demerit/unit, or Du, control chart.

Essential operations for the Du control chart include:

1. Develop a list of defects that are categorized into classes of severity.
2. Collect historical data to characterize the process.
3. Determine the average demerits/unit and the associated standard deviation.
4. Calculate control limits based on the average ±3 standard deviations.
5. Construct the control chart with upper and lower limits, and plot the data.
6. Continue to monitor the process, looking for evidence of process changes.

The following illustration will serve as a demonstration for the construction of the Du control chart.

Step 1. Develop defect list and classifications.

Hotel rooms are audited to check compliance to a set of standard requirements. Nonconformances to the requirement are classified as critical, major, and minor. Each defect is assigned demerits according to the classification. A critical defect counts as 100 demerits, a major as 25 demerits, and a minor as 5 demerits. The number of demerits per classification is a management decision related to factors such as potential liability, cost to repair, or impact on the ability of the service to be rendered adequately or the product to perform its intended mission.

Using the scale of critical = 100, major = 25, and minor = 5, we are allowing a critical defect to have 20× the impact on the control as a minor and 4× the impact as a major. More categories of classification may be used, but three is a reasonable number to use in most cases.

Specific defects are as follows:

Critical	*Minor*	*Major*
Ground fault detector malfunction	Missing mint on bed in evening	No towels
TV not working	Dust on furniture	No telephone directory
No hot water	Drawer moderately difficult to open	No shampoo
	No motel stationery	Drapery drawstring broken
	Motel phone number missing from room phone	

Step 2. Collect historical data for process characterization.

Approximately 50 sets of samples as a minimum are taken to characterize the process for the period to be used as a reference time period. Each day, 40 rooms are audited after the housekeeping department has prepared the rooms for new guests.

A tally sheet is maintained for the audit results summary. This record sheet will contain a count of the total critical, major, and minor defects; the computed total defects for each sample of size *n;* and the computed number of demerits per unit of inspection. The demerits per unit are determined for each sample by dividing the total demerits for that sample by the total units inspected for that sample.

The total demerits are calculated for Sample #1 as follows:

$$(1 \text{ critical} \times 100 \text{ demerits} + (2 \text{ major} \times 25 \text{ demerits}) + (6 \text{ minor} \times 5 \text{ demerits})$$

$$\text{Total demerits for sample one} = 180$$

The demerits per unit are equal to the total demerits divided by the total units inspected:

$$180/40 = 4.50 \text{ demerits/unit}$$

Table 1. Completed Tally Record.

Sample number:	Size n	Date	Criticals	Majors	Minors	Total demerits	Demerits/ unit
1	40	7/7/94	1	2	6	180	4.50
2	40	7/8/94	0	3	2	85	2.13
3	40	7/9/94	1	3	9	220	5.50
4	40	7/10/94	1	2	4	170	4.25
5	40	7/11/94	2	0	3	215	5.75
6	40	7/12/94	0	4	6	130	3.25
7	40	7/13/94	0	3	5	100	2.50
8	40	7/14/94	0	1	6	55	1.38
9	40	7/15/94	1	2	5	175	4.38
10	40	7/16/94	0	1	4	45	1.13
11	40	7/17/94	0	3	4	95	2.38
12	40	7/18/94	1	4	6	230	5.75
13	40	7/19/94	1	0	8	140	3.50
14	40	7/20/94	0	3	7	110	2.75
15	40	7/21/94	0	1	5	50	1.25
Totals:	600		8	32	80	2000	3.33

Step 3. Calculate the average demerits/unit and three standard deviations.

The binomial distribution is used to model the Du control chart. The standard deviation for the binomial is given by $\sqrt{npq}$, where n = sample size, p = fractional defective, and $q = 1 - p$.

Rearranging this equation based on a proportional defective rather than a number defective, the standard deviation is equal to $\sqrt{\dfrac{\bar{p}(1-\bar{p})}{n}}$.

The average fractional defective term $\bar{p}$ varies from one type of defect to another. Of the 600 samples inspected (rooms audited), there are 8 defects that are critical, 32 that are major, and 80 that are minor. In order to combine the effects of these defects and allow a relative weighting in accordance with the demerit assignment, we must weigh the calculation of the standard deviation. This is accomplished by calculating a weighted average standard deviation, σDu:

$$\sigma_{Du} = W_c \sqrt{\frac{p_c(1-p_c)}{n}} + W_{ma}\sqrt{\frac{p_{ma}(1-p_{ma})}{n}} + W_{mi}\sqrt{\frac{p_{mi}(1-p_{mi})}{n}}$$

where: W = demerit assignment for classified defects
p = fractional defects

Rearranging, we have

$$\sigma_{Du} = \sqrt{\frac{W_c^2 p_c (1 - p_c) + W_{ma}^2 p_{ma} (1 - p_{ma}) + W_{mi}^2 p_{mi} (1 - p_{mi})}{n}}$$

p_c = the fractional contribution of the critical defects and is equal to the total number of critical defects found divided by the total number of units inspected:

$$p_c = \frac{8}{600} = 0.013$$

In a similar manner, we calculated the fractional contribution of the major and minor defects:

$$p_{ma} = \frac{32}{600} = 0.053 \quad \text{and} \quad p_{mi} = \frac{80}{600} = 0.133$$

Using the following relationship, we compute the standard deviation:

$$\sigma_{Du} = \sqrt{\frac{W_c^2 p_c (1 - p_c) + W_{ma}^2 p_{ma} (1 - p_{ma}) + W_{mi}^2 p_{mi} (1 - p_{mi})}{n}}$$

$$W_c = 100 \qquad W_{ma} = 25 \qquad W_{mi} = 5$$
$$p_c = 0.013 \qquad p_{ma} = 0.053 \qquad p_{mi} = 0.133$$

$$\sigma_{Du} = \sqrt{\frac{128.31 + 31.37 + 2.44}{40}} = \sqrt{\frac{162.12}{40}} = 2.01$$

Three standard deviations = $3 \times 2.01 = 6.03$

$$\text{The average number of demerits / unit} = \frac{\text{Total demerits}}{\text{Total units inspected}} = \frac{2000}{600} = 3.33$$

Step 4. Determine control limits.

All Shewhart control chart limits are based on an average response ±3 standard deviations. We expect to find approximately 99.7 percent of the data points within this range.

Upper Control Limit (UCL) = $3.33 + 6.03 = 9.36$

Lower Control Limit (LCL) = $3.33 - 6.03 = -2.70$

A negative control limit is meaningless; therefore, we default to 0.00 for the LCL.

Step 5. Construct the control chart with upper and lower control limits and plot data.

Locate the control limits and label them. Draw a broken line for the control limits, and extend it halfway between the vertical lines that represent the 15th sample and the 16th sample. Draw a solid line for the average, and extend it to the end of the chart.

Record all information in the data area, including date; sample size; total demerits; demerits/unit; and number of critical, major, and minor defects found in the sample.

A wavy vertical line is drawn between sample #15 and sample #16 to separate the historical data that were used to develop the control chart average and control limits from future values. Label the left portion of the control chart as *History*.

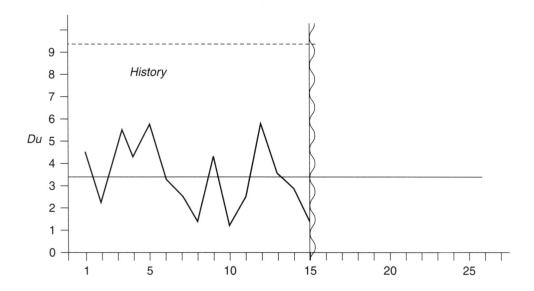

Step 6. Continue to monitor the process, looking for changes in the future.

The process appears to be statistically stable in that approximately half of the points are distributed above the average and half below, and runs of length seven are not present. Data collected in the future can be used to detect a shift in the process average. Evidence of a process change are indicated when:

1. There is a run of seven points in a row on one side of the average
2. A single point is present outside the control limits
3. Seven points are steadily increasing or decreasing

These three indicators or rules are statistically rare with respect to their probability of occurrence for stable process. They are sufficiently rare enough that when these patterns do occur, we assume that a process change has taken place. Using these rules and the following data, determine if there has been a process change. Do the data indicate that the performance has gotten better or worse when compared to the historical average of 3.33 demerits per unit?

Note that the sample size has been increased to $n = 120$. A recalculation of the control limits is required anytime the sample size changes. The historical values for Du, p_c, p_{ma}, and p_{mi} are used in the recalculation. Only the sample size n is changed in the computation. The new value for three standard deviations is calculated by

$$3\sigma \text{ for } Du = 3\sqrt{\frac{162.12}{120}} = 3.49$$

The new upper and lower control limits using $n = 120$ are:

UCL $= 3.33 + 3.49 = 6.82$

LCL $= 3.33 - 3.49 = -0.16 = 0.00$

Future data from the process are as follows:

Sample number	Size n	Date	Criticals	Majors	Minors	Total demerits	Demerits/ unit
16	120	7/22/94	2	4	12		
17	120	7/23/94	1	12	16		
18	120	7/24/94	3	10	21		
19	120	7/25/94	3	2	9		
20	120	7/26/94	2	4	2		
21	120	7/27/94	5	3	13		
22	120	7/28/94	7	10	13		
23	120	7/29/94	5	9	3		
24	120	7/30/94	3	2	1		
25	120	8/1/94	6	13	3		
26	120	8/2/94	4	0	14		

The total demerits and demerits/unit have been left blank intentionally for the reader to complete. The finished control chart can be seen in the following figure. It appears that the process average demerits/unit has shifted positively.

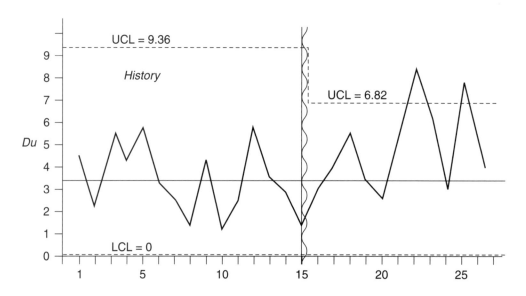

Optional problem:

A complex electronic device goes through a comprehensive final audit. Defects are categorized into four classifications and are assigned demerit weighting factors accordingly:

1. Critical $C = 100$
2. Severe $S = 50$
3. Major $M = 10$
4. Incidental $I = 1$

Construct a Du control chart based on the following data:

Lot no.	Sample size n	Defects distribution			
		C	S	M	I
1	100	1	0	4	9
2	100	0	1	5	3
3	100	0	2	1	5
4	100	1	3	2	8
5	100	0	1	5	6
6	400	3	2	6	24
7	400	2	5	14	30
8	400	1	8	10	18
9	250	5	2	11	12
10	250	2	4	7	16
10	250	5	2	4	11

How would this control chart compare to a traditional u chart?

The demerit/unit control chart is sometimes referred to as a D chart.

An alternative method given by Dale H. Besterfield (1994, 271–272) relies on an average number of nonconformities per unit rather than a weighted binomial relationship.

The plot point is given by the formula

$$D = W_c U_c + W_{ma} + W_{mi} U_{mi}$$

where: D = demerits/unit
W_c = weighting factor for criticals
W_{ma} = weighting factor for majors
W_{mi} = weighting factor for minors

and

U_c = counts for criticals per unit
U_{ma} = counts for majors per unit
U_{mi} = counts for minors per unit

If the weighting factors for the critical, major, and minors are 12, 9, and 4 respectively, the demerits per unit would be calculated

$$D = 12U_c + 9U_{ma} + 4U_{mi}$$

The central or average line is calculated from the formula

$$\bar{D}u = 12\bar{U}_c + 9U_{ma} + 4U_{mi}$$

and the standard deviation for Du is given by

$$\sigma = \sqrt{\frac{(12)^2\,\bar{U}_c + (9)^2\,\bar{U}_{mi} + (4)^2\,\bar{U}_{mi}}{n}}$$

Example:
The sample size is $n = 40$, and the average defects/unit for the critical, major, and minor defects are 0.07, 0.38, and 1.59 respectively. What is the centerline, UCL, and LCL for this chart?

The process average demerits/unit $\bar{D}u$ is

$$\bar{D}u = 12\bar{U}_c + 9\bar{U}_{ma} + 4\bar{U}_{mi}$$

$$\bar{D}u = (12)(.07) + (9)(0.38) + (4)(1.59) = 10.62$$

$$\sigma = \sqrt{\frac{(12)^2\,\bar{U}_c + (9)\bar{U}_{mi} + (4)^2\,\bar{U}_{mi}}{n}}$$

$$\sigma = \sqrt{\frac{(12)^2\,(0.07) + (9)^2\,(0.38) + (4)^2\,(1.59)}{40}}$$

$$\sigma = 1.29 \text{ and } 3\sigma = 3.87$$

UCL = 10.62 + 3.87 = 14.49

LCL = 10.62 − 3.87 = 6.75

A sample of 40 is taken where 6 criticals, 7 majors, and 12 minors are found. Calculate Du for this point. Does the calculated value indicate an out-of-control condition?

$$Du = (12)\left(\frac{6}{40}\right) + (9)\left(\frac{15}{40}\right) + (4)\left(\frac{18}{40}\right)$$

$$Du = 6.98$$

This point is not outside the control limits, and out of control is contraindicated.

BIBLIOGRAPHY

Besterfield, D. H. 1994. *Quality Control.* 4th edition. Englewood Cliffs, NJ: Prentice Hall.

Grant, E. L., and R. S. Leavenworth. 1996. *Statistical Quality Control.* 7th edition. New York: McGraw-Hill.

Hayes, G. E., and H. G. Romig. 1982. *Modern Quality Control.* 3rd edition. Encino, CA: Glencoe.

Montgomery, D. C. 1996. *Introduction to Statistical Quality Control.* 3rd edition. New York: John Wiley and Sons.

DESCRIPTIVE STATISTICS

Processes can be thought of as defined activities that lead to a change. Processes are dynamic and always add value to services or manufactured devices. As quality practitioners, we are all interested in processes. Processes can be measured in a numerical sense, and all observations regarding processes can be ultimately converted into numerical data. Examples of measurements include service-related and manufacturing-related measurements.

Service-related
Length of time required to answer customer questions
Proportion of customers that are not satisfied with a service
Number of errors found on expense reports
Units of production per hour
Error on shipping papers
Nonconformities on a service audit

Manufacturing-related
Total defects on 45 printed circuit boards
Surface finish
Diameter of a part
pH of a solution

In most cases, we have an overabundance of both the opportunity as well as available data for measurements of a characteristic.

All of the opportunities for a process to be measured can be considered as a universe from which we can develop a characterization. This entire opportunity can be considered as the universe or population of a process. In magnitude this opportunity is unlimited.

For our discussion, we will equate process = population = universe. The abbreviation for the population is N. N literally consists of *all* of the observations that can be made on a process. In most cases, N is exceedingly large.

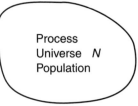

Process
Universe N
Population

Since this population is so large, we cannot from a practical point measure all of it. The process might be one of answering calls at a phone help line where hundreds of calls are received per hour. We are, nevertheless, interested in the length of time required to answer each call. The size of the opportunity (population) does not diminish our interest in the process. If we cannot measure all of the calls, we can always take a sample of the population. Samples are subsets of the larger population or universe.

We can act upon data from samples to calculate such information as averages, proportions, and standard deviations. The following are examples of this population/sample relationship as used in quality:

1. You are interested in determining the percentage of customers that use your service who would consider using it again. You have sampled 500 and found that six would not use your service again. This proportion is 6/500, or 0.012.
2. A shipment of 1200 bolts has been received. The average weight of a random sample of 25 is determined to be 12.68 grams each.
3. Approximately 45 patients a day are admitted to a local hospital. A daily sample of 20 are chosen each day to determine the standard deviation for the time required to complete the admission process.

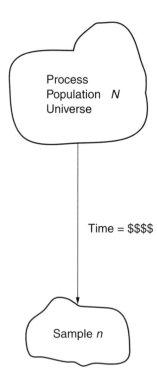

Information calculated from samples are called *statistics*. Examples of statistics frequently used in quality are:

Proportion (or percentage) P

Average $\bar{X}$

Standard deviation S

Traditional statistics such as the average are abbreviated using English letters or symbols made up of English letters. The average proportion is written with a letter P and a bar over it. Any statistic with a bar over it refers to the average of that statistic. For example, $\bar{R}$ represents the average range. The abbreviations may or may not be capitalized.

Statistics by their very nature are subject to error. Statistics are estimates of the true description of the process. If we could and did use all of the data in the population or universe to calculate a defining characteristic, then there would be no error. Such numbers

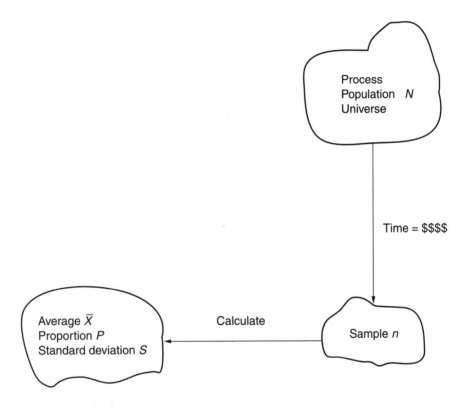

derived from *all* of the data are no longer call statistics but rather *parameters.* The true proportion is still called the proportion; however, it is abbreviated with a lowercase Greek letter for the English letter P—pi or π. The true standard deviation is abbreviated using the Greek lowercase letter for the English letter S—sigma or σ. The case for the average is unique in that the name of the true average is the mean, and the abbreviation for the mean is the lowercase Greek letter for the English letter M—mu or μ.

This parameter is estimated by the corresponding statistic, and it is the parameter that actually describes the process without error.

The following final statement summarizes the relationship of a process to a statistic: "We are all interested in the process. But due to the vastness of it, we are required to obtain a sample from it. From the sample, we calculate a statistic. The statistic is an estimate of the truth or parameter. The truth describes the process without error."

The use of statistics is critical to the science of quality. There is no substitute for the value of numerical data, as was so well stated by Lord Kelvin in 1883: "When you can

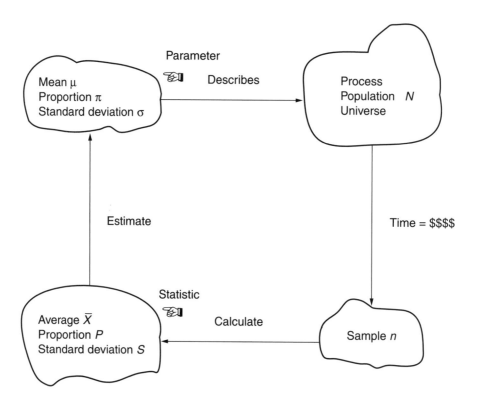

measure what you are speaking about and express it in numbers you know something about it; but when you cannot express it in numbers your knowledge is of a meager and unsatisfactory kind."

The degree to which a statistic represents the truth is a function of the sample size, error, and level of confidence.

Statistics used to describe sets of data are called descriptive statistics and generally fall into the two catagories of location and variation.

Location statistics give an indication of the central tendency of a set of data, and variation statistics give an indication of the variation or dispersion of a set of data. Both these

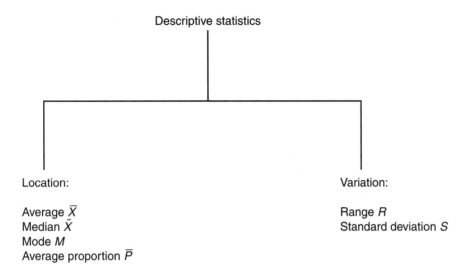

Descriptive statistics

Location:

Average $\bar{X}$
Median $\tilde{X}$
Mode M
Average proportion $\bar{P}$

Variation:

Range R
Standard deviation S

numbers used in conjunction with each other give a better overall description of data sets and processes.

The nature of the actual data collected can be defined as *variables* or *attributes*. A variable is any data collected initially as a direct number and can take on any value within the limits that it can exist. Examples of variables data would be temperature, pressure, time, weight, and volume.

Attribute data are data that evolve from a direct observation where the response is *yes* or *no*. The actual attribute data can and must, however, be translated into a number so that mathematical operations can be made. Examples of attribute data would be the characteristic of meeting a specification requirement. The specification is either met (*yes* response) or it is not (*no* response).

Variables data can be converted to attribute data by comparing to a specification requirement.

Example:
The weight of a product is 12.36 pounds—a variable. The specification for the weight is 12.00 ± 0.50 pounds. Observation: the product meets the specification requirement, so the response is yes—an attribute.

BIBLIOGRAPHY

Hayes, E., and H. G. Romig. 1982. *Modern Quality Control.* 3rd edition. Encino, CA: Glencoe.

Juran, J. M., and F. M. Gryna. 1980. *Quality Planning and Analysis.* 3rd edition. New York, NY: McGraw-Hill.

Petruccelli, J. D., B. Nandram, and M. Chen. 1999. *Applied Statistics for Engineers and Scientists.* Upper Saddle River, NJ: Prentice Hall.

Sternstein, M. 1996. *Statistics.* Hauppauge, NY: Barron's Educational Series.

DESIGNED EXPERIMENTS

Designed experiments provide a statistical tool to allow for the efficient examination of several input *factors* and to determine their effect on one or more *response* variables.

Input factors are those characteristics or operating conditions over which we may or may not have direct control but that can and do affect a process. Response variables or factors are the observations or measurements that result from a process.

Examples of this input factor/response variable relationship follow.

Process: Wood gluing operation

Input factors	*Response variables*
Amount of glue	Tensile strength, pounds/inch^2
Type of glue	
Drying temperature	
Drying time	
Moisture content of wood	
Type of wood	
Relative humidity of environment	

For all of these input factors, there is an optimum level of setting that will maximize the bond strength. Some of these factors may be of more importance than others.

Designed experiments (frequently referred to as design of experiments, or DOE) can assist in determining which factors play a role in affecting the level of response.

There are essentially the following five steps in a DOE:

1. Brainstorming to identify potential input factors (or factors) and output responses (or responses) and establishing *levels* for the factors and the measure for the response(s)
2. Designing the experimental design or *matrix*
3. Performing the experiment
4. Analyzing the results
5. Performing a validation run to test the results

Example A:
A manufacturer of plastic/paper laminate wants to investigate the lamination process to see if improvements to the process can be made.

Step 1. Brainstorm.

During this initial stage, individuals gather to discuss and define input factors and output responses. The results of this meeting yield the following:

Input factors	*Response variables*
Top roll tension setting, lb.	Amount of curl
Rewind tension setting, lb.	Number of wrinkles per 100 ft.
Bottom roll tension setting, lb.	Peel strength, lb./in.
Take-up speed, ft./min.	
Type of paper	
Thickness of plastic film, mils	
Type of plastic film	

For this initial experiment, three factors and one response variable are selected. The input factors are identified as *A, B,* and *C.* The response variable is the amount of curl, and the three input factors are

Input factor	*Description*
A	Top roll tension
B	Bottom roll tension
C	Rewind tension

Response variable	*Description*
Curl, inches	A 36-inch strip of paper/plastic film is hung against a vertical plane. The distance from the end of the laminate to the vertical plane is measured in millimeters.

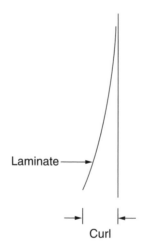

Laminate

Curl

The objective of this experiment is to better understand the effects of the three chosen factors on the amount of curl in the laminate. The current method calls for these factors to be set as follows:

Factor	Setting
A	22
B	22
C	9

For this experiment, we will set each of the factors to a "+" level and a "–" level. The notation for this three-factor, two-level experiment is

$$2^3$$

where: 2 = number of levels
3 = number of factors

The total number of experimental conditions is given by one $2^3 = 8$.

The general case for any two-level experiment, where all possible experimental conditions are run is given by

$$2^n$$

where: n = the number of factors

Each of the factors will be set at a + and – level around the traditional value at which they are run.

Note: Some text on DOE will use *1* for one level and *2* for the second level.

Factor	– Setting	+ Setting
A	16	28
B	16	28
C	6	12

The new conditions are set at such a level as to give a change in the response variable if the factor is a contributor to the response variable. The change in the factor setting should be large enough to change the response variable but not enough to critically affect the process.

Step 2. Design the experiment.

This experiment is a full 2^3 and will have eight experimental setups. Start by having eight *runs* and three columns for the factors *A, B,* and *C.*

	Factors		
Run	**A**	**B**	**C**
1			
2			
3			
4			
5			
6			
7			
8			

Start by placing a "–" for the first run of factor A and then alternating + and – for the first column.

	Factors		
Run	**A**	**B**	**C**
1	–		
2	+		
3	–		
4	+		
5	–		
6	+		
7	–		
8	+		

The second column is treated in a similar manner, except there are pairs of –'s and +'s. The third column is completed with groups of four –'s and four +'s. This completed table represents the design or matrix for the two-level, three-factor experiment, or the 2^3 design.

	Factors		
Run	**A**	**B**	**C**
1	–	–	–
2	+	–	–
3	–	+	–
4	+	+	–
5	–	–	+
6	+	–	+
7	–	+	+
8	+	+	+

The order of one through eight listed in numerical sequence is called the standard order. The actual sequence for the eight experiments should be a random order. The – and + signs determine the setting for the particular run.

For example, run #3 would have:

Factor *A* (Top roll tension) set at 16

Factor *B* (Bottom roll tension) set at 28

Factor *C* (Rewind tension) set at 6

Step 3. Perform the experiment.

Each of the eight experimental runs is made in a random order. The results of each run are recorded, and a single measurement observation will be made for each run.

Step 4. Analyze the data.

Main Effects:

For each of the main factors *A, B,* and *C,* an *effect* will be determined. The effect will be the average effect obtained when the factor under consideration is changed from a – setting to a + setting. The signs of the column to which a factor has been assigned will be used to determine the effect.

Step 5. Validation.

The original objective was to minimize the curl response. From the graphical analysis of the effects we have concluded that factors *A* and *C* are major factors with respect to their effect on curl. Since we are minimizing the response of curl, we must set the factors *A* and *C* to the + and – settings respectively. The top roll tension should be set to 28, the re-wind tension set to 6, and the remaining factor *B* to the most economical setting as it has no significant effect on tension.

The process should now be run for a more extended period of time to collect data on the curl response. This data from the improved process can now be compared to the historical data using traditional methods of hypothesis testing. Essentially we want to know if there has been a measured, statistically significant reduction in the amount of curl.

	Factors			
Run	**A**	**B**	**C**	**Response, curl**
1	–	–	–	87
2	+	–	–	76
3	–	+	–	90
4	+	+	–	83
5	–	–	+	101
6	+	–	+	92
7	–	+	+	100
8	+	+	+	92

Effects are determined by taking the difference between the average response when the factor is set + and the average response when the factor is set −.

Main Effect A:

$$\left(\frac{76+83+92+92}{4}\right)-\left(\frac{87+90+101+100}{4}\right)=85.75-94.50=-8.75$$

The average effect in going from a + setting for factor A to a − setting for factor A is −8.75 units of curl.

Main Effect B:

$$\left(\frac{90+83+100+92}{4}\right)-\left(\frac{87+76+101+92}{4}\right)=91.25-89.00=+2.25$$

Main Effect C:

$$\left(\frac{101+92+100+92}{4}\right)-\left(\frac{87+76+90+83}{4}\right)=96.25-84.00=+12.25$$

The main effects can be visualized by drawing a cube plot of the responses.

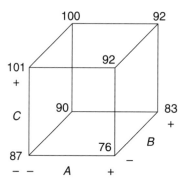

It can be seen that in order to minimize the curl response, the following setting for the factors should be made:

Factor	Setting
Factor A (Top roll tension)	$A+=28$
Factor B (Bottom roll tension)	$B-=16$
Factor C (Rewind tension)	$C-=6$

In addition to the main effects, there can be interaction effects. Interactions result when the effect of one main effect depends on or is related to the effect of another effect. For example, the rate at which a chemical reaction occurs can be influenced by a catalyst. The selection of the catalyst can be a determining factor. Catalyst A might perform well but only at a higher temperature, where catalyst B would perform as well as catalyst A but only at a lower temperature. The catalyst and temperature factors interact with each other; in other words, there is a catalyst-temperature interaction.

In order to evaluate interactions of the three factors *A*, *B*, and *C*, the interactions will be determined. There are three interactions that involve two factors:

AB = the interaction of top roll tension and bottom roll tension

BC = the interaction of bottom roll tension and rewind tension

AC = the interaction of top roll tension and rewind tension

These interactions may also be written as $A \times B$, $B \times C$, and $A \times C$. In addition, there is a single three-factor interaction expressed as *ABC*.

The computation of these interactions is performed using the signs of the columns as was the case with the main effects. The signs for the interaction columns are determined by multiplying the signs of the main effects used in the interaction.

Run	A	B	C	AB	AC	BC	ABC	Response, curl
					Factors			
1	−	−	−	+	+	+	−	87
2	+	−	−	−	−	+	+	76
3	−	+	−	−	+	−	+	90
4	+	+	−	+	−	−	−	83
5	−	−	+	+	−	−	+	101
6	+	−	+	−	+	−	−	92
7	−	+	+	−	−	+	−	100
8	+	+	+	+	+	+	+	92

AB effect:

$$\left(\frac{87+83+101+92}{4}\right)-\left(\frac{76+90+92+100}{4}\right)=90.75-89.50=+1.25$$

AC effect:

$$\left(\frac{87+90+92+92}{4}\right)-\left(\frac{76+83+101+100}{4}\right)=90.25-90.00=+0.25$$

BC effect:

$$\left(\frac{87+76+100+92}{4}\right)-\left(\frac{90+83+101+92}{4}\right)=88.75-91.50=-0.75$$

ABC effect:

$$\left(\frac{76+90+101+92}{4}\right)-\left(\frac{87+83+92+100}{4}\right)=89.75-90.50=-0.75$$

Are these effects significant, or are the differences due to random, experimental error?

There are two ways to determine if these effects are statistically significant. One way is to repeat the experiment and determine the actual experimental error, and the other is utilized when only one observation is made for each run (as is this case). When only one observation per run is available, the use of a normal probability plot can be helpful.

NORMAL PROBABILITY PLOT

Normal probability plotting paper is used. The effect is located on the horizontal axis, and the relative rank of the effect is expressed as a percent median rank (%MR).

Step 1. Arrange all of the effects in ascending order (starting with the lowest at the top). The order of this ranking is expressed as i.

Order i	Effect value	Effect
1	−8.75	A
2	−2.75	BC
3	−0.75	ABC
4	0.25	AC
5	1.25	AB
6	2.25	B
7	12.25	C

Note that there is one less effect than there are experimental runs.

Step 2. Calculate the percent median rank values.

$$\%MR = \frac{i - .3}{n + .4} \times 100$$

where: i = the order of the effect
n = the total number of effects

This median rank is called the Benard Median Rank and follows the general expression of $MR = \frac{i - c}{n - 2c + 1}$ where $c = 0.3$. Frequently another form of the median rank is the Hazen, where $c = 0.5$.

$$\text{First \% median rank} = \%MR = \frac{i - .3}{n + .4} \times 100 = \%MR = \frac{1 - .3}{7 + .4} \times 100 = \frac{0.7}{7.4} \times 100 = 9.5$$

$$\text{Second \% median rank} = \frac{2 - .3}{7 + .4} \times 100 = 23.0$$

$$\text{Third \% median rank} = \frac{3 - .3}{7 + .4} \times 100 = 36.5$$

Order i	Effect value	Effect	Percent median rank
1	−8.75	A	9.5
2	−2.75	BC	23.0
3	−0.75	ABC	36.5
4	0.25	AC	50.0
5	1.25	AB	63.5
6	2.25	B	77.0
7	12.25	C	90.5

Step 3. Plot the effects as a function of the percent median rank on the normal probability paper.

Draw a straight line through the points, concentrating the emphasis around the points nearest the zero effect. Effects falling off of this straight line are judged to be significant. To maximize a response variable, set the main effect to the same sign as the effect. To minimize the response variable, set the factor to the reverse of the effect. In this example, factors A and C are statistically significant.

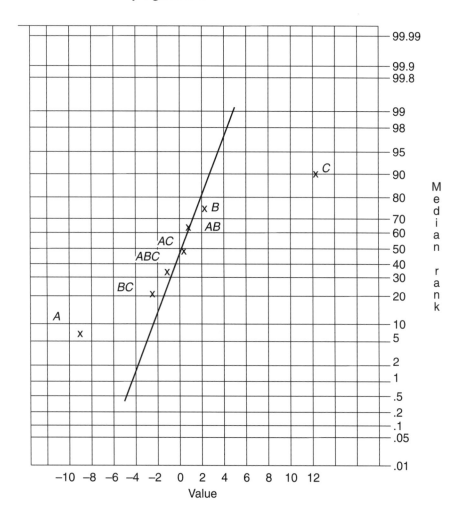

VISUALIZATION OF TWO-FACTOR INTERACTIONS

Main effects can be visualized using the cube plot. Two-factor interactions can be visualized using linear plots.

The following demonstrates how to develop the two-factor interaction for the *AB* interaction.

Step 1. Draw a two-axis graph.

The response will be plotted on the vertical axis, and one of the two main effects involved in the two-factor interaction will be plotted on the horizontal axis. The horizontal value is attribute in nature. Only the designations of − and + are used for the selected factor.

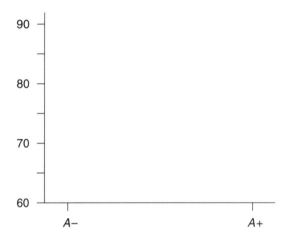

Since factor *A* was chosen for the horizontal axis, the remaining factor *B* will be plotted as a pair of linear lines—one for *B*− and one for *B*+.

The *B*− line will be plotted first. There are two experimental runs where *A* is set to *A*− and factor *B* is set to *B*−. These two runs are run #1 and run #5. The average response for these two is:

$$\frac{87+101}{2} = 94.0$$

When factor *A* is set to *A*− and factor *B* is set to *B*−, the average response is 94.0.

The other end of the *B*− line is determined by calculating the average response when *A* is set at *A*+ and *B* is set at *B*−. *A*+ and *B*− are found in two experimental runs—run #2 and run #6. The average response for these two is

$$\frac{76+92}{2} = 84.0$$

These two values define the *B*− line on the graph.

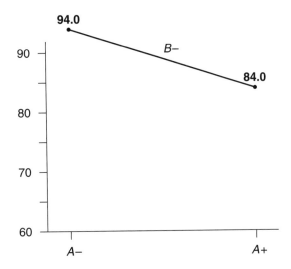

The remaining line is the B+ line. There are two runs where B is set + and A is set −. These are run #3 and run #7. The average of these two responses is

$$\frac{90+100}{2} = 95.0$$

The other end of the B+ line is at the A+ position. There are two runs where B is set + and A is set +. These are run #4 and run #8. The average of these two responses is:

$$\frac{83.0+92.0}{2} = 87.5$$

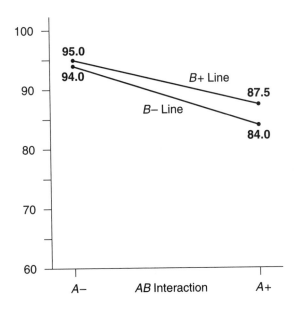

Plotting the *B+* line completes the two-factor interaction plot.

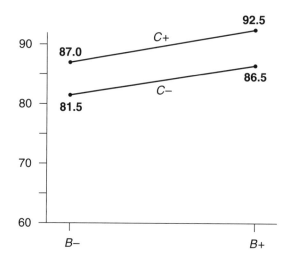

The linear graph for the *BC* interaction follows:

The C– line:

$$C - B - \text{point} = \frac{87.0 + 76.0}{2} = 81.5$$

$$C - B + \text{point} = \frac{90.0 + 83.0}{2} = 86.5$$

The C+ line:

$$C + B - \text{point} = \frac{93.0 + 81.0}{2} = 87.0$$

$$C + B + \text{point} = \frac{95.0 + 90.0}{2} = 92.5$$

In the original experiment, only one observation was made for each experiment run. The only technique to determine if the effects were statistically significant was to perform a normal probability plot and make a subjective judgment if any of the points fell off of the straight line (with emphasis on the zero point). This method is subject to errors of judgment, but it is the best technique available when only one observation is available.

In order to determine experimental error, the entire experiment must be replicated at least one more time in order to calculate a measure of variation—the standard deviation.

The entire set of eight runs is replicated. The new data from the first and second replications yield an average and standard deviation for each of the eight runs:

	Factors							1st	2nd	Avg.	Std. dev.
Run	A	B	C	AB	AC	BC	ABC				
1	−	−	−	+	+	+	−	87	88	87.5	0.707
2	+	−	−	−	−	+	+	76	78	77.0	1.414
3	−	+	−	−	+	−	+	90	92	91.0	1.414
4	+	+	−	+	−	−	−	83	80	81.5	2.121
5	−	−	+	+	−	−	+	101	96	98.5	3.535
6	+	−	+	−	+	−	−	92	91	91.5	0.707
7	−	+	+	−	−	+	−	100	104	102.0	2.828
8	+	+	+	+	+	+	+	92	91	91.5	0.707

The new calculated effects, in decreasing order, based on the average of two observations/run are:

Factor	Effect
A	−9.38
ABC	−1.13
BC	−1.13
AB	−0.63
AC	0.63
B	2.88
C	11.63

While a normal probability plot would yield the same conclusion that factor A and factor B are probably statistically significant, again, the conclusion is somewhat subjective.

By using the t-distribution and determining the confidence interval for each of the effects, a better judgment can be made.

The following method can test if the effect is within the normal variation of the experimental error or if the effect is statistically removed from the limits of normal variation. That is, the effect is real and not just a random value.

A confidence interval will be determined for each effect. If zero is found to be within the confidence interval limits, the conclusion will be that the observed effect is simply random variation within the expectation for the calculated experimental error. If, however, the confidence interval does not contain zero, the conclusion will be made that the effect is real and is statistically significant.

Calculate the pooled standard deviation for all the observations.

$$Sp = \sqrt{\frac{V_1 S_1^2 + V_2 S_2^2 + \ldots V_n S_n^2}{V_1 + V_2 + \ldots V_n}}$$

where: V_1 = number of observations in run #1
S^2_1 = the variance (standard deviation squared) for sample run #1
V_n = number of observations in the nth run
S^2_n = variance of the nth sample run

$$Sp = \sqrt{\frac{(2)(0.707)^2 + (2)(1.414)^2 + (2)(1.414)^2 + \ldots + (2)(2.828)^2 + (2)(0.707)^2}{2 + 2 + 2 + 2 + 2 + 2 + 2 + 2}}$$

$$Sp = \sqrt{\frac{0.9997 + 3.9988 + 3.9988 + 8.9973 + 24.9925 + 0.9997 + 15.9952 + 0.9997}{16}}$$

$$Sp = 1.952$$

Note: If the number of replicates is the same for all runs, then the pooled standard deviation is simply the square root of the average variances.

Each of the calculated effects is actually a point estimate. The confidence interval for each of these estimated effects is determined by

$$\text{Effect} \pm \text{Error}$$

The error is calculated as:

$$t_{\alpha/2,(r-1)2^{k-f}} \left(Sp \sqrt{\frac{2}{pr}} \right)$$

where: α = risk (Confidence = 1 − Risk)
p = number of +'s per effect column
r = number of replicates
Sp = pooled standard deviation
f = degree of fractionation

The error of the estimated effect at a level of confidence of 95 percent is

$$E = t_{\alpha/2,(r-1)2^{k-f}} \left(Sp \sqrt{\frac{2}{pr}} \right)$$

$$t_{\alpha/2,(r-1)2^{k-f}} = t_{0.025,8} = 2.306$$

$$E = 2.306 \left(1.952 \sqrt{\frac{2}{(4)(2)}} \right) \qquad E = 2.25$$

Any effect that is contained within the limits of 0 ± 2.25 is considered to be not statistically significant.

The following effects are significant:

Effect	95 percent confidence interval
A	-11.63 to -7.13
B	0.63 to 5.13
C	8.75 to 13.88

Note: Any effect whose absolute value is greater than the error is statistically significant.

The main effect *B* is not deemed to be statistically significant at the 95 percent level of confidence. This was not evident using the normal probability plot method.

FRACTIONAL FACTORIAL EXPERIMENTS

If four factors are to be investigated at two levels each, then the total number of experiments required would be $2^4 = 16$, and in order to establish an experimental error, two replicates will be required for a total of 32 observations. This number of observations and experiments can be time consuming and cost prohibitive in many cases.

The same number of factors can be evaluated using half the normal runs by running a one-half fractional factorial experiment.

Consider the requirement for a full four-factor experiment. This design can be created by extending the eight runs to 16 and adding a fourth column for the next factor *D*.

		Factors		
Run	**A**	**B**	**C**	**D**
1	−	−	−	−
2	+	−	−	−
3	−	+	−	−
4	+	+	−	−
5	−	−	+	−
6	+	−	+	−
7	−	+	+	−
8	+	+	+	−
9	−	−	−	+
10	+	−	−	+
11	−	+	−	+
12	+	+	−	+
13	−	−	+	+
14	+	−	+	+
15	−	+	+	+
16	+	+	+	+

Three-level interactions are very rare. If the assumption is made that any three-level interaction is actually due to experimental error, then by letting the ABC interaction of this 2^4 experiment equal the D main effect, the full 16 runs for a full 2^4 can be cut in half. In other words, let $D = ABC$. $D = ABC$ is called a *generator*, and the *identity* is $I = ABCD$.

By mixing the effect of main effect D with the three-level interaction ABC, we have confounded the two effects. D and ABC are called *aliases*. Since D and ABC are aliases, we cannot differentiate between D and ABC.

			Factors		
Run	A	B	C	D	ABC
1	−	−	−	−	−
2	+	−	−	−	+
3	−	+	−	−	+
4	+	+	−	−	−
5	−	−	+	−	+
6	+	−	+	−	−
7	−	+	+	−	−
8	+	+	+	−	+
9	−	−	−	+	−
10	+	−	−	+	+
11	−	+	−	+	+
12	+	+	−	+	−
13	−	−	+	+	+
14	+	−	+	+	−
15	−	+	+	+	−
16	+	+	+	+	+

Let $D = ABC$.

		Factors		D
Run	A	B	C	ABC
1	−	−	−	−
2	+	−	−	+
3	−	+	−	+
4	+	+	−	−
5	−	−	+	+
6	+	−	+	−
7	−	+	+	−
8	+	+	+	+
9	−	−	−	−
10	+	−	−	+
11	−	+	−	+
12	+	+	−	−
13	−	−	+	+
14	+	−	+	−
15	−	+	+	−
16	+	+	+	+

Notice that run #1 through run #8 are now duplicated by run #9 through run #16. Run #9 through run #16 can be eliminated. Four factors can be run using eight experimental runs. This experimental design is called a 2^{4-1} design.

Fractional factorial designs can be designated by the general form of

$$2^{k-f}$$

where: k = number of factors 2^{k-f}
f = degree of fractionation
f = 1½ fractional design
f = 2¼ fractional design
f = 3⅛ fractional design

The number of experimental runs is determined by the value of 2^{k-f}. In a 2^{7-3}, there are 16 runs—2^4 = 16.

The *resolution* of a design determines the degree to which confounding is present. Resolutions are expressed as roman numerals III, IV, or V and are named by the number of letters in the shortest word of the defining relationship. In the example of a 2^{4-1}, the defining relationship was $D = ABC$. The identity element is $I = ABCD$. There are four letters in the word; therefore, the resolution is IV.

The proper designation for this design is:

$$2_{IV}^{4-1}$$

Possible experimental design resolutions are as follows:

R_{III}: Main effects and two-factor interactions are confounded, so be careful.

R_{IV}: Main effects are clear from any two-factor interactions, and two-factor interactions are confounded.

R_V: Main effects are clear, and two-factor interactions are clear.

Example:
In a 2^{5-2}, the defining relationships are: $D = AB$ and $E = AC$.

This is a 1/4 fractional design. The number of factors is five, the number of runs is eight, and the resolution is III ($I = ABD$ and $I = ACE$), three letters make up the identity element.

This is a 2_{III}^{5-2} experiment.

Table 1. Selected Fractional Factorial Designs.

Number of factors k	Designation	Number of runs	Design generator
3	2_{III}^{3-1}	4	$A = BC$
4	2_{IV}^{4-1}	8	$D = ABC$
5	2_{V}^{5-1}	16	$E = ABCD$
	2_{III}^{5-2}	8	$D = AB$
			$E = AC$
6	2_{VI}^{6-1}	32	$F = ABCDE$
	2_{IV}^{6-2}	16	$E = ABC$
			$F = BCD$
7	2_{VII}^{7-1}	64	$G = ABCDEF$
	2_{IV}^{7-2}	32	$F = ABCD$
			$G = ABDE$
	2_{IV}^{7-3}	16	$E = ABC$
			$F = BCD$
			$G = ACD$
	2_{III}^{7-4}	8	$D = AB$
			$E = AC$
			$F = BC$
			$G = ABC$

Case study:

The AstroSol company manufactures solar panels used to generate electricity. The manufacturing process consists of deposition of alternating layers of tin (Sn) and cadmium telluride (CdTe) and then baking these deposited layers in an oven. The power output is then measured and reported in units of watts/ft^2.

The R&D department wants to maximize the output of the cells. After considerable review and discussion, the design team decides to look at five factors in a 1/2 fractional factorial design, 2_{V}^{5-1}, as this design will isolate a main and two factorial interactions.

The five factors to be considered are:

Thickness of Sn in microns

Thickness of CdTe in microns

Bake temperature, °F

Baking time, minutes

Source of CdTe

All other factors in the process will remain fixed at their current levels.

A relative – and + for each of the factors around the current operating conditions is determined for each factor.

Factor	Description	−	+
A	Thickness of Sn, μ	20	40
B	Thickness of CdTe, μ	30	50
C	Bake temperature, °F	350	400
D	Bake time, minutes	40	50
E	Source of CdTe	United States	China

For a 2_V^{5-1} fractional design, the base design will be that of a 2^4 full factorial where the fifth factor will be generated by $E = ABCD$ and $I = ABCDE$.

Interactions:

$AB = CDE$	$AE = BCD$	$BE = ACD$	$DE = ABC$
$AC = BDE$	$BC = ADE$	$CD = ABE$	
$AD = BCE$	$BD = ACE$	$CE = ABD$	

Since this is a resolution V, all main effects are clear and all two-factor interactions are clear.

The 16 runs were replicated three times so that experimental error and statistical significance can be determined. The individual responses, response average, and standard deviation are recorded in Table 2.

Calculation of Main Effects

A effect:

$$\left(\frac{9.23+9.76+8.27+9.83+8.23+10.17+8.30+9.50}{8}\right)$$
$$-\left(\frac{9.30+10.73+9.30+9.73+10.10+10.97+8.30+10.77}{8}\right) = -0.74$$

B effect:

$$\left(\frac{10.73+9.76+9.73+9.83+10.97+10.17+10.77+9.50}{8}\right)$$
$$-\left(\frac{9.30+9.23+9.30+8.27+10.10+8.23+8.50+8.30}{8}\right) = 1.28$$

C effect:

$$\left(\frac{9.30+8.27+9.73+9.83+8.50+8.30+10.77+9.50}{8}\right)$$
$$-\left(\frac{9.30+9.23+10.73+9.76+10.10+8.23+10.97+10.17}{8}\right) = -0.54$$

Table 2. Design Matrix.

					ABCD	CDE	BDE	BCE	BCD	ADE	ACE	ACD	ABE	ABD	ABC	Response				
Run	A	B	C	D	E	AB	AC	AD	AE	BC	BD	BE	CD	CE	DE	Y_1	Y_2	Y_3	Avg.	S
1	−	−	−	−	+	+	+	+	−	+	+	−	+	−	−	9.3	9.4	9.2	9.30	0.10
2	+	−	−	−	−	−	−	−	−	+	+	+	+	+	+	9.1	9.4	9.2	9.23	0.15
3	−	+	−	−	−	−	+	+	+	−	−	−	+	+	+	10.8	10.8	10.6	10.73	0.12
4	+	+	−	−	+	+	−	−	+	−	−	+	+	−	−	9.8	9.6	9.9	9.76	0.15
5	−	−	+	−	+	+	−	+	−	−	+	−	−	+	−	9.5	9.0	9.4	9.30	0.26
6	+	−	+	−	−	−	+	−	−	−	+	+	−	−	+	8.0	8.4	8.4	8.27	0.23
7	−	+	+	−	−	−	−	+	+	+	−	−	−	−	+	9.9	9.6	9.7	9.73	0.15
8	+	+	+	−	+	+	+	−	+	+	−	+	−	+	−	9.8	9.9	9.8	9.83	0.06
9	−	−	−	+	+	+	+	−	−	+	−	−	−	−	+	10.4	10.5	9.4	10.10	0.61
10	+	−	−	+	−	−	−	+	−	+	−	+	−	+	−	8.5	8.0	8.2	8.23	0.25
11	−	+	−	+	−	−	+	−	+	−	+	−	−	+	−	10.8	11.1	11.0	10.97	0.15
12	+	+	−	+	+	+	−	+	+	−	+	+	−	−	+	10.2	9.7	10.6	10.17	0.45
13	−	−	+	+	+	+	−	−	−	−	−	−	+	+	+	8.6	8.2	8.7	8.50	0.26
14	+	−	+	+	−	−	+	+	−	−	−	+	+	−	−	7.8	8.4	8.7	8.30	0.46
15	−	+	+	+	−	−	−	−	+	+	+	−	+	−	−	11.0	10.6	10.7	10.77	0.21
16	+	+	+	+	+	+	+	+	+	+	+	+	+	+	+	9.2	9.9	9.4	9.50	0.36

D effect:

$$\left(\frac{10.10+8.23+10.97+10.17+8.50+8.30+10.77+9.50}{8}\right)$$
$$-\left(\frac{9.30+9.23+10.73+9.76+9.30+8.27+9.73+9.83}{8}\right)=0.05$$

E effect:

$$\left(\frac{9.30+9.76+8.27+9.73+8.23+10.97+8.50+9.50}{8}\right)$$
$$-\left(\frac{9.23+10.73+9.30+9.83+10.10+10.17+8.30+10.77}{8}\right)=-0.52$$

Two-Factor Interactions

AB interaction:

$$\left(\frac{9.30+9.76+9.30+9.83+10.10+10.17+8.50+9.50}{8}\right)$$
$$-\left(\frac{9.23+10.73+8.27+9.73+8.23+10.97+8.30+10.77}{8}\right)=0.07$$

AC interaction:

$$\left(\frac{9.30+10.73+8.27+9.83+10.10+10.97+8.30+9.50}{8}\right)$$
$$-\left(\frac{9.23+9.76+9.30+9.73+8.23+10.17+8.50+10.77}{8}\right)=0.16$$

AD interaction:

$$\left(\frac{9.30+10.73+9.30+9.73+8.23+10.17+8.30+9.50}{8}\right)$$
$$-\left(\frac{9.23+9.76+8.27+9.83+10.10+10.97+8.50+10.77}{8}\right)=-0.27$$

AE interaction:

$$\left(\frac{10.73+9.76+9.30+8.27+10.10+8.23+10.77+9.50}{8}\right)$$
$$-\left(\frac{9.30+9.23+9.73+9.83+10.97+10.17+8.50+8.30}{8}\right)=0.08$$

BC **interaction:**

$$\left(\frac{9.30+9.23+9.73+9.83+10.10+8.23+10.77+9.50}{8}\right)$$
$$-\left(\frac{10.73+9.76+9.30+8.27+10.97+10.17+8.50+8.30}{8}\right)=0.11$$

BD **interaction:**

$$\left(\frac{9.30+9.23+9.30+8.27+10.97+10.17+10.77+9.50}{8}\right)$$
$$-\left(\frac{10.73+9.76+9.73+9.83+10.10+8.23+8.50+8.30}{8}\right)=0.29$$

BE **interaction:**

$$\left(\frac{9.23+9.76+9.30+9.73+10.10+10.97+8.30+9.50}{8}\right)$$
$$-\left(\frac{9.30+10.73+8.27+9.83+8.23+10.17+8.50+10.77}{8}\right)=0.14$$

CD **interaction:**

$$\left(\frac{9.30+9.23+10.73+9.76+8.50+8.30+10.77+9.50}{8}\right)$$
$$-\left(\frac{9.30+8.27+9.73+9.83+10.10+8.23+10.97+10.17}{8}\right)=-0.06$$

CE **interaction:**

$$\left(\frac{9.23+10.73+8.27+9.73+10.10+10.17+8.50+9.50}{8}\right)$$
$$-\left(\frac{9.30+9.76+9.30+9.83+8.23+10.97+8.30+10.77}{8}\right)=-0.03$$

DE **interaction:**

$$\left(\frac{9.23+10.73+9.30+9.83+8.23+10.97+8.50+9.50}{8}\right)$$
$$-\left(\frac{9.30+9.76+8.27+9.73+10.10+10.17+8.30+10.77}{8}\right)=-0.01$$

A summary of all main and two-factor interaction effects are ranked in descending order and percent median rank values are calculated as follows:

Order i	Effect value	Effect	Percent median rank
1	−0.74	A	4.5
2	−0.54	C	11.0
3	−0.52	E	17.5
4	−0.27	AD	24.0
5	−0.06	CD	30.5
6	−0.03	CE	37.0
7	−0.01	DE	43.5
8	0.05	D	50.0
9	0.07	AB	56.5
10	0.08	AE	63.0
11	0.11	BC	69.5
12	0.14	BE	76.0
13	0.16	AC	82.5
14	0.29	BD	89.0
15	1.28	B	95.5

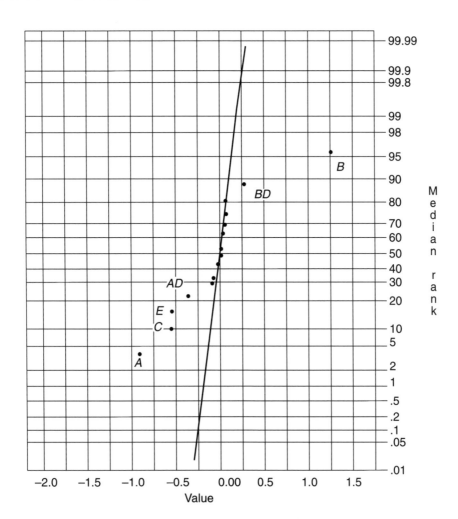

The effects are plotted on normal probability paper to estimate which effects and interactions are significant.

A normal probability plot of effects is shown in the following figure:

Based on the analysis of the normal probability plot, it appears that main effects A, C, E, and B are statistically significant. In addition, consideration should be given to the BD and AD interactions.

In order to maximize the power output factors, A, C, and E should be set at the $-$ level and factor B at the $+$ level.

Statistical Significance Based on t test

Step 1. Calculate the pooled standard deviation.

$$Sp = \sqrt{\frac{V_1 S_1^2 + V_2 S_2^2 + \ldots\ldots V_n S_n^2}{V_1 V_2 + \ldots\ldots V_n}}$$

$$Sp = \sqrt{\frac{(3)(0.10)^2 + (3)(0.15)^2 + (3)(0.12)^2 + \ldots\ldots(3)(0.21)^2 + (3)(0.36)^2}{3 + 3 + 3 + \ldots\ldots 3 + 3}}$$

$$Sp = \sqrt{\frac{3.9855}{48}} \quad Sp = 0.288$$

Step 2. Determine the statistical significance at the 90 percent level of confidence.

$$\text{Error, } E = t_{\alpha/2,(r-1)2^{k-f}}\left(Sp\sqrt{\frac{2}{pr}}\right)$$

$$\alpha = 0.10 \quad p = 8$$

$$f = 1 \quad r = 3 \quad k = 5$$

$$t_{\alpha/2,(r-1)2^{k-f}} = t_{.05,32} = 1.698$$

$$E = 1.698\left(0.288\sqrt{\frac{2}{(8)(3)}}\right) E = 0.14$$

Any effect ± 0.14 that contains zero is judged not statistically significant at the 90 percent confidence level.

Order *i*	Effect value	Effect	90% confidence interval	Statistically significant?
1	−0.74	A	−0.88 to −0.60	Yes
2	−0.54	C	−0.68 to −0.40	Yes
3	−0.52	E	−0.66 to −0.38	Yes
4	−0.27	AD	−0.41 to −0.13	Yes
5	−0.06	CD	−0.20 to 0.08	No
6	−0.03	CE	−0.17 to 0.11	No
7	−0.01	DE	−0.15 to 0.13	No
8	0.05	D	−0.09 to 0.19	No
9	0.07	AB	−0.07 to 0.21	No
10	0.08	AE	−0.06 to 0.22	No
11	0.11	BC	−0.03 to 0.25	No
12	0.14	BE	0.00 to 0.28	No
13	0.16	AC	0.02 to 0.30	Yes
14	0.29	BD	0.15 to 0.43	Yes
15	1.28	B	1.14 to 1.42	Yes

Conclusion:

Factors *A, B, C,* and *E* appear to be statistically significant at the 90 percent confidence level. The following factor setting will maximize the output.

Set *A, C,* and *E* at the − level, and set factor *B* at the + level.

Factor	Description	Optimum setting
A	Thickness of Sn	20 microns
B	Thickness of CdTe	50 microns
C	Bake temperature	350°
E	Source of CdTe	United States
D	Bake time	Does not significantly affect power output; therefore, set a lower time of 40 minutes to reduce cycle time.

VARIATION REDUCTION

Variation reduction is accomplished when we reduce the standard deviation; however, standard deviations are not normally distributed and cannot be used directly as a response to minimize. One approach is to take multiple observations during an experimental run, calculate the standard deviation, and take the log, or ln. The resulting transformed response can then be treated as a normally distributed response and the effects calculated in the traditional manner. The objective of the experiment is to reduce the log *S*, or ln *S*. Dr. George Box (1978) of the University of Wisconsin, Madison, and his colleagues have published a series of papers suggesting the use of $-\log 10(s)$ as a response variable when the objective is to reduce variation.

An alternative is to simply calculate the variance (standard deviation squared) and compare the average variance for all of the − factor settings to the average of all of the

+ settings using an F test. If the calculated F value exceeds a critical F value, we conclude that the factor under consideration is statistically significant.

Example:
The objective is to minimize the variation of compressive strength of a cast concrete material. A full three-factor experiment (2^3) was run. The experiment was replicated three times, and the standard deviation was determined using the three observations for each of the eight runs.

Run	A	B	C	AB	AC	BC	ABC	Average	S	S²
1	−	−	−	+	+	+	−	1900	3.9	15.21
2	+	−	−	−	−	+	+	1250	5.4	29.16
3	−	+	−	−	+	−	+	1875	15.1	228.01
4	+	+	−	+	−	−	−	1160	14.8	219.04
5	−	−	+	+	−	−	+	2100	10.3	106.09
6	+	−	+	−	+	−	−	1740	8.0	64.0
7	−	+	+	−	−	+	−	1940	13.6	184.96
8	+	+	+	+	+	+	+	1350	11.1	123.21

We now calculate an F value for each of the effects.

For effect A:

$$(\bar{S})^2{}_+ = \frac{29.16 + 219.04 + 64.0 + 123.21}{4} = 108.85$$

$$(\bar{S})^2{}_- = \frac{15.21 + 228.01 + 106.09 + 184.96}{4} = 133.57$$

$$F = \frac{(\bar{S})^2{}_{\text{larger}}}{(\bar{S})^2{}_{\text{smaller}}} \qquad F = \frac{133.57}{108.85} = 1.23$$

For effect B:

$$(\bar{S})^2{}_+ = \frac{228.01 + 219.04 + 184.96 + 123.21}{4} = 188.81$$

$$(\bar{S})^2{}_+ = \frac{15.21 + 29.16 + 106.09 + 64.0}{4} = 53.62$$

$$F = \frac{(\bar{S})^2{}_{\text{larger}}}{(\bar{S})^2{}_{\text{smaller}}} \qquad F = \frac{188.81}{53.62} = 3.52$$

The remaining F values are calculated in a similar manner.

Effect	*F* value	
A	1.23	
B	3.52	(Statistically significant at a confidence level of 90 percent)
C	1.03	
AB	1.09	
AC	1.25	
BC	1.75	
ABC	1.01	

A critical *F* value is determined. This value is looked up in an *F* table using one-half the degrees of freedom as calculated previously for the *t* value used in testing for statistical significance.

$$df = \frac{(r-1)2^{k-f}}{2}$$

$$df = 8$$

where: r = number of replicates = 3
k = number of factors = 3
f = degree of fractionation = 0 (this is a full factorial)

At a level of confidence of 0.90, the risk is 0.10; dividing the risk between the two levels, we have an $\alpha/2$ of 0.05.

The appropriate critical *F* value is found for $F_{.05,8,8} = 3.44$. Any effects that have a calculated *F* value greater than this are considered statistically significant. We can see that factor *B* is significant and that the average variance was smaller when factor *B* was set to the – level; therefore, we should set *B* at the – level to minimize variation. The other factors are not statistically significant contributors to variation.

In the previous example if the objective had been to maximize the compressive strength, the response would have been the average compressive strength. Calculation of all of these effects using the average compressive strength yields the following:

Factor	Effect
A	+579
B	−373
C	+236
AB	−74
AC	+104
BC	−109
ABC	−41

The calculated experimental error is determined to be

$$\text{Experimental error, } E = t_{\alpha/2,(r-1)2^{k-f}} \left(Sp \sqrt{\frac{2}{pr}} \right)$$

$$Sp = \text{the average standard deviation}$$

$$Sp = \sqrt{\frac{\Sigma S_i^2}{n}} = 10.38$$

Experimental error at 90 percent confidence = 7.4

$$t_{\alpha/2,(r-1)2^{k-f}} \left(Sp \sqrt{\frac{2}{pr}} \right) = t_{0.05,16} \left(10.38 \sqrt{\frac{2}{(4)(3)}} \right) = (1.746)(10.38)(0.41) = 7.4$$

At the 90 percent level of confidence, all of the factors in their interactions are statistically significant. This is due to the fact that the magnitude of the effects is relatively large compared to the experimental error of ±7.4.

In this case to maximize the compressive strength, we would set factors A and C to the + level and factor B to the − level. Setting B to the − level also minimizes the variation.

RESPONSE MODELING

Having determined the factor effects, we can establish a model to predict responses.

The expected response is determined by taking the grand average of all responses and adding the contribution of the various effects.

$$\text{Expected response, } X = \text{Grand average} + \left(\frac{\text{Effect } A}{2} \right) + \left(\frac{\text{Effect } B}{2} \right) + \left(\frac{\text{Effect } C}{2} \right) + \ldots$$

Consider the following case. An experiment was run using a full-factorial, three-factor experiment. The effects were as follows:

Factor	Effect	The average of all responses was 125.
A	+60	
B	−44	
C	+30	
AB	−12	
AC	+8	
BC	+10	
ABC	−4	

What is the expected maximum given these effects?

In order to maximize the response, we would set the factors at $A+$, $B-$, and $C+$.

$$\text{Maximum expected value} = \text{Grand average} + \left(\frac{\text{Effect } A}{2}\right) + \left(\frac{\text{Effect } B}{2}\right) + \left(\frac{\text{Effect } C}{2}\right) + \ldots$$

$$\text{Maximum expected value} = 125 + \left(\frac{60}{2}\right) + \left(\frac{44}{2}\right) + \left(\frac{30}{2}\right) = 192$$

This model only features the main-effects contribution. If we include all of the interactions, we simply include their effects with each one multiplied by the appropriate signs for the settings of the main effects.

$$\text{Maximum expected value} = 125 + \left(\frac{60}{2}\right) + \left(\frac{44}{2}\right) + \left(\frac{30}{2}\right) + \left((A)(B)\frac{\text{Effect } AB}{2}\right)$$

$$+ \left((B)(C)\frac{\text{Effect } BC}{2}\right) + \text{other terms}$$

where: (A) = setting for factor A = $(+1)$
          (B) = setting for factor B = (-1)
          (C) = setting for factor C = $(+1)$

The complete expression for the expected maximum value when the parameter settings are $A+$, $B-$, and $C+$ is

$$\text{Maximum expected} = 125 + 30 + 22 + 15 + (1)(-1)(-6) + (1)(1)(4) + (-1)(1)(5) + (1)(-1)(1)(2)$$
$$\qquad\qquad\qquad\qquad\qquad\qquad (AB) \qquad\quad (AC) \qquad\quad (BC) \qquad\qquad (ABC)$$

$$\text{Maximum expected} = 125 + 30 + 22 + 15 + 6 + 4 - 5 - 2 = 195$$

Note: Only those effects that are statistically significant should be used in the model. For the previous example, it is assumed that all effects are significant.

The same approach is used when minimization is the objective. The signs for the effects are reversed in those cases. For example, if the main effects for the factors were $A-$, $B+$, and $C-$, then setting A to the + level, B to the – level, and C to the + level would minimize the response.

We may use this modeling approach to determine where to set parameters in order to achieve a target value.

Example:
A manufacturer of ice cream desires to achieve a target fill weight of 2.50 pounds. Two factors will be examined to determine where they should be set to achieve the target weight. These factors are the filling temperature and the percent overfill. Overfill is the percentage of air incorporated into the ice cream.

Factors and settings are as follows:

Factor	Description	+	−
A	Fill temperature, °F	25	20
B	Overfill, %	110	90

Experimental response: Fill weight in pounds

This experiment will be a full 2^2 factorial (two factors at two levels) and will require four runs. No replicates will be run.

The design matrix and responses are shown in the following table.

Run	A	B	AB	Response
1	−	−	+	2.31
2	+	−	−	2.82
3	−	+	−	2.16
4	+	+	+	2.38

Effects are calculated as follows:

A main effect:

$$\left(\frac{2.82+2.38}{2}\right)-\left(\frac{2.31+2.16}{2}\right)=2.60-2.24=+0.36$$

B main effect:

$$\left(\frac{2.16+2.38}{2}\right)-\left(\frac{2.31+2.82}{2}\right)=2.27-2.57=-0.30$$

AB interaction effect:

$$\left(\frac{2.31+2.38}{2}\right)-\left(\frac{2.82+2.16}{2}\right)=2.35-2.49=-0.14$$

The general prediction equation is given by

$$\hat{Y}=\text{Grand average}+\left(\frac{\text{Effect }A}{2}\right)(A)+\left(\frac{\text{Effect }B}{2}\right)(B)+\left(\frac{\text{Effect }AB}{2}\right)(A)(B)$$

where: $\hat{Y}$ = the expected value of the response

For our specific case where the target value is $y = 2.50$, this expression becomes

$$2.50 = 2.42 + 0.18A - 0.15B - 0.07AB$$

Since we have one prediction equation with two variables, we must fix one. Suppose we set factor A to the + setting, or 25°. The resulting prediction equation becomes

$$\hat{Y} = 2.42 + 0.18(+1) - 0.15(B) - 0.07(+1)(B)$$
$$2.50 = 2.42 + 0.18 - 0.15(B) - 0.07(B)$$
$$2.50 = 2.60 - 0.22(B)$$
$$-0.10 = -0.22(B)$$
$$B = +0.45$$

In order for $B = +0.45$ to be utilized in setting the process factor B (percent overrun), we must translate the coded factor level to a process setting. This can be visualized in the following diagram:

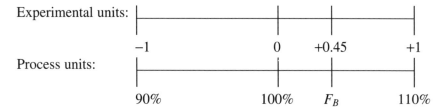

The process setting F_B can be determined by

$$\frac{0.45 - 0.00}{F_B - 100} = \frac{1.0 - 0.0}{110 - 100}$$
$$(F_B - 100)(1) = 0.45(110 - 100)$$
$$F_B - 100 = 4.5$$
$$F_B = 104.5\%$$

The proper process conditions to achieve a fill weight of 2.50 pounds are setting the temperature to 25°F and the overfill to 104.5 percent.

AN ALTERNATIVE METHOD FOR EVALUATING EFFECTS BASED ON STANDARD DEVIATION

Montgomery has shown that $\ln S^2_+ / S^2_-$ has an approximate standard normal distribution. Using the standard deviation as a response and calculating the variance of the average standard deviations where the factor is $+$ and $-$ and then calculating the natural log of the $+$ term divided by the $-$ term yields a test statistic. If the absolute value of this test statistic is greater than $Z_{\alpha/2}$, we may conclude that the effect is statistically significant. We are essentially performing a hypothesis test.

1. $Ho: \sigma_+^2 = \sigma_-^2$
   $Ha: \sigma_+^2 \neq \sigma_-^2$
2. Compute the test statistic.

If $\left| \ln \dfrac{(\overline{S}_+)^2}{(\overline{S}_-)^2} \right| > Z_{\alpha/2}$, reject *Ho* and accept *Ha*. Conclude that the effect is significant.

For a 95 percent level of confidence, $Z_{\alpha/2} = 1.96$.

Example:

Consider the 2^2 experiment where the objective is to minimize the variation in the percent shrinkage of pressure-treated wood. Note that we are not minimizing the shrinkage but rather, the variation in the shrinkage. Two factors will be evaluated. Factor *A* is pressure, where + = 150 psi and − = 100 psi. Factor *B* is time, where + = 4 hours and − = 2 hours. Four experimental runs are planned. Three replicates are run, and the percent shrinkage is determined for each run.

Run	A	B	AB	Responses	Average	Standard deviation
1	−	−	+	2.31, 2.42, 2.46	2.397	0.0777
2	+	−	−	2.46, 2.51, 2.53	2.500	0.0361
3	−	+	−	4.86, 4.77, 4.28	4.637	0.3121
4	+	+	+	3.75, 3.87, 4.61	4.077	0.4658

Effects based on the standard deviation are as follows:

A main effect:

$$\left(\frac{0.0361 + 0.4658}{2} \right) - \left(\frac{0.0777 + 0.3121}{2} \right) = 0.2510 - 0.1949 = 0.0561$$

B main effect:

$$\left(\frac{0.3121 + 0.4658}{2} \right) - \left(\frac{0.0777 + 0.361}{2} \right) = 0.3890 - 0.0569 = 0.3321$$

AB interaction:

$$\left(\frac{0.0777 + 0.4658}{2} \right) - \left(\frac{0.0361 + 0.3121}{2} \right) = 0.2718 - 0.1741 = 0.0977$$

Test for statistical significance as follows:

Effect A:

$$\overline{S}_+^2 = \left(\frac{0.0361 + 0.4658}{2} \right)^2 = 0.0630$$

$$\overline{S}_-^2 = \left(\frac{0.0777 + 0.3121}{2} \right)^2 = 0.0380$$

$$\left| \ln \frac{(\overline{S}_+)^2}{(\overline{S}_-)^2} \right| = \left| \ln \frac{0.0630}{0.0380} \right| = 0.51 \qquad 0.51 \ngtr 1.96; \text{ therefore, factor } A \text{ is not significant.}$$

Effect B:

$$\bar{S}_+^2 = \left(\frac{0.3121 + 0.4658}{2}\right)^2 = 0.1513$$

$$\bar{S}_-^2 = \left(\frac{0.0777 + 0.0361}{2}\right)^2 = 0.0032$$

$$\left|\ln\frac{(\bar{S}_+)^2}{(\bar{S}_-)^2}\right| = \left|\ln\frac{0.1513}{0.0361}\right| = 3.86 \quad 3.86 > 1.96; \text{ therefore, factor } A \text{ is significant.}$$

Effect AB interaction:

$$\bar{S}_+^2 = \left(\frac{0.0777 + 0.4658}{2}\right)^2 = 0.0738$$

$$\bar{S}_-^2 = \left(\frac{0.0361 + 0.3121}{2}\right)^2 = 0.0303$$

$$\left|\ln\frac{(\bar{S}_+)^2}{(\bar{S}_-)^2}\right| = \left|\ln\frac{0.0738}{0.0303}\right| = 0.89 \quad 0.89 \ngtr 1.96; \text{ therefore, the } AB \text{ interaction is not significant}$$

Conclusion:

To minimize the variation in the actual percent shrinkage, the time of treatment should be two hours. The pressure is not a statistically significant factor with respect to minimizing variation. Note that we reverse the signs of the effect to determine the setting for the factor in order to minimize the response. Since the sign for the main effect for B is $+$, we chose to set factor B to the $-$ setting, or two hours.

No calculated effects were made using the shrinkage response; however, it is obvious that setting $B-$ would minimize the actual percent shrinkage as well as the variation in the shrinkage.

PLACKETT-BURMAN SCREENING DESIGNS

These series of designs are modestly useful for screening but should be used with great caution. At best, the Plackett-Burman designs are a resolution III. Confounding of main effects with two- and three-factor interactions are significant.

The design columns for several Plackett-Burman designs follow:

$n = 8$ $+ + + - + - -$

$n = 12$ $+ + - + + + - - - + -$

$n = 16$ $+ + + + - + - + + - - + - - -$

$n = 20$ $+ + - - + + + + - + - + - - - - + + -$

$n = 24$ $+ + + + + - + - + + - - + + - - + - + - - - -$

To generate the design matrix, the appropriate generating column is selected. The first column is generated by placing the signs in a vertical column. The last sign of the first column becomes the first sign in the second column, and the remainder of the second column follows the sequence designated by the generating column. After $n - 1$ columns have been developed, the final row is made by using all −'s.

An example is shown for a seven-factor, eight-run Plackett-Burman design.

Run	A	B	C	D	E	F	G
1	+	−	−	+	−	+	+
2	+	+	−	−	+	−	+
3	+	+	+	−	−	+	−
4	−	+	+	+	−	−	+
5	+	−	+	+	+	−	−
6	−	+	−	+	+	+	−
7	−	−	+	−	+	+	+
8	−	−	−	−	−	−	−

In the event that less than eight factors are to be screened, simply drop the required number of columns (factors).

Case study:

You have been asked to examine a process with respect to three factors that are suspected to influence a response. The objective is for the process to yield a response of 55.0 with a minimum of variation. The factors and their settings for this experiment are as follows:

The design matrix and data are as follows:

			Setting	
Factor	Description	−	+
A	Feed rate, lbs/hr.	40	60
B	Temperature, °C	30	70
C	Percent filler	22	30

Run	A	B	C	AB	AC	BC	ABC	I	II	Average	Standard deviation
1	−	−	−	+	+	+	−	40.48	45.52	43.0	3.56
2	+	−	−	−	−	+	+	30.01	27.99	29.0	1.43
3	−	+	−	−	+	−	+	50.92	67.08	59.0	5.71
4	+	+	−	+	−	−	−	42.05	31.95	37.0	7.14
5	−	−	+	+	−	−	+	69.69	72.31	71.0	1.85
6	+	−	+	−	+	−	−	47.95	50.05	49.0	1.48
7	−	+	+	−	−	+	−	69.54	64.46	67.0	3.59
8	+	+	+	+	+	+	+	42.46	47.54	45.0	9.98

Effects based on average response are as follows:

Main effect A:

$$\left(\frac{29+37+49+45}{4}\right)-\left(\frac{43+59+71+67}{4}\right)=-10$$

Main effect C:

$$\left(\frac{71+49+67+45}{4}\right)-\left(\frac{43+29+59+37}{4}\right)=+16$$

Interaction effect AC:

$$\left(\frac{43+59+49+45}{4}\right)-\left(\frac{29+37+71+67}{4}\right)=-2$$

Interaction ABC:

$$\left(\frac{29+59+71+45}{4}\right)-\left(\frac{43+37+49+67}{4}\right)=2$$

Main effect B:

$$\left(\frac{59+37+67+45}{4}\right)-\left(\frac{43+29+71+49}{4}\right)=+4$$

Interaction effect AB:

$$\left(\frac{43+37+71+45}{4}\right)-\left(\frac{29+59+49+67}{4}\right)=-2$$

Interaction BC:

$$\left(\frac{43+29+67+45}{4}\right)-\left(\frac{59+37+71+49}{4}\right)=-8$$

Determine the statistical significance for effects based on average responses. The level of confidence is chosen to be 95 percent. The error for the calculated effects is given by

$$\text{Error, } E = t_{\alpha/2,(r-1)2^{k-f}}\left(Sp\sqrt{\frac{2}{pr}}\right)$$

where: $\alpha = 0.05$
$\quad\quad\quad r = 2$
$\quad\quad\quad f = 0$
$\quad\quad\quad p = 4$

$$Sp = \sqrt{\frac{(3.56)^2 + (1.43)^2 + \ldots + (9.98)^2}{8}} = 5.20 \qquad \sqrt{\frac{2}{(4)(2)}} = 0.50$$

$$t_{0.025,8} = 2.303$$

Error, $E = (2.303)(5.20)(0.50) = 5.99$

Any effects greater than the experimental error 5.99 are deemed statistically significant at the 95 percent level. The effects that are significant are main effects A, C, and the two-factor interaction effect BC.

The model for predicting the response using only those effects that are significant is

$Y_{expected}$ = grand average – ½(effect A)(A) + ½(effect C)(C) – ½(effect BC)(B)(C)

$Y_{expected}$ = 50 – 5(A) + 8(C) – 4(B)(C)

Effects based on standard deviation are as follows:

Main effect A:

$$\left(\frac{1.43 + 7.14 + 1.48 + 9.98}{4}\right) - \left(\frac{3.56 + 5.71 + 1.85 + 3.59}{4}\right) = 0.62$$

Main effect B:

$$\left(\frac{5.71 + 7.14 + 3.59 + 9.98}{4}\right) - \left(\frac{3.56 + 1.43 + 1.85 + 1.48}{4}\right) = 2.31$$

Main effect C:

$$\left(\frac{1.85 + 1.48 + 3.59 + 9.98}{4}\right) - \left(\frac{3.56 + 1.43 + 5.71 + 7.14}{4}\right) = 0.11$$

Interaction AB effect:

$$\left(\frac{3.56 + 7.14 + 1.85 + 9.98}{4}\right) - \left(\frac{1.43 + 5.71 + 1.48 + 3.59}{4}\right) = 1.23$$

Interaction AC effect:

$$\left(\frac{3.56 + 5.71 + 1.48 + 9.98}{4}\right) - \left(\frac{1.43 + 7.14 + 1.85 + 3.59}{4}\right) = 0.78$$

Interaction BC effect:

$$\left(\frac{3.56 + 1.43 + 3.59 + 9.98}{4}\right) - \left(\frac{5.71 + 7.14 + 1.85 + 1.48}{4}\right) = 0.27$$

Interaction *ABC* effect:

$$\left(\frac{1.43+5.71+1.85+9.98}{4}\right)-\left(\frac{3.56+7.14+1.48+3.59}{4}\right)=0.37$$

Determine the statistical significance for effects based on standard deviation as a response. Each one of the effects based on standard deviation will be tested independently for statistical significance. The level of significance chosen is 90 percent.

The test statistic will be the natural log of the ratio of the square of the average standard deviation of the −'s to the +'s.

$$\text{Test statistic: } \left|\ln\frac{(\bar{S}_+)^2}{(\bar{S}_-)^2}\right|$$

If the test is greater than $Z_{\alpha/2}$, we will reject the hypothesis that the effect is not statistically significant at a level of confidence of $1-\alpha$. In this case, the test statistic must be greater than 1.645.

Calculate the test statistic for the effects as follows:

Main effect *A*:

$$(\bar{S}_+)^2 = \left(\frac{1.43+7.14+1.48+9.98}{4}\right)^2 = 5.01$$

$$(\bar{S}_-)^2 = \left(\frac{3.56+5.71+1.85+3.59}{4}\right)^2 = 3.68$$

$$\left|\ln\frac{(\bar{S}_+)^2}{(\bar{S}_-)^2}\right| = \left|\ln\frac{5.01}{3.68}\right| = \left|\ln\frac{25.10}{13.54}\right| = |\ln 1.85| = 0.62$$

Since the test statistic is less than the critical value of 1.645, we cannot reject the null hypothesis that the variation from a positive setting of *A* is any different than the variation from a negative setting of *A*. We conclude that factor *A* is not a significant factor affecting variation.

Similar calculations for all of the factors and interactions give the following test statistics:

Factor	Test statistic	
A	0.62	
B	2.31	(Statistically significant at a confidence level of 90%)
C	0.11	
AB	1.23	
AC	0.78	
BC	0.27	
ABC	0.37	

In order to minimize variation, factor *B* must be set at the − setting, or 30°C.

ACHIEVING THE FINAL OBJECTIVE

The target value for the primary response is to obtain a value of 55.0 with a minimum of variation. Minimum variation is obtained with factor B set at the $-$ setting, or 30° C. The expected value for Y now becomes

$$Y_{\text{expected}} = 50 - 5(A) + 8(C) - 4(B)(C) \text{ since } B = -1$$

$$Y_{\text{expected}} = 50 - 5(A) + 8(C) - 4(-1)(C)$$

$$Y_{\text{expected}} = 50 - 5(A) + 12(C)$$

Setting $Y = 55$, we may adjust factor A or C to achieve the desired target value. Of the two factors, factor A had more of an affect on variation, even though it was not statistically significant. We will set factor A to its negative setting (40 lb./hr. feed rate).

This leaves only factor C to adjust to target.

$$Y_{\text{expected}} = 55 - (5)(-1) + 12(C)$$

$$Y_{\text{expected}} = 60 + 12(C)$$

$$55 = 60 + 12(C)$$

$$C = -0.42$$

This setting is in standardized units of the experimental design, where $-1 = 22\%$, $0 = 26\%$, and $+1 = 30\%$.

Setting the percent filler to 24.4 percent will achieve a desired response of 55.0.

BIBLIOGRAPHY

Barrentine, L. B. 1999. *An Introduction to Design of Experiments.* Milwaukee, WI: ASQ Quality Press.

Box, G. E. P., W. G. Hunter, and J. S. Hunter. 1978. *Statistics for Experimenters.* New York: John Wiley and Sons.

Montgomery, D. C. 1997. *Design and Analysis of Experiments.* 4th edition. New York: John Wiley and Sons.

Schmidt, S. R., and R. G. Launsby. 1994. *Understanding Industrial Designed Experiments.* 4th edition. Colorado Springs, CO: Air Academy Press.

DISCRETE DISTRIBUTIONS

When the characteristic being measured can only take on integer values such as 0, 1, 2, . . . , the probability distribution is a *discrete distribution*. The number of nonconforming units in a sample would be a discrete distribution.

HYPERGEOMETRIC DISTRIBUTION

The hypergeometric distribution is the only discrete distribution that has the lot or population size as an element. The hypergeometric distribution is appropriately used when randomly sampling n items without replacement from a lot of N items of which D items are defective. The probability of finding exactly x defective items is given by

$$P(x) = \frac{\binom{D}{x}\binom{N-D}{n-x}}{\binom{N}{n}}$$

where: D = Nonconforming units in the population or lot
N = Size of the population or lot
n = Sample size
x = Number of nonconforming units in the sample

Note: The expression $\binom{a}{b} \approx \frac{a!}{b!(a-b)!}$

The mean and standard deviation of the hypergeometric distribution are

$$Mean,\ \mu = \frac{nD}{N}$$

and

$$Standard\ deviation,\ \sigma = \sqrt{\frac{nD}{N}\left(1 - \frac{D}{N}\right)\left(\frac{N-n}{N-1}\right)}$$

Example application of the hypergeometric distribution:
A collection (lot) of 60 parts is known to have 15 nonconforming. If a sample of 20 is randomly chosen from this lot, what is the probability that exactly three nonconforming parts will be found?

Lot size $N = 60$

Sample size $n = 20$

Number defective in lot $D = 14$

$$P(x) = \frac{\binom{15}{3}\binom{60-15}{20-3}}{\binom{60}{20}} \quad P(x) = \frac{(455)(1.103 \times 10^{12})}{4.1918 \times 10^{15}}$$

$$P(x) = 0.1197$$

If we want to know the probability of getting three or less nonconforming units $P(£3)$, we must sum the probabilities of getting exactly zero, one, two, and three.

$$P(x \leq 3) = P(x = 0) + P(x = 1)\ P(x = 2) + P(x = 3)$$

Using the hypergeometeric distribution function, we determine these individual probabilities and the sum as

$$P(x \leq 3) = 0.0008 + 0.0087 + 0.043 + 0.1197$$

$$P(x \leq 3) = 0.1722$$

Problem:

A commuter airplane has 28 passengers; nine of them are womem. Assuming a random order of deplaning, what is the probability that at least one woman will be in the first five that deplane? What is the probability that the first five deplaning will be all women?
Answer: 0.4732 and 0.0013

BINOMIAL DISTRIBUTION

The binomial distribution is independent of the population size and assumes the following:

1. Only two outcomes are possible for each trial. In quality applications, the only outcomes resulting from an inspection are that the items are either conforming or nonconforming to a requirement. Only two possibilities are available—good or bad.
2. Independence of outcome means that the result of one trial does not influence the outcome of another trial. For example, tossing a coin and getting a head does not influence the probability of getting a head on the next toss. The probability of getting a head in a fair toss will always be 50 percent.

The standard deviation of the binomial is given by

$$\sigma = \sqrt{npq}$$

where: n = sample size
p = proportion defective
$q = 1 - p$

The mean is given by

$$\mu = np$$

The binomial equation or probability distribution function is

$$P(x) = \binom{n}{x} P^x Q^{n-x}$$

Example 1:
A process is known to perform at 7.0 percent nonconforming. If a sample of 68 is chosen at random, what is the probability of getting exactly three nonconforming units?

$n = 68$

$p = 0.07$

$q = 0.93$

$x = 3$

$$P(x) = \left(\frac{n}{x}\right) P^x Q^{n-x}$$

$$P(x = 3) = \left(\frac{68}{3}\right)(0.07)^3 (0.93)^{65} = (50116)(1.00034)(0.0089) = 0.1517$$

Example 2:
A process is running approximately 25 percent defective. What is the probability that three or less defective will be found in a random sample of 20 units?

The probabilities of getting exactly zero, one, two, and three must be determined and these individual probabilities summed in order to determine the probability of getting three or less.

$$P(x \leq 3) = P(x = 0) + P(x = 1) + P(x = 2) + P(x = 3)$$

The probability of getting exactly zero is a unique case for the binomial equation. The binomial equation for this case reduces to

$$P(x = 0) = (1 - p)^n$$

or

$$P(x) = Q^n$$

In this example, $P(x = 0) = (0.75)^{20} = 0.0032$.

The probability of getting exactly one defective is:

$$P(x=1) = \binom{20}{1}(0.25)^1 (0.75)^{19} = (20)(0.25)(0.0042) = 0.0211$$

In a similar manner, we calculate the probability of getting exactly two and three defective.

$$P(x=2) = \binom{20}{2}(.25)^2 (.75)^{18} = 0.0669$$

$$P(x=3) = \binom{20}{3}(.25)^3 (.75)^{17} = 0.1339$$

The probability of getting three or less defective is the sum of the probabilities of getting zero, one, two, and three defective.

$$P(x \leq 3) = 0.0032 + 0.0211 + 0.0669 + 0.1339 = 0.2281$$

POISSON DISTRIBUTION

The Poisson distribution can be thought of as a limiting form of the binomial distribution. If the factor p in the binomial approaches zero and the sample n approaches infinity such that the term np approaches a constant λ, the binomial becomes the Poisson distribution.

The Poisson distribution has only one parameter λ, which is the product of the process proportion defective p and the sample size n chosen from the process or population.

The Poisson distribution is given by

$$P(x) = \frac{e^{-\lambda}\lambda^x}{x!}$$

where: $\lambda = np$

The Poisson is unique in that both the variance and mean are equal to λ.

$$\text{Mean, } \mu = \lambda = np$$

$$\text{Standard deviation, } \sigma = \sqrt{np}$$

A typical application of the Poisson distribution is that of the U and C control charts where defects per unit or defects are monitored with respect to time. Any randomly occurring event that occurs on a per unit basis (such as defects per hour, defects per meter, or defects per square yard) represents a suitable application of the Poisson distribution.

Example 1:
A textbook has 450 pages on which exactly nine typographical errors are found. These errors are randomly distributed throughout the text. What is the probability of not finding an error on 40 randomly selected pages?

$$p = \text{proportion nonconforming} = \frac{9}{450} = 0.02$$

$$np = \lambda = (40)(0.02) = 0.89$$

$$x = 0$$

$$P(x) = \frac{e^{-\lambda}\lambda^x}{x!} \quad P(x = 0) = \frac{e^{-0.8}\lambda^0}{0!} = \frac{(0.4493)(1)}{1} = 0.4493$$

Example 2:
During a period of one week, 126,000 garments have been manufactured in which 4536 defects were found. If the process continues to perform at this level, what is the probability of finding no more than two defects in a sample of 60 garments?

$$p = \text{proportion defective} = \frac{4536}{126,000} = 0.036$$

$$n = \text{sample size} = 60$$

$$np = \lambda = (60)(0.063) = 2.16$$

The probability of finding no more than two is the probability of finding two or less defective. This probability is determined by summing the probabilities of finding exactly zero, one, and two defects.

$$P(x \le 2) = P(x = 0) + P(x = 1) + P(x = 2)$$

$$P(x = 0) = \frac{e^{-\lambda}\lambda^x}{x!} = \frac{e^{-2.16}\lambda^0}{0!} = \frac{(0.1153)(1)}{1} = 0.1153$$

$$P(x = 1) = \frac{e^{-2.16}\lambda^1}{1!} = \frac{(0.1153)(2.16)^1}{1!} = 0.2490$$

$$P(x = 2) = \frac{e^{-2.16}\lambda^2}{2!} = \frac{(0.1153)(2.16)^2}{2!} = 0.2690$$

The probability of getting two or less defective is the sum of the probabilities of getting zero, one, and two defective.

$$P(x \le 2) = 0.1153 + 0.2490 + 0.2690 + 0.6333$$

THE NORMAL DISTRIBUTION AS AN APPROXIMATION OF THE BINOMIAL

The binomial distribution is appropriate for independent trials, each of which has a probability p and a number of trials n. If the number of trials n is large, then the central limit theorem may be used to justify the normal distribution with a mean of np and a variance of $np(1-p)$ as an approximation to the binomial.

$$P(x = a) = \frac{1}{\sqrt{2\pi np(1-p)}} e^{\frac{1}{2}[(a-np)^2 / np(1-p)]}$$

The binomial distribution is discrete, whereas the normal distribution is continuous. The normal distribution deals with variables that can take on any value, not necessarily only integers. It is appropriate to use a *continuity correction factor* in the use of the normal distribution.

$$P(x) \cong \Theta\left(\frac{x + 0.5 - np}{\sqrt{2\pi np(1-p)}}\right) - \Theta\left(\frac{x - 0.5 - np}{\sqrt{2\pi np(1-p)}}\right)$$

where: Θ = the cumulative normal distribution function

The normal distribution is appropriate and acceptable for p of approximately 0.50 and $n > 10$. For larger values of p, larger sample sizes are needed. The normal distribution is not appropriate for $p < 1/(n + 1)$ or $p > n/(n + 1)$.

Example 1:
A sample of $n = 80$ is chosen from a population characterized by $p = 0.15$. What is the probability of getting exactly two defective using the normal distribution?

$$np > 10 \text{ and } p < .50$$

$n = 80$

$p = 0.15$

$np = 12.0$

$$P(x = 2) = \Theta\left(\frac{2 + 0.5 - 12}{\sqrt{2(3.14)(12)(1 - .30)}}\right) - \Theta\left(\frac{2 - 0.5 - 12}{\sqrt{2(3.14)(12)(1 - .3)}}\right)$$

$$P(x = 2) = \Theta(-1.31) - \Theta(-1.45)$$

$$P(x = 2) = 0.09509 - 0.07353$$

$$P(x = 2) = 0.02156$$

Example 2:
A sample of $n = 60$ items is chosen from a process where $p = 0.15$. What is the probability of getting exactly two defective using the normal, binomial, and Poisson distribution as a model? Which distribution would be the best for this calculation?

Using the normal distribution:

$n = 60$

$p = 0.15$

$np = (60)(0.15) = 9.0$

$x = 2$

$$P(x=2) = \Theta\left(\frac{x+0.5-np}{\sqrt{2\pi np(1-p)}}\right) - \Theta\left(\frac{x-0.5-np}{\sqrt{2\pi np(1-p)}}\right)$$

$$P(x=2) = \Theta\left(\frac{2+0.5-9}{\sqrt{2(3.14)(9)(1-.15)}}\right) - \Theta\left(\frac{2-0.5-9}{\sqrt{2(3.14)(9)(1-.15)}}\right)$$

$$P(x=2) = \Theta(-0.94) - \Theta(1.08)$$

$$P(x=2) = 0.17361 - 0.14007$$

$$P(x=2) = 0.03354$$

Using the binomial distribution:

$n = 60$

$p = 0.15$

$q = 0.85$

$x = 2$

$$P(x=2) = \binom{n}{x} p^x q^{n-x}$$

$$P(x=2) = \binom{60}{2}(0.15)^2 (0.85)^{60-2}$$

$$P(x=2) = (1770)(0.0225)(0.000081) = 0.0032$$

Using the Poisson distribution:

$$P(x) = \frac{e^{-\lambda}\lambda^x}{x!}$$

$$P(x=2) = \frac{e^{-9.0}(9.0)^2}{2!}$$

$$P(x=2) = \frac{(0.000123)(81)}{2}$$

$$P(x=2) = 0.0050$$

Summary:

Distribution	P(x = 2)
Normal	0.0335
Binomial	0.0032
Poisson	0.0050

PASCAL DISTRIBUTION

The Pascal distribution is similar to the binomial in that it is based on independent trials with the probability of success for a given trial not being influenced by the result of other trials. If we have a series of independent trials with a probability of success of p, then x will be the trial on which the rth success will occur. This value of x is a Pascal random variable. The probability distribution function defining this function is given by

$$P(x) = \binom{x-1}{r-1} p^r (1-p)^{x-r} \quad x = r,\ r+1,\ r+2,\ \ldots$$

A special case of the Pascal distribution is when $r = 1$. In this case, we are interested in determining the probability of the number of trials required until the first success.

This particular case where we fix the number of successes to one and determine the probability of the number of trials is the **geometric distribution.** Other values of $r > 0$ (and r is not necessarily an integer) define the **negative binomial distribution.**

Example 1:
Given a process performing at $p = 0.08$, what is the probability of getting a total of two defective on the fifteenth sample?

$p = 0.08$

$x = 15$

$r = 2$

$$P(x) = \binom{x-1}{r-1} p^r (1-p)^{x-r}$$

$$P(x = 15) = \binom{14}{1} (0.08)^2 (0.92)^{13}$$

$$P(x = 15) = (14)(0.0064)(0.3383)$$

$$P(x = 15) = 0.0303$$

CUMULATIVE DISTRIBUTION TABLES

In order to facilitate expeditious calculations of Poisson probabilities, cumulative probability tables have been developed. A vertical column provides for various values for x, and a horizontal heading provides for values of np or λ. The intersection of the appropriate row and column will give the probability of getting x or less defective.

The following illustrates the use of a cumulative Poisson distribution table.

What is the probability of finding two or less defective in a sample of $n = 60$ chosen from a process where $p = 0.0833$?

$x = 2$

$n = 60$

$p = 0.083$

$\lambda = (60)(0.083) = 5.0$

Moving across the horizontal, we look up the value for λ (5.00). We then move down the vertical axis until the value for x (2) is found. At the intercept, the probability of 0.125 is given. This is the cumulative probability of getting two or less where $\lambda = 5.00$.

$P(x \leq 2) = P(x = 0) + P(x = 1) + P(x = 2)$

$P(x \leq 2) = 0.0067 + 0.0337 + .0842$

$P(x \leq 2) = 0.1246$

					λ					
x:	4.1	4.2	4.3	4.4	4.5	4.6	4.7	4.8	4.9	5.0
0	0.017	0.015	0.014	0.012	0.011	0.010	0.009	0.008	0.007	0.007
1	0.085	0.078	0.072	0.066	0.061	0.056	0.052	0.048	0.044	0.040
2	0.224	0.210	0.197	0.185	0.174	0.163	0.152	0.143	0.133	0.125
3	0.414	0.395	0.377	0.360	0.342	0.326	0.310	0.294	0.279	0.265
4	0.609	0.590	0.570	0.551	0.532	0.513	0.495	0.476	0.458	0.441
5	0.769	0.753	0.737	0.720	0.703	0.686	0.668	0.651	0.634	0.616
6	0.879	0.868	0.856	0.844	0.831	0.818	0.805	0.791	0.777	0.762
7	0.943	0.936	0.929	0.921	0.913	0.905	0.896	0.887	0.877	0.867

The following figure shows the selection of distributions for the approximation of the hypergeometric:

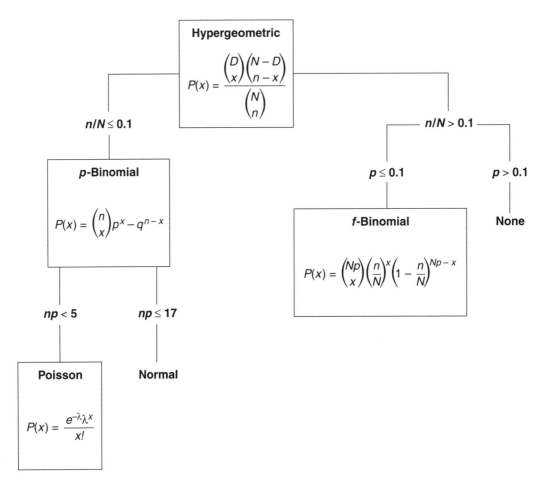

BIBLIOGRAPHY

Besterfield, D. H. 1994. *Quality Control.* 4th edition. Englewood Cliffs, NJ: Prentice Hall.

Grant, E. L, and R. S. Leavenworth. 1996. *Statistical Quality Control.* 7th edition. New York: McGraw Hill.

Hayes G. E, and H. G. Romig. 1982. *Modern Quality Control.* 3rd edition. Encino, CA: Glencoe.

Juran J. M., and F. M. Gryna. 1980. *Quality Planning and Analysis.* 3rd edition. New York, NY: McGraw-Hill.

Montgomery, D. C. 1996. *Introduction to Statistical Quality Control.* 3rd edition. New York: John Wiley and Sons.

Petruccelli, J. D., B. Nandram, and M. Chen. 1999. *Applied Statistics for Engineers and Scientists.* Upper Saddle River, NJ: Prentice Hall.

Sternstein, M. 1996. *Statistics.* Hauppauge, NY: Barron's Educational Series.

Walpole, R. E., and R. H. Myers. 1993. *Probability and Statistics for Engineers and Scientists.* 5th edition. Englewood Cliffs, NJ: Prentice Hall.

EVOLUTIONARY OPERATION, EVOP

An alternative approach to process optimization is the use of evolutionary operation, or EVOP. First introduced by George Box in 1957, the technique of EVOP is closely related to Sequential Simplex Optimization. It can be performed for any number of process variables but from a practical perspective is limited to three process variables. The original method developed by Box utilized the range of observations to estimate the standard deviation, which was used to measure experimental error and serve as a measure of statistical significance. The method presented here will use the standard deviation calculated directly and the statistical significance of effects evaluated using the *t*-distribution.

The underlying principle of EVOP is to make small process changes during the phase and replicate the conditions (cycles) until there is positive evidence that one of the conditions moves the process into an area of improvement. The following two-factor EVOP will serve as an example of the EVOP process.

The current operating conditions are used to produce a pressure-sensitive adhesive. Polymer concentration is 54.0 percent, and the coating thickness is 0.8 mils. The response variable is peel strength in pounds per linear inch. We will vary the operating parameters slightly to evaluate the response. After deciding on the amount of change, we will obtain peel strength for each new condition. This initial result of one response per setup is called *cycle 1*.

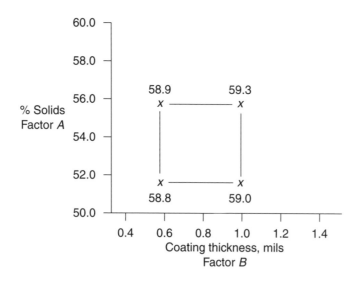

The initial trial change will be to set the percentage of total solids to 52.0 percent and 56.0 percent (±2 percent) around the normal operating condition and to set the coating thickness to 0.6 and 1.0 mils.

The operating conditions and the responses are treated as a traditional 2^2, two-level experiment.

Factor	Description	−	+
A	% Solids	52.0	56.0
B	Coating thickness	0.6	1.0

Phase 1, cycle 1:

Run	A	B	Response
1	−	−	58.8
2	+	−	58.9
3	−	+	59.0
4	+	+	59.3

The effects will not be calculated, because with only a single observation, we cannot determine the experimental error. An additional set of runs (cycle 2) will be made using the same set of operating conditions (phase 1).

Phase 1, cycle 2:

Run	A	B	Cycle 1	Cycle 2	Average	Standard deviation, S	Variance, S^2
1	−	−	58.8	59.1	58.95	0.212	0.0449
2	+	−	58.9	58.8	58.85	0.071	0.0050
3	−	+	59.0	58.4	58.70	0.424	0.1798
4	+	+	59.3	60.0	59.65	0.495	0.2450

Main effects:

$$A: \left(\frac{58.85+59.65}{2}\right)-\left(\frac{58.95+58.70}{2}\right)=59.25-58.83=0.42$$

$$B: \left(\frac{58.70+59.65}{2}\right)-\left(\frac{58.95+58.85}{2}\right)=59.18-58.90=0.28$$

Error of effects:

$$t_{\alpha/2,(r-1)2^{k-f}}\left(Sp\sqrt{\frac{2}{pr}}\right)$$

where: α = risk
   k = number of factors
   f = number of times the experiment is fractionated
   r = number of replicates or cycles
   Sp = pooled standard deviation
   p = number of +'s in a column

For this example, we will use a 95 percent confidence level ($\alpha = 0.05$).

$\alpha = 0.05$

$k = 2$

$f = 0$

$r = 2$

$p = 2$

$Sp = 0.345$

$$E = (2.776)(0.345)(0.707)$$

$$E = \pm 0.68$$

Since the effects are smaller than the error, we cannot conclude that the effects are statistically significant at the 95 percent level of confidence. We will replicate the experiment again by running cycle 3.

Phase 1, cycle 3:

Run	A	B	Cycle 1	Cycle 2	Cycle 3	Average	Standard deviation, S	Variance, S²
1	−	−	58.8	59.1	58.7	58.87	0.208	0.0433
2	+	−	58.9	58.8	59.5	59.07	0.379	0.1436
3	−	+	59.0	58.4	59.6	59.00	0.600	0.3600
4	+	+	59.3	60.0	61.0	60.10	0.854	0.7293

$$A: \left(\frac{59.07+60.10}{2}\right)-\left(\frac{58.87+59.00}{2}\right) = 59.59 - 58.94 = 0.65$$

$$B: \left(\frac{59.00+60.10}{2}\right)-\left(\frac{58.87+59.07}{2}\right) = 59.55 - 58.97 = 0.58$$

Error of effects:

$$t_{\alpha/2,(r-1)2^{k-f}}\left(Sp\sqrt{\frac{2}{pr}}\right)$$

$\alpha = 0.05$

$k = 2$

$f = 0$

$r = 3$

$p = 2$

$Sp = 0.565$

$$E = (2.306)(0.565)(0.58) = 0.756$$

Since the error is still larger than the effects, cycle 4 will be run.

Phase 1, cycle 4:

Run	A	B	Cycle 1	Cycle 2	Cycle 3	Cycle 4	Average	Standard deviation, S	Variance, S^2
1	−	−	58.8	59.1	58.7	58.9	58.88	0.170	0.0289
2	+	−	58.9	58.8	59.5	59.1	59.08	0.310	0.0961
3	−	+	59.0	58.4	59.6	59.0	59.00	0.490	0.2401
4	+	+	59.3	60.0	61.0	59.9	60.05	0.705	0.4970

$$A: \left(\frac{59.08+60.05}{2}\right)-\left(\frac{58.88+58.00}{2}\right)=59.66-58.94=0.72$$

$$B: \left(\frac{59.00+60.05}{2}\right)-\left(\frac{58.88+59.08}{2}\right)=59.53-58.98=0.55$$

Error of effects:

$$t_{\alpha/2,(r-1)2^{k-f}}\left(Sp\sqrt{\frac{2}{pr}}\right)$$

$\alpha = 0.05$

$k = 2$

$f = 0$

$r = 4$

$p = 2$

$Sp = 0.46$

$$E = (2.179)(0.46)(0.50) = 0.50$$

The error is now less than the effect, so we conclude that the effects are statistically significant. Factor A and factor B should both be set positive (+) in order to maximize the peel strength.

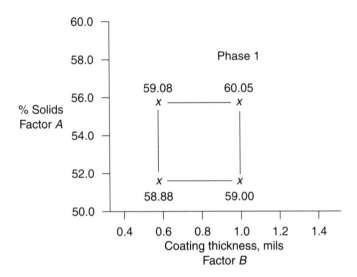

A new set of conditions will be established (phase 2) around the new settings, where A is + (percentage of solids = 56.0 percent) and where B is + (coating thickness = 1.0).

Factor	Description	–	+
A	% Solids	54.0	58.0
B	Coating thickness	0.8	1.2

Phase 2, cycle 1:

Run	A	B	Response
1	–	–	58.9
2	+	–	59.3
3	–	+	59.5
4	+	+	60.6

Cycle 2 will be run to obtain a standard deviation for the responses.

Phase 2, cycle 2:

Run	A	B	Cycle 1	Cycle 2	Average	Standard deviation, S	Variance, S^2
1	−	−	58.9	59.1	59.00	0.141	0.0199
2	+	−	59.3	59.5	59.40	0.141	0.0199
3	−	+	59.5	59.4	59.45	0.071	0.0050
4	+	+	60.6	60.9	60.75	0.212	0.0449

Main effects:

$$\text{A:}\ \left(\frac{59.40+60.75}{2}\right)-\left(\frac{59.0+59.45}{2}\right)=60.08-59.23=0.085$$

$$\text{B:}\ \left(\frac{59.45+60.75}{2}\right)-\left(\frac{59.00+59.40}{2}\right)=60.10-59.20=0.90$$

Error of effects:

$$t_{\alpha/2,(r-1)2^{k-f}}\left(Sp\sqrt{\frac{2}{pr}}\right)$$

$\alpha = 0.05$

$k = 2$

$f = 0$

$r = 2$

$p = 2$

$Sp = 0.150$

$$E = (2.776)(0.150)(0.707) = 0.817$$

The error for both effects is statistically significant.

The following figure shows process averages after phase 2, cycle 2:

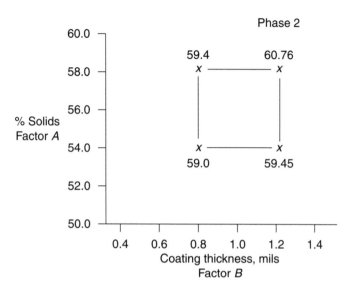

After phase 2, we would set another set of process conditions (phase 3) to continue our process optimization. The suggested conditions for phase 3 are as follows:

Phase 3, cycle 1:

Factor	Description	–	+
A	% Solids	56.0	60.0
B	Coating thickness	1.0	1.4

The EVOP is continued until no additional gain is achievable at the designated level of confidence.

BIBLIOGRAPHY

Box, G. E. P., and N. R. Draper. 1969. *Evolutionary Operation.* New York: John Wiley and Sons.

Montgomery, D. C. 1996. *Introduction to Statistical Quality Control.* 3rd edition. New York: John Wiley and Sons.

Montgomery, D. C. 1997. *Design and Analysis of Experiments.* 4th edition. New York: John Wiley and Sons.

EXPONENTIALLY WEIGHTED MOVING AVERAGE, EWMA

The individual/moving range control chart has the following two modest disadvantages:

1. It is relatively weak in detecting a small process change in a timely manner.
2. It is potentially affected by nonnormal distributions.

An alternative to the individual/moving range control chart (and the average/range control chart) is the exponentially weighted moving average control chart (EWMA).

An EWMA is a moving average of past data where each of the data points is assigned a weight. These weights decrease in an exponential decay manner from the present point value to subsequent points. The EWMA tends to reflect the current condition biased with, to varying degrees, the values of the immediate historical data points. The amount of decrease of the weights is an exponential function of the weighting factor λ, which can assume any value between zero and one.

$$W_{t-1} = \lambda(1 - \lambda)^1$$

where: W_{t-1} = weight associated with observation X_{t-1}, with X_t being the most recent observation
λ = weighting factor 0 to 1

When λ is very small, the moving average at any time t carries with it a great deal of inertia from the previous data; therefore, it is relatively insensitive to short-lived process changes.

In selecting the value for λ, use the following relationship between the weighting factor λ and the sample size n for a Shewhart $\overline{X}/R$ control chart.

$$\lambda = \frac{2}{n+1}$$

If $\lambda = 1$ is selected for the EWMA, all of the weight is given to the current sample, and in effect it is an individual Shewhart control chart.

The weights $\lambda (1 - \lambda)^1$ decrease geometrically with the age of the sample. If $\lambda = 0.2$, then the weight assigned to the current sample (used in calculating EWMA) is 0.20, and the weights given to the preceding samples are 0.16, 0.128, 0.1024, and so on. Because these decline geometrically, the EWMA is sometimes referred to as a geometric moving average (GMA). Since the EWMA is a weighted average of the current sample and all of the past data, it is very insensitive to the assumption of normalcy for the distribution of

173

observations. It is, therefore, ideal for monitoring individual values. The EWMA can also be used with averages.

If the $\bar{X}_i$ are independent random variables with variance $\dfrac{\sigma^2}{n}$, then the variance of Z_t is

$$\sigma^2{}_{z_t} = \frac{\sigma^2}{n}\left(\frac{\lambda}{2-\lambda}\right)[1-(1-\lambda)^{2t}]$$

As t increases, $\sigma^2_{z_t}$ increases to the limiting value

$$\sigma^2{}_z = \frac{\sigma^2}{n}\left(\frac{\lambda}{2-\lambda}\right)$$

The upper and lower control limits for the EWMA are

$$\text{UCL} = \bar{\bar{X}} + 3\sigma\sqrt{\frac{\lambda}{2-\lambda}}$$

and

$$\text{LCL} = \bar{\bar{X}} - 3\sigma\sqrt{\frac{\lambda}{2-\lambda}}$$

The first few samples have the quantity $\sqrt{\dfrac{\lambda}{2-\lambda}}$ adjusted by the amount of $[1-(1-\lambda)^{2t}]$. The upper and lower control limits are

$$\bar{\bar{X}} \pm 3\frac{\sigma}{\sqrt{n}}\sqrt{\left(\frac{\lambda}{2-\lambda}\right)[1-(1-\lambda)^{2t}]}$$

These control limits will increase rapidly to their limiting value.

The following example will be used to illustrate the construction of the EWMA control chart.

Step 1. Collect data from which the baseline EWMA will be constructed.

At least $k = 25$ samples should be collected and, where possible, 40 to 50 should be collected if available. For this example, however, we will use $k = 15$. The data will be used to provide a historical statistical characterization of the process being monitored.

The following data represent individual observations of waiting times to be served in a branch of the Midwest National Bank. Times are recorded in minutes.

Sample t	X_t
1	60
2	91
3	132
4	100
5	80
6	120
7	70
8	95
9	140
10	80
11	110
12	100
13	82
14	105
15	65

$\bar{X} = 95.3$

$S = 23.5$

Note: the sample standard deviation S is used as an estimate of σ. Any appropriate estimate based on SPC relationships may be used.

Step 2. Calculate the EWMAs.

The EWMAs are calculated for individual observations as

$$Z_t = \lambda X_t + (1 - \lambda)Z_{t-1}$$

where: Z_t = EWMA at time t
 λ = weighting factor ($\lambda = 0.2$ for this example)
 Z_{t-1} = EWMA for the immediate preceding point

The initial Z_0 = the process average $\bar{X} = 95.3$.

Z_1 is

$$Z_t = \lambda X_t + (1 - \lambda)Z_{t-1} \quad Z_1 = \lambda X_1 + (1 - \lambda)Z_0 \quad Z_1 = 0.2(60.0) + (1 - 0.2)95.3 \quad Z_1 = 88.2$$

Z_2 is

$$Z_2 = \lambda X_2 + (1 - \lambda)Z_1 \quad Z_2 = 0.2(91.0) + (1 - 0.2)88.2 \quad Z_2 = 88.8$$

Z_3 is

$$Z_3 = \lambda X_3 + (1 - \lambda)Z_2 \quad Z_3 = 0.2(132.0) + (1 - 2)88.8 \quad Z_3 = 97.4$$

The remaining EWMAs are calculated and tallied in the following table.

Summary of individual observations X_t and EWMAs Z_t:

Sample t	X_t	Z_t
1	60	88.2
2	91	88.8
3	132	97.4
4	100	97.9
5	80	94.3
6	120	99.4
7	70	93.5
8	95	93.8
9	140	103.0
10	80	98.4
11	110	100.7
12	100	100.6
13	82	96.9
14	105	98.5
15	65	91.8

Step 3: Calculate the control limits.

For EWMAs based on individual observations ($n = 1$), the control limits are based on

$$\bar{\bar{X}} \pm 3S \sqrt{\left(\frac{\lambda}{2-\lambda}\right)[1-(1-\lambda)^{2t}]}$$

For this example, $\bar{X} = 95.3$ and $S = 23.5$

Limits for sample #1:

$$UCL_1 = 95.3 + 3(23.5) \sqrt{\left(\frac{0.2}{2-0.2}\right)[1-(1-0.2)^2]} = 95.3 + 14.1 = 109.4$$

$$LCL_1 = 95.3 + 3(23.5) \sqrt{\left(\frac{0.2}{2-0.2}\right)[1-(1-0.2)^2]} = 95.3 + 14.1 = 81.2$$

Limits for sample #2:

$$UCL_2 = 95.3 + 3(23.5) \sqrt{\left(\frac{0.2}{2-0.2}\right)[1-(1-0.2)^4]} = 95.3 + 18.1 = 113.4$$

$$LCL_2 = 95.3 + 3(23.5) \sqrt{\left(\frac{0.2}{2-0.2}\right)[1-(1-0.2)^4]} = 95.3 + 18.1 = 77.2$$

Limits for sample #3:

$$UCL_3 = 95.3 + 3(23.5)\sqrt{\left(\frac{0.2}{2-0.2}\right)[1-(1-0.2)^6]} = 95.3 + 20.2 = 115.5$$

$$LCL_3 = 95.3 + 3(23.5)\sqrt{\left(\frac{0.2}{2-0.2}\right)[1-(1-0.2)^6]} = 95.3 + 20.2 = 75.1$$

Limits for sample #4:

$$UCL_4 = 95.3 + 3(23.5)\sqrt{\left(\frac{0.2}{2-0.2}\right)[1-(1-0.2)^8]} = 95.3 + 21.4 = 116.7$$

$$LCL_4 = 95.3 + 3(23.5)\sqrt{\left(\frac{0.2}{2-0.2}\right)[1-(1-0.2)^8]} = 95.3 + 21.4 = 73.9$$

Limits for sample #5:

$$UCL_5 = 95.3 + 3(23.5)\sqrt{\left(\frac{0.2}{2-0.2}\right)[1-(1-0.2)^{10}]} = 95.3 + 22.2 = 117.5$$

$$LCL_5 = 95.3 + 3(23.5)\sqrt{\left(\frac{0.2}{2-0.2}\right)[1-(1-0.2)^{10}]} = 95.3 + 22.2 = 73.1$$

The remaining control limits are calculated and listed in the following summary. Notice that the control limits stabilize soon to a constant limit of $X \pm S$.

Sample			Control limits For Z_t	
t	X_t	Z_t	LCL	UCL
1	60	88.2	81.2	109.4
2	91	88.8	77.2	113.4
3	132	97.4	75.1	115.5
4	100	97.9	73.9	116.7
5	80	94.3	73.1	117.5
6	120	99.4	72.6	118.0
7	70	93.5	72.3	118.3
8	95	93.8	72.1	118.5
9	140	103.0	72.0	118.6
10	80	98.4	.	.
11	110	100.7	.	.
12	100	100.6	.	.
13	82	96.9	.	.
14	105	98.5	.	.
15	65	91.8	.	.

Step 4. Construct the chart, and plot the data.

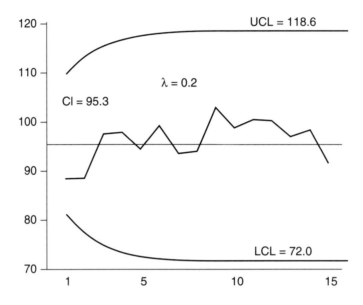

Step 5. Continue to collect data, plotting and looking for signs of a change.

Ten new data values are obtained. Calculate the EWMA values, and plot the results.

Sample #	X_i	Z_t
(15	65	91.9)
16	110	95.4
17	115	99.4
18	111	101.7
19	118	104.9
20	126	109.2
21	117	110.7
22	122	112.3
23	131	116.6
24	124	118.1
25	129	120.2

The completed EWMA chart with all data points is as follows:

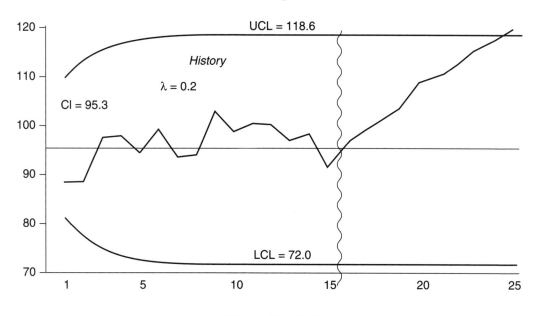

BIBLIOGRAPHY

Montgomery, D. C. 1996. *Introduction to Statistical Quality Control.* 3rd edition. New York: John Wiley and Sons.

Sweet, A. L. 1986. Control charts using coupled exponentially weighted moving averages, *Transactions of IIE* 18, No.1: 26–33.

Wheeler, D. J. 1995. *Advanced Topics in Statistical Process Control.* Knoxville, TN: SPC Press.

F-TEST

One of the main objectives in the pursuit of quality in manufacturing (and service for that matter) is to reduce variability. This is consistent with the Taguchi philosophy of performing at the target with minimum variation. One of the best methods to assess whether or not variation reduction has been accomplished is to compare the variances (or square of the standard deviation).

Consider a process variable 1 that has a standard deviation of σ_1 and a process variable 2 that has a standard deviation σ_2. For example, σ_1 might represent the standard deviation of the tablet weight of a pharmaceutical product before attempts were made to reduce variability and σ_2 the standard deviation after efforts were made to reduce the variation. A sample of seven tablets is taken from the production line prior to the improvement, and five tablets are taken from the production line after implementation of the modification. By comparing the variances of the two samples, we may test the hypothesis that one of the variances is greater than the other variance. This test relies upon the F statistic named after Ronald A. Fisher (1890–1962) who developed the probability distribution in the 1920s.

The weights of tablets prior to process improvement:	*The weights of tablets after process improvement:*
12.3	12.6
12.7	12.5
12.5	12.6
12.6	12.5
12.4	12.4
12.9	
12.2	
$S_1 = 0.24$	$S_2 = 0.08$
$S_1^2 = 0.0576$	$S_2^2 = 0.0064$

The F statistic or ratio is determined by dividing the larger variance by the smaller variance.

$$F = \frac{S_1^2}{S_2^2} = \frac{0.0576}{0.0064} = 9.00$$

With this F-test, we are testing the hypothesis that σ_1^2 is greater than σ_2^2 (the alternative hypothesis) as opposed to σ_1^2 equal to σ_2^2 (the null hypothesis).

$$H_0: \sigma_1^2 = \sigma_2^2$$

$$H_a: \sigma_1^2 > \sigma_2^2$$

We will reject the null hypothesis and accept the alternative hypothesis if our F calculated value is greater than the critical F value. A level of significance must be chosen. For this example, a confidence level of 95 percent is selected meaning that a risk, $\alpha = 5\%$ is accepted.

The critical F value is found in the F table in the appendix. The degrees of freedom are determined by the sample size minus one. The horizontal rows represent the degrees of freedom for the sample giving the smaller variance, v_2. The vertical columns represent the degrees of freedom for the sample giving the larger variance, v_1. Each row of v_2 has four levels of α corresponding to 10, 5, 2.5, and 1%.

For this example $v_1 = 7 - 1 = 6$ (the sample size of the larger variance) and $v_2 = 5 - 1 = 4$ for the smaller variance.

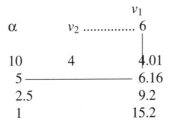

α	v_2	v_1 6
10	4	4.01
5		6.16
2.5		9.2
1		15.2

The critical F value for $v_1 = 6$, $v_2 = 4$, and a risk of $\alpha = 0.05$ is 6.16. Since our calculated F ratio is 9.0 and exceeds 6.16 we will reject the null hypothesis ($\sigma_1^2 \neq \sigma_2^2$) and accept the alternative hypothesis ($\sigma_1^2 > \sigma_2^2$) concluding that a reduction in variation has been demonstrated.

Single-sided vs. double-sided tests

In the previous example, we were concerned whether or not a decrease in variability was evident. We were interested only in a one-way change. Test of this type are called single sided or one-tailed test and the F-table is utilized as it is presented. There are, however, cases in which a double sided or two-tailed test are appropriate. Such a two-tailed test would be the case where the null hypothesis is $\sigma_1^2 = \sigma_2^2$ and the alternative hypothesis is $\sigma_1^2 \neq \sigma_2^2$. For these cases, the tabled values are doubled; in other words, the 5 percent level of significance becomes the 10 percent level.

Example 1:
Two different methods of training have been evaluated. One test group consists of 16 participants, and the other consists of 11. The respective standard deviations are 14.0 and 7. Assume a risk, $\alpha = 0.05$.

The F-ratio is:

$$F = \frac{196}{49} = 4.00 \quad \alpha / 2 = 0.025$$

The critical F-ratio, $F_{15,10,0.025} = 3.52$. Since F calculated is greater than F critical, we reject the null hypothesis, $H_a: \sigma_1^2 = \sigma_2^2$ and accept the alternative hypothesis, $H_a: \sigma_1^2 \neq \sigma_2^2$.

Example 2:
A method of testing (X), has been used extensively to measure the water content in a latex co-polymer. Based upon a very large number of test results, this method has a standard deviation of $S_X = 0.071$ units. A new method (Y) has been developed in an effort to reduce the variation of tests results. Eight independent tests gave a standard deviation of $S_y = 0.038$. It would be an improvement if the test variation were to be reduced as a result of using the new method. Do the data support the hypothesis that a reduction in variation is evident?

$$H_0: \sigma_X^2 = \sigma_Y^2$$

$$H_a: \sigma_X^2 \neq \sigma_Y^2 \text{ (for a two-tailed test)}$$

$$F = \frac{S_X^2}{S_Y^2} = \frac{(0.071)^2}{(0.038)^2} = 3.49$$

Selecting a risk of $\alpha = 0.10$, we have $\alpha/2 = 0.05$.

The critical F value is $F_{\infty,7,0.05} = 3.23$. Since the calculated F is greater than F critical, we will reject H_0 in favor of H_a, concluding that there is a difference in the variation in the two methods

HISTOGRAMS

Histograms or frequency histograms are used to present a picture or graphical representation of data. The frequency histogram consists of a vertical axis that corresponds to the frequency at which a value or group of values occurs and a horizontal axis giving the value of the data or group of data. The shape of the histogram can give insight to the nature of the distribution of data. A frequency histogram shaped like a bell is indicative of a normal distribution.

There are numerous techniques for developing a frequency histogram. All of these techniques involve the following common set of steps:

1. Collection of a set of data, typically 50 or more
2. Determination of a number of class intervals or cells
3. Determination of the class interval width
4. Calculation of all class intervals and categorization of the data into the appropriate class intervals
5. Construction of the frequency histogram

EXAMPLE OF A FREQUENCY HISTOGRAM

Step 1. Collect data.

For this example, the resistance in ohms for 126 coils is measured. The data for these measurements follow:

42.4	39.9	42.2	40.5	44.2	40.5
39.1	42.1	34.6	41.4	46.0	38.5
43.3	38.5	41.2	43.7	31.0	36.1
34.2	38.7	37.0	42.0	39.6	40.6
37.4	35.1	36.4	34.3	47.3	45.9
41.3	42.2	37.9	37.5	41.3	44.8
41.9	31.4	36.4	42.9	41.2	41.9
39.4	35.6	40.9	33.9	39.5	46.6
43.6	40.8	37.8	37.7	40.5	43.6
40.6	47.3	36.5	45.1	37.3	39.2
40.1	41.2	35.7	29.5	42.4	38.5
38.4	38.3	38.0	40.2	42.1	42.6
46.7	38.0	39.4	38.9	40.7	38.9
40.1	37.9	43.6	40.8	40.4	36.9

35.5	47.7	43.0	44.6	47.2	40.6
34.8	33.4	35.8	39.2	41.5	41.5
43.0	36.8	40.1	40.0	41.7	39.2
43.1	33.0	39.2	36.3	32.8	41.0
41.5	43.5	40.6	38.9	37.7	44.8
41.8	40.7	37.4	47.3	36.4	45.1
42.0	42.5	38.1	36.1	42.0	41.3

Step 2. Determine the number of class intervals or cells K into which the data will be sorted.

The number of class intervals K depends on the total number of data values or observations N. Juran's *Quality Control Handbook* (1980, 23.13) gives the following table:

Number of observations N	Recommended number of intervals K
20–50	6
51–100	7
101–200	8
201–500	9
501–1000	10
> 1000	11–20

Juran (1980) suggests that the following equation may be used to estimate the number of class intervals. The resulting value should be rounded to the nearest integer. This relationship will give values approximately equal to those given in Juran's handbook.

$$K = 1.5 \ln N + 0.5$$

For our example, the number of observations $N = 126$, and the number of class intervals K is given by

$$K = 1.5(4.84) + 0.5$$
$$K = 7.76 = 8$$

Step 3. Determine the class interval width.

The class interval width is determined by dividing the difference in the largest and smallest observations by the number of class intervals.

$$\text{Interval width} = \frac{47.7 - 29.5}{8} = 2.2$$

Step 4. Calculate all class intervals, and catagorize the data into the appropriate class interval.

There will be eight intervals, each with a class interval width of 2.00 units.

The beginning of the first interval will be at the smallest observation minus one-half the interval width (1.00). The end of the first interval will be at the beginning point plus the full interval width (2.00).

First interval bounds: 29.5 − 1.1 to 29.5 + 2.2 or 28.4 to 31.7

Each of the following intervals is calculated by taking the ending point of the previous and adding the interval width, 2.2.

1st   interval = 28.4 to 31.7

2nd   interval = 31.7 to 33.9

3rd   interval = 33.9 to 36.1

4th   interval = 36.1 to 38.3

5th   interval = 38.3 to 40.5

6th   interval = 40.5 to 42.7

7th   interval = 42.7 to 44.9

8th   interval = 44.9 to 47.1

9th   interval = 47.1 to 49.3   ←   This additional interval is required to accommodate the maximum observed value of 47.7.

These intervals will hold observations from the beginning point up to but not including the ending point. For example, the fifth interval will contain all of the observations from 39.4 up to but not including 41.4. If we encounter an observation of 41.4, it will be assigned to the sixth interval.

A table of the intervals and a tally count of all of the observations is made, assigning each observation to its appropriate interval:

Interval	Frequency or tally count
28.4 to 31.7	3
31.7 to 33.9	3
33.9 to 36.1	10
36.1 to 38.3	22
38.3 to 40.5	25
40.5 to 42.7	38
42.7 to 44.9	14
44.9 to 47.1	6
47.1 to 49.3	5

Step 5. Construct the frequency histogram.

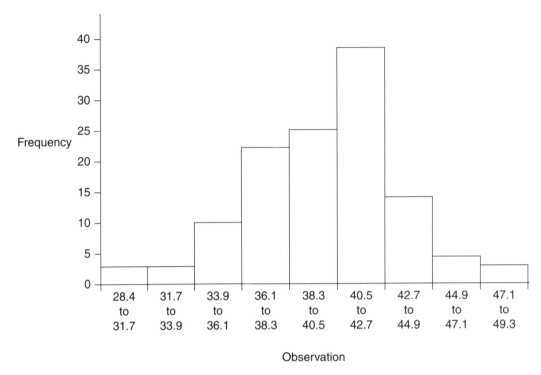

Observation

An alternative to the traditional frequency histogram is the application of a technique called the *leaf-stem plot*. This method is a quick and relatively simple way to present data in a graphical format. Start by determining the largest and smallest observations and recording these to one decimal place less than that of the data.

In the previous example, the smallest and largest values were 29.5 and 47.7. These values will be recorded as 29 and 47.

A vertical table is made beginning with 29 and ending with 47. A vertical line is drawn to the right of the values.

A		B		C	
29		29		29	
30		30		30	
31		31		31	
32		32		32	
33		33		33	
34		34		34	
35		35		35	
36		36		36	
37		37		37	
38		38		38	
39		39	1	39	1

```
40|          40|          40|
41|          41|          41|
42|4         42|4         42|4
43|          43|          43|3
44|          44|          44|
45|          45|          45|
46|          46|          46|
47|          47|          47|
```

1. The first observation from the 126 values is 42.4. Go down the table until reaching 42, then place a "4" to the right of the vertical line. This will represent the location of the observation 42.4.
2. The second observation from the 126 values is 39.1. Go down the table until reaching 39, then place a "1" to the right of the vertical line. This will represent the location of the observation 39.1.
3. The third observation from the 126 values is 43.3. Go down the table until reaching 43, then place a "3" to the right of the vertical line. This will represent the location of the observation 43.3.

Continue until all 126 observations have been posted. The resulting leaf-stem plot will aid in visualizing the distribution of the observations.

Completed leaf-stem plot:

```
29 | 5
30 |
31 | 4 0
32 | 8
33 | 4 0 9
34 | 2 8 6 3
35 | 5 1 6 7 8
36 | 8 4 4 5 3 1 4 1 9
37 | 4 9 0 9 8 4 5 7 3 7
38 | 4 5 7 3 0 0 1 9 9 5 5 9
39 | 1 4 9 4 2 2 6 5 2 2
40 | 6 1 1 8 7 9 1 6 5 2 8 0 5 7 4 5 6 6
41 | 3 9 5 8 2 2 4 3 2 5 7 9 5 0 3
42 | 4 0 1 2 5 2 0 9 4 1 0 6
43 | 3 6 0 1 5 6 0 7 6
44 | 6 2 8 8
45 | 1 9 1
46 | 1 7 0 6
47 | 3 7 3 3 2
```

Optional problem:

Construct a leaf-stem plot for the following 76 weights of employees. Is this distribution normally distributed? What could contribute to the shape of this distribution?

136	92	139	132
198	184	216	133
106	181	240	232
190	138	204	207
94	112	152	146
150	163	148	192
142	211	111	193
116	183	147	135
206	81	222	213
144	162	196	206
136	207	146	193
156	205	144	133
124	194	122	212
108	168	194	204
195	225	236	223
231	107	205	180
172	111	153	120
151	196	128	117

BIBLIOGRAPHY

Betteley, G, N. Mettrick, E. Sweeney, and D. Wilson. 1994. *Using Statistics in Industry.* New York: Prentice Hall.

Ishikawa, K. 1993. *Guide to Quality Control.* 2nd edition. White Plains, NY: Quality Resources.

Juran, J. M., and F. M. Gryna. 1980. *Quality Planning and Analysis.* 3rd edition. New York, NY: Mc Graw-Hill.

Tague, N. R. 1995. *The Quality Toolbox.* Milwaukee, WI: ASQC Quality Press.

HYPOTHESIS TESTING

Hypothesis tests are used to derive statistically sound judgements regarding evaluating sample data compared to population parameters or other external criteria. A hypothesis is a systematic approach to assessing beliefs about reality; it is confronting a belief with evidence and then deciding, in light of this evidence, whether the initial belief (or hypothesis) can be maintained as acceptable or must be rejected as untenable.

FORMULATION OF THE HYPOTHESIS

A researcher in any field sets out to test a new method or a new theory and first must formulate a hypothesis, or claim, that is felt to be true. For example, we might establish that the true average starting salary for college graduates with a B.S. degree is greater than $38,000 per year. The hypothesis that the researcher wants to establish is called the *alternative hypothesis.* To be paired with this alternative hypothesis that the researcher believes to be true is the *null hypothesis,* which is opposite of the alternative hypothesis. These two hypotheses are sometimes referred to as *opposing hypotheses.* When they are both stated in terms of the appropriate population parameter, the null hypothesis and alternative hypothesis describe two states of being that cannot simultaneously be true. In hypothesis testing, we will take an indirect approach to obtaining evidence to support the alternative hypothesis (what we are trying to prove). Instead of trying to prove that the alternative hypothesis is true, we attempt to produce evidence to show that the null hypothesis is false.

> The hypothesis that we hope to disprove is called the *null hypothesis* or *Ho.*

> The hypothesis for which we want to develop supporting evidence and that we want to prove is called the *alternative hypothesis* or *Ha.*

Formulate the appropriate null hypothesis and alternative hypothesis for testing that the starting salary for graduates with a B.S. degree in electrical engineering is greater than $38,000 per year.

The hypothesis is stated in terms of the population parameter.

Ho: $\mu = \$38,000$

Ha: $\mu > \$38,000$

In this example, we attempt to gather evidence that the starting salary is greater than $38,000 per year rather than not equal to $38,000 per year. This is an example of a one-tail hypothesis test.

CONCLUSIONS AND CONSEQUENCES FOR A HYPOTHESIS TEST

The objective of any hypothesis testing is to make a decision as to whether to reject the null hypothesis *Ho* in favor of the alternative hypothesis *Ha*. While we would like to make a correct decision all of the time, we cannot because our decision is based on a statistic that is derived from a sample chosen from the population. As with all statistics, there is an associated error. Two types of error can occur: type I and type II. These two errors are depicted in Table 1.

A type I error occurs when we reject a null hypothesis that is, in fact, true. The probability of a type I error is α. α is also the level of significance for the hypothesis test.

A type II error occurs if we fail to reject a null hypothesis when it is, in fact, false. The probability of a type II error is β.

Table 1. Test Result.

True state of the world	Null hypothesis is accepted	Null hypothesis is rejected
Null hypothesis *Ho* is, in fact, true	(a) Correct result, O.K. $p = 1 - \alpha$	(b) Type I error $p = \alpha$ risk
Null hypothesis *Ho* is, in fact, false	(c) Type II error $p = \beta$ risk	(d) Correct result, O.K. $p = 1 - \beta$

The level of significance α must be chosen for the hypothesis. This significance, or risk, is the probability that we will reject the null hypothesis when, in reality, the null hypothesis is, in fact, true. Sometimes the significance is expressed as a level of confidence of the hypothesis. Confidence is $1 - \alpha$. Typical levels of confidence are 0.99, 0.95 and 0.90.

The level of significance, one of the easily controlled requirements, is the percentage of time over an extended period that the statistical test will make a type I error.

For this example, we will set the significance, or α, risk at $\alpha = 0.05$.

SELECTING A TEST STATISTIC AND REJECTION CRITERIA

For our example, we want to test the following hypothesis:

Ho: $\mu = \$38,000$ (The hypothesized mean or μ_0)

Ha: $\mu > \$38,000$

The first step is to sample the population in order to determine a sample average $\overline{X}$.

In this step of the hypothesis test, we select a test statistic, which is the type of statistic to be computed from a random sample taken from the population of interest. This test statistic will be used to establish the probability of truth or falsity of the null hypothesis. Hypothesis tests are used to test statistical inferences regarding

1. The sample mean $\overline{X}$
2. The difference between two sample means, $\overline{X}_A - \overline{X}_B$
3. The sample proportion P
4. The difference between two sample proportions, $P_A - P_B$

Each of these has its own sampling distribution and an associated test statistic. For large samples where $n \geq 30$ and $n < 0.05N$, we will use the normal distribution as a model. For small samples of $n < 30$, we will use the *t*-distribution as a model. The *t*-distribution is actually used when the standard deviation σ is unknown. When the sample size $n > 30$, we assume that while the true standard is still unknown, the sample standard deviation is an adequate estimation of σ, and we use the normal distribution.

For our example, we will calculate the test statistic using $n = 75$; therefore, a normal distribution will be used.

The sampling distribution for averages where $n \geq 30$ is such that

$$Z = \frac{\overline{X} - \mu}{\sigma_{\overline{X}}}$$

where: $\mu =$ the true but unknown average

$\sigma_{\overline{X}} =$ standard deviation of averages of n observations

$$\sigma_{\overline{X}} = \frac{S}{\sqrt{n}}$$

$S =$ sample standard deviation

In this example, we will sample $n = 45$ and determine the average $\overline{X}$.

Assume from the sample data that we find that $\overline{X} = \$39,750$ and $S = \$300$.

Calculation of the test statistic Z yields

$$Z = \frac{39,000 - 38,000}{\dfrac{1695}{\sqrt{50}}}$$

$$Z = 4.17$$

We will compare the test statistic to the defined rejection region.

For this example of a one-tail hypothesis for a sample average and a large (greater than 30) sample size, the rejection region is

$$Z > Z_\alpha$$

If our calculated Z test statistic is greater than the Z_α, we will reject the null hypothesis and accept the alternative. Z_α for $\alpha = 0.05$ can be determined by using a standard t-distribution table where the degrees of freedom are equal to ∞ or by using a standard normal distribution in a reverse manner (that is, to determine the Z score that will yield a table value equal to the α value).

The Z_α for $\alpha = 0.05$ is 1.645.

Since Z is, in fact, greater than Z_α, we will reject the null hypothesis and accept the alternative hypothesis that the true average starting salary for electrical engineers is greater than $38,000 per year. To summarize, hypothesis testing requires the following operations:

1. Formulate the null and alternative hypotheses.
2. Select the significance level for the test.
3. Compute the value for the test statistic.
4. Compare the test statistic to a critical value for the appropriate probability distribution, and observe if the test statistic falls within the region of acceptance or rejection.

The general formula for computing a test statistic for making an inference is

$$\text{test statistic} = \frac{\text{observed sample statistic} - \text{tested value}}{\text{standard error}}$$

where the *observed sample statistic* is the statistic of interest from the sample, *tested value* is the hypothesized population parameter, and *standard error* is the standard deviation of the sampling population divided by the square root of the sample size.

Hypothesis tests can be divided into several types depending on the following:

1. Large vs. small sample sizes (large $\geq$ 30 and small $<$ 30)
2. Hypothesis test about a population mean μ
   Hypothesis test about a population proportion π
3. Hypothesis test about the difference between two population means, $\mu_1 - \mu_2$
   Hypothesis test about the difference between two population proportions, $\pi_1 - \pi_2$

In the following example, we have illustrated the hypothesis for a small-sample test about a population mean.

Hypothesis Test about a Population Mean: Small-Sample Case

a. One-tail test   b. Two-tail test

$Ho:$ $\mu = \mu_0$ $Ho:$ $\mu = \mu_0$
$Ha:$ $\mu > \mu_0$ $Ha:$ $\mu \neq \mu_0$
(or $Ha:$ $\mu < \mu_0$)

Test statistic: Test statistic:

$$t = \frac{\overline{X} - \mu_0}{\dfrac{s}{\sqrt{n}}}$$ $$t = \frac{\overline{X} - \mu_0}{\dfrac{s}{\sqrt{n}}}$$

Reject Ho if   Reject Ho if

$t > t_\alpha$ $(or\ t < -t_\alpha)$ $t < -t_{\alpha/2}$ or $t > t_{\alpha/2})$

Example 1. Small-sample, one-tail test:

A claim has been made that the actual amount of mustard in Acme brand mustard is less than the advertised amount of 15 ounces. You have sampled 25 units and determined that the average is $\overline{X} = 14.2$ ounces and the sample standard deviation is $S = 1.3$. What is your conclusion?

Step 1. Formulate two opposing hypotheses.

  $Ho:$ $\mu = 15$

  $Ha:$ $\mu < 15$

Step 2. Select a test statistic.

$$t = \frac{\overline{X} - \mu_0}{\dfrac{s}{\sqrt{n}}}$$

Step 3. Derive a rejection rule.

We will set the level of significance at $\alpha = 0.10$. If the calculated test statistic is less than $-t_\alpha$, we will reject the null hypothesis and accept the alternative that the true population average is less than 15.0 ounces.

Looking up the $t_{\alpha,\,n-1}$ in the t table, we find 1.318.

If our test statistic is less than -1.318, we will reject the null hypothesis and accept the alternative hypothesis (that the true average weight is, in fact, equal to or greater than 15.0 ounces).

Step 4. Select a sample, calculate the test statistic, and accept or reject the null hypothesis.

$n = 25$ $\bar{X} = 14.2$ $S = 1.3$

$$t = \frac{14.2 - 15.0}{\dfrac{1.3}{\sqrt{25}}} = -3.08$$

Since the test statistic is less than -1.318, we will reject the null hypothesis and conclude that the true average weight is less than 15.0 ounces.

Note: It is assumed that the variances, while unknown, are equal. If the variances of the two populations are assumed not to be equal, we modify the degrees of freedom when looking up the critical t-score. The degrees of freedom for these cases are determined by

$$df = \frac{\left(\dfrac{S_1^{\,2}}{n_1} + \dfrac{S_2^{\,2}}{n_2}\right)^2}{\dfrac{\left(\dfrac{S_1^{\,2}}{n_1}\right)^2}{n_1} + \dfrac{\left(\dfrac{S_2^{\,2}}{n_2}\right)^2}{n_2}}$$

Example 2. Small-sample, two-tail test:

A manufacturing manager claims that on average it takes 22.0 minutes to build a unit. How can you verify the claim having secured test data for 15 production times?

Step 1. Formulate two opposing hypotheses.

Ho: $\mu = 22.0$

Ha: $\mu \neq 22.0$

This is a two-tail test, as we are not testing that the true time is greater than or less than but that the true average is not equal to 22.0 minutes.

Step 2. Select a test statistic.

$$t = \frac{\bar{X} - \mu_0}{\dfrac{S}{\sqrt{n}}}$$

Step 3. Derive a rejection rule.

We will set the level of significance at $\alpha = 0.05$. However, since we are performing a two-tail test, we must look up one-half the level of significance when using the t-distribution table. For this example, we will look up $t_{\alpha/2, n-1}$. The correct t value for $n - 1 = 14$ and $\alpha/2 = 0.025$ is $t_{0.025,14} = 2.145$.

Step 4. Select a sample, calculate the test statistic, and accept or reject the null hypothesis.

From the 15 data values, we find the average $\bar{X} = 25.8$ and the standard deviation $S = 4.70$. The resulting test statistic is

$$t = \frac{\bar{X} - \mu_0}{\dfrac{s}{\sqrt{n}}} \qquad t = \frac{25.8 - 22.0}{\dfrac{4.70}{\sqrt{15}}} \qquad t = 3.14$$

Since the test statistic is greater than the rejection region of $t > 3.14$, we will reject the null hypothesis and conclude that the true average build time is not equal to 22.0 minutes.

Hypothesis Test about a Population Mean: Large-Sample Case

a. One-tail test	b. Two-tail test
$Ho: \mu = \mu_0$	$Ho: \mu = \mu_0$
$Ha: \mu > \mu_0$	$Ha: \mu \neq \mu_0$
(or $Ha: \mu < \mu_0$)	
Test statistic:	Test statistic:
$Z = \dfrac{\bar{X} - \mu_0}{\dfrac{s}{\sqrt{n}}}$	$Z = \dfrac{\bar{X} - \mu_0}{\dfrac{s}{\sqrt{n}}}$
Reject Ho if	Reject Ho if
$Z > Z_\alpha$ (or $Z < -Z_\alpha$)	$Z < -Z_{\alpha/2}$ (or $Z > Z_{\alpha/2}$)

Example 1. Large-sample, one-tail test:

A software company wants to verify that the technical service help line answers customers' inquiries in less than two minutes. The results of the data are summarized as follows:

Sample size, $n = 120$ Average, $\bar{X} = 1.93$ Standard deviation, $S = 0.28$

What is your conclusion?

Step 1. Formulate two opposing hypotheses.

 Ho: $\mu = 2.00$

 Ha: $\mu < 2.00$

Step 2. Select a test statistic.

$$Z = \frac{\overline{X} - X_0}{\frac{S}{\sqrt{n}}} \quad Z = \frac{1.93 - 2.00}{\frac{0.28}{\sqrt{120}}} \quad Z = -2.73$$

Step 3. Derive a rejection rule.

We will set the level of significance at $\alpha = 0.10$.

 If the test statistic Z is less than the critical value of $-Z_\alpha$, we will reject the null hypothesis that the true average response time is equal to 2.0 minutes in favor of the alternative that the true average response time is, in fact, less than 2.0 minutes. Locating 0.10 in the main body of the Standard Normal Distribution table, we find the corresponding Z value to be 1.282 and $-Z_\alpha = -1.282$.

Step 4. Apply the rejection criteria and form a conclusion.

The value for our test statistic is not less than −1.282; therefore, we cannot reject the null hypothesis and conclude that the true average length of time is not less than 2.0 minutes.

Example 2. Large-sample, two-tail test:

A diet plan states that, on average, participants will lose 12 pounds in four weeks. As a statistician for a competing organization, you want to test the claim. You sample 70 participants who have been on the subject diet plan for four weeks. You find that the average weight loss has been 11.4 pounds with a sample standard deviation of 1.6 pounds. What is your conclusion based on the data using a 5 percent level of significance?

Step 1. Formulate two opposing hypotheses.

 Ho: $\mu = 12.0$

 Ha: $\mu \neq 12.0$

Step 2. Select a test statistic.

$$Z = \frac{\overline{X} - u_0}{\frac{S}{\sqrt{n}}} \quad Z = -3.14$$

Step 3. Derive a rejection rule.

If the value of our test statistic is less than $-Z_{\alpha/2}$ or greater than $Z_{\alpha/2}$, we will reject the null hypothesis in favor of the alternative hypothesis.

Step 4. Select a sample, calculate a test statistic, and accept or reject the null hypothesis.

The value for $Z_{\alpha/2}$ with $\alpha = 0.05$ is 1.96, and the test statistic is -3.14; therefore, we reject the null hypothesis and conclude that the true average weight loss is not equal to 12 pounds.

Small-Sample Test of Hypothesis about $(\mu_1 - \mu_2)$

a. One-tail test   b. Two-tail test

$Ho: (\mu_1 - \mu_2) = D_0$ $Ho: (\mu_1 - \mu_2) = D_0$
$Ha: (\mu_1 - \mu_2) > D_0$ $Ha: (\mu_1 - \mu_2) \neq D_0$
(or $Ha:(\mu_1 - \mu_2) = D_0$

Test statistic: Test statistic:

$$t = \frac{(\overline{X}_1 - \overline{X}_2) - D_0}{\sqrt{S_p^2 \left(\dfrac{1}{n_1} + \dfrac{1}{n_2} \right)}} \qquad t = \frac{(\overline{X}_1 - \overline{X}_2) - D_0}{\sqrt{S_p^2 \left(\dfrac{1}{n_1} + \dfrac{1}{n_2} \right)}}$$

Reject H_o if: Reject H_o if:

$t > t_\alpha$ or $(t < -t_\alpha)$ $t > t_{\alpha/2}$ or $(t < -t_{\alpha/2})$

where:

$$S_p^2 = \frac{(n_1 - 1)S_1^2 + (n_2 - 1)S_2^2}{n_1 + n_2 - 2}$$

D_0 = the hypothesis difference; frequently $D_0 = 0$

The following assumptions are made:

1. The population variances are equal.
2. The populations have the same distribution, assumed to be normal.
3. Random samples are selected in an independent manner from the two populations.

Example 1. Small-sample, two-tail test:

Two brands of walking shoes are compared with respect to sole wear on a treadmill-type testing machine. Shoes were tested until 10 mils of sole were removed. Fifteen samples of brand A and 18 samples of brand B were evaluated. We want to test the hypothesis that there is no difference in the sole wear rate between the two brands.

Step 1. Formulate the two opposing hypotheses.

$Ho:(\mu_1 - \mu_2) = 0$ (no difference in mileage)

$Ha:(\mu_1 - \mu_2) \neq 0$ (a difference in mileage)

Step 2. Calculate the test statistic.

We will use the test statistic of

$$t - \frac{(\overline{X}_1 - \overline{X}_2) - D_0}{\sqrt{S_p^2 \left(\dfrac{1}{n_1} + \dfrac{1}{n_2}\right)}}$$

We will sample the wear rate for 15 samples of brand A and 18 samples of brand B and will calculate the average, standard deviation for each and determine the pooled variance S_p^2.

$n_1 = 15$	$n_2 = 18$
$\overline{X}_1 = 18.4$	$\overline{X}_2 = 20.6$
$S_1 = 3.4$	$S_2 = 3.8$

The pooled variance is

$$S_p^2 = \frac{(n_1 - 1)S_1^2 + (n_2 - 1)S_2^2}{n_1 + n_2 - 2} \qquad S_p^2 = \frac{(15-1)(3.4)^2 + (18-1)(3.8)^2}{15 + 18 - 2}$$

$$S_p^2 = 13.14$$

Using this pooled variance of 13.14, we calculate the test statistic.

$$t = \frac{(\overline{X}_1 - \overline{X}_2) - D_0}{\sqrt{S_p^2 \left(\dfrac{1}{n_1} + \dfrac{1}{n_2}\right)}} \qquad t = \frac{(18.4 - 20.6) - 0}{\sqrt{13.14\left(\dfrac{1}{15} + \dfrac{1}{18}\right)}} \qquad t = -1.74$$

Step 3. Derive a rejection criteria.

We will test our hypothesis at the 10 percent significance level. We are performing a two-tail test that the true difference between the two brands can be either less than or greater than zero.

Looking up the t value for a 5 percent risk ($t_{\alpha/2}$) and $n_1 + n_2 - 2$ degrees of freedom, we find the critical t value to be ± 1.697 (we use degrees of freedom of $n = 30$, because the exact degrees of $n = 31$, which would have been correct, was not available from our t table).

Step 4. State a conclusion.

Since our test statistic t exceeds the critical region, we reject the null hypothesis and conclude that the wear rate is different.

Hypothesis test for proportions:

Large-Sample Test of Hypothesis about a Population Proportion

a. One-tail test   b. Two-tail test

Ho: $\pi = \pi_0$ *Ho:* $\pi = \pi_0$

Ha: $\pi > \pi_0$ *Ha:* $\pi \neq \pi_0$

(or *Ha:* $\pi < \pi_0$)

Test statistic: Test statistic:

$$Z = \frac{\pi - \pi_0}{\sqrt{\dfrac{\pi_0(1-\pi_0)}{n}}}$$ $$Z = \frac{\pi - \pi_0}{\sqrt{\dfrac{\pi_0(1-\pi_0)}{n}}}$$

Reject H_0 if: Reject H_0 if:

$Z > Z_\alpha$ (or $Z < -Z_\alpha$) $Z < -Z_{\alpha/2}$ (or $Z > Z_{\alpha/2}$)

Example 1: Large-sample, one-sided test (testing to a hypothesized proportion):
During a recent business meeting, the human resource manager made a remark that 75 percent of the employees would prefer to work four 10-hour work days as opposed to the normal five 8-hour work days. 78.5 percent of a random sample of 85 employees agreed that the 10-hour schedule would be better. At the 10 percent significance level, was the human resource manager correct in his assertion?

Step 1. Formulate the opposing hypotheses.

Ho: $\pi = 0.75$

Ha: $\pi > 0.75$

Step 2. Calculate the test statistic.

$$Z = \frac{\pi - \pi_0}{\sqrt{\dfrac{\pi_0(1-\pi_0)}{n}}}$$

$$Z = \frac{.785 - .750}{\sqrt{\dfrac{.750(1-.750)}{125}}}$$

$$Z = 0.90$$

Step 3. Derive the rejection rule or criteria.

We are testing at a significance level of $\alpha = 10\%$ (or a level of confidence of 90 percent). The Z_α for a 10 percent risk is 1.282.

Step 4. We have already selected the sample, calculated the test statistic, and established the rejection criteria. We now compare the test statistic Z to the critical Z_α of 1.282.

Since our calculated Z test statistic is not greater than the critical Z of 1.282, we cannot reject the null hypothesis. We cannot conclude that over 75 percent of the employees support the notion that the 10-hour work day would be better than the 8-hour work day.

Up to this point, we have been performing hypothesis tests based on a single statistic, mean, or proportion and a single hypothesized expectation. For example, is the hypothesized mean equal to or greater than a specific value?

We may use hypothesis tests to compare two population means or proportions to test if they are equal or not. This type of test is referred to as a hypothesis test about the difference between two statistics. These tests are also one- or two-tail tests, can be done for population means and proportions, and are available for both small and large samples.

The following examples will be given for comparisons of means for small samples (less than 30) and large samples and comparisons of proportions for large samples. Comparisons of population proportions are generally not performed with samples less than 30.

Large-Sample Test of Hypothesis for $\pi_1 - \pi_2$ or the Difference in Two Population Proportions

a. One-tail test   b. Two-tail test

$Ho: (\pi_- - \pi_2) = 0*$ $Ho: (\pi_- - \pi_2) = 0$

$Ha: (\pi_- - \mu_2) < 0$ $Ha: (\pi_- - \pi_2) \neq 0$

      or

    $(\pi_- - \pi_2) > 0$

Test statistic (both cases)

$$Z = \frac{P_1 - P_2}{\sigma_{(P_1 - P_2)}}$$

Reject Ho if: Reject Ho if:

$Z < -Z_\alpha$ or $Z < -Z_{\alpha/2}$

$Z > Z_\alpha$ for $Z > Z_{\alpha/2}$

$Ha: (\pi_- - \pi_2) > 0$

*This test can be applied to differences other than zero.

Example 1:
Over the past decade it has been suspected that the percentage of adults living in the south-eastern region of the United States that eat grits (a gourmet delicacy indigenous to this area) has declined. A sample survey by the Grain Research Institute of Technology (GRIT) in 1986 found that from a random sample of 1500 adults, 525 ate grits on a regular basis. The same survey taken in 1996 with a sample of 1850 found that 685 adults ate grits. Based on this data, has the consumption of grits increased or has it remained the same?

Step 1. Formulate the opposing hypotheses.

$Ho: (\pi_- - \pi_2) = 0$

$Ha: (\pi_- - \pi_2) > 0$

π_1 = true proportion of population that ate grits in 1996

π_2 = true proportion of population that ate grits in 1986

Step 2. Calculate the test statistic.
Set the risk at $\alpha = 0.05$.

$$Z = \frac{P_1 - P_2}{\sigma_{(P_1 - P_2)}}$$

$$P_1 = \frac{(np)_1}{n_1}$$

where: $(np)_1$ = number eating grits in sample from 1996 (n_1)

$$n_1 = 1850$$

$$P_1 = \frac{685}{1850} = 0.370$$

$$P_2 = \frac{(np)_2}{n_2}$$

where: $(np)_2$ = number eating grits in sample from 1986 (n_2)

$$n_2 = 1500$$

$$P_2 = \frac{528}{1500} = 0.352$$

$$\sigma_{(P_1 - P_2)} = \sqrt{P(1-P)\left(\frac{1}{n_1} + \frac{1}{n_2}\right)} \quad \text{or} \quad \sigma_{(P_1 - P_2)} = \sqrt{\frac{P_1 q_1}{n_1} + \frac{P_2 q_2}{n_1}}$$

$$P = \text{combined proportion} = \left(\frac{(np)_1 + (np)_2}{n_1 + n_2}\right) = \frac{685 + 528}{1850 + 1500} = 0.362$$

$$\sigma_{(P_1-P_2)} = \sqrt{P(1-P)\left(\frac{1}{n_1}+\frac{1}{n_2}\right)} \quad \rightarrow \quad \sigma_{(P_1-P_2)} = \sqrt{(0.362)(0.638)\left(\frac{1}{1850}+\frac{1}{1500}\right)}$$

$$\sigma_{(P_1-P_2)} = \sqrt{(0.362)(0.638)(0.0012)} \quad \sigma_{(P_1-P_2)} = 0.017$$

Test statistic:

$$Z = \frac{P_1-P_2}{\sigma_{(P_1-P_2)}} \quad Z = \frac{0.370-0.352}{0.017} = 1.06$$

Step 3. Compare the test statistic to the rejection criteria.

Reject the null hypothesis *Ho:* $\pi_1 = \pi_2$ in favor of the alternative hypothesis *Ha:* $\pi_1 > \pi_2$ if the test statistic (1.06) is greater than Z_α.

For this example, the confidence level is 95 percent with a risk or significance level of 5 percent. The appropriate Z-score for $Z_{.05}$ is 1.645.

The test statistic *is not greater than* the Z_α value of 1.645; therefore, the null hypothesis cannot be rejected. While there is a positive difference between the *sample* data from 1986 and 1996, this difference is not large enough to indicate a difference in the total *population*. The proportion of the population that eat grits has not increased.

Example 2:

The proportion defective of a sample of 350 printed circuit boards sampled during March was found to be 0.075. During April, the process was modified in an effort to lower the defect rate. A sample of 280 printed circuit boards yielded a proportion defective of 0.050. Does the evidence support the conclusion that a process improvement has been made?

Step 1. Formulate the opposing hypotheses.

Ho: $\pi_1 = \pi_2$

Ha: $\pi_1 > \pi_2$

where: π_1 = old process
π_2 = improved process

Step 2. Calculate the test statistic.

$$Z = \frac{p_1 - p_2}{\dfrac{p_1 q_1}{n_1} + \dfrac{p_2 q_2}{n_2}} \quad Z = \frac{0.075 - 0.050}{\sqrt{\dfrac{(0.075)(0.925)}{350} + \dfrac{(0.050)(0.95)}{250}}} = 1.29$$

Step 3. Compare the test statistic (Z calculated) to the critical Z-score.

If our calculated test Z-score is greater than the critical Z-score, we will reject the null hypothesis (that both processes are the same) and accept the alternative hypothesis that the old process is worse than the improved process. For this example, we will set the risk $\alpha = 0.050$, in which case the critical Z-score is 1.645. Z calculated is not greater than Z critical; therefore, we cannot conclude that an improvement has, in fact, been made.

Hypothesis about the Variance

A manager of the reservations center for a major airline has supported the decision to upgrade the computer system for handling flight reservations. His critics, however, claim that the new system, while more powerful than the old system, is too difficult to learn and will lead to more variation in the time to check in passengers. The old system has a documented standard deviation of 7 minutes per passenger. A hypothesis test is designed to settle this issue. The level of significance is set at 2 percent. It will be based on a random sample of 30 customers using the new system.

Step 1. Formulate two opposing hypotheses.

$Ho: \sigma^2 = 49$

$Ha: \sigma^2 < 49$

Note: The hypothesis test must be in terms of the variance, which is the square of the standard deviation.

Step 2. Calculate the test statistic.
The test statistic will be chi-square χ^2.

$$\chi^2 = \frac{s^2(n-1)}{\sigma^2}$$

The sample standard deviation S from the 30 sample observations is 3.2 minutes.

$$\chi^2 = \frac{(3.2)^2(29)}{(7)^2} = 6.06$$

Step 3. Compare the calculated χ^2 to a critical χ^2.

Rejection criteria:

If χ^2 calculated is less than χ^2 critical, reject Ho and accept Ha.
   The calculated $\chi^2 = 6.06$, and chi-square for a risk of 2 percent and a sample size of $n = 30$ is $\chi^2_{1-a,\ n-1} = \chi^2_{.98,29} = 15.574$.

$$\chi^2 \text{ calc.} < \chi^2 \text{ crit.}\quad 6.06 < 15.574$$

We reject the null hypothesis that the true standard deviation is equal to 7 minutes and conclude that the new system does, in fact, have a lower standard deviation.

BIBLIOGRAPHY

Dixon, W. L., and F. J. Massey, Jr. 1969. *Introduction to Statistical Analysis.* 3rd edition. New York: McGraw-Hill.

Montgomery, D. C. 1996. *Introduction to Statistical Quality Control.* 3rd edition. New York: John Wiley and Sons.

Petruccelli, J. D., B. Nandram, and M. Chen. 1999. *Applied Statistics for Engineers and Scientists.* Upper Saddle River, NJ: Prentice Hall.

Sternstein, M. 1996. *Statistics.* Hauppauge, NY: Barron's Educational Series.

Walpole, R. E., and R. H. Myers. 1993. *Probability and Statistics for Engineers and Scientists.* 5th edition. Englewood Cliffs, NJ: Prentice Hall.

INDIVIDUAL-MEDIAN/RANGE CHART

The selection of control charts for variables characteristics are traditionally the individual/moving range, average/range, average/standard deviation and the median/range. In all of these cases, there is an assumption that the characteristic is originating from a single population. In some cases, there is a macro contributing or influencing process that is dominating a set of smaller microprocesses. An example of such a case would be a multi-cavity mold operation. Each of the individual cavities would represent several microprocesses, and the injection pressure common to all of the cavities would be the macroprocess. Another example would be a filling line where several filling heads would represent the individual microprocesses, and the master pump supplying all of the individual filling heads would represent the macroprocess. If in the latter example a twelve-head filling line were to be monitored using an average/range control chart where five samples were randomly chosen every hour and the average weight were charted, there would be a 67 percent probability of not including a specific filling head in the sample. Even with hourly sampling, production could continue for a full day without sampling one of the filling heads.

If a faulty filling head were to develop a state of short filling which were to lead to an out-of-control condition, the usual response would be to adjust the primary pump. This action would lead to an increased overall average causing all of the other positions of the filling heads to be increased, when our original intention was to increase only the filling head in question.

An optional method to monitor the overall process and address the performance of the individual microprocesses is found in the application of the median-individual/range control chart. The larger macroprocess will be monitored by the movement of the individual medians around the average median. The microprocesses will be monitored by the movement of the individuals around the average. These two charts will be constructed on a common chart. The process variation will be monitored using a traditional range control chart.

The steps for the construction for this chart, as with all variable control charts, involve:

1. Collection of historical data to characterize the current performance of the process
2. Calculation of a location and variation statistic
3. Calculation of statistical control limits for the location and variation statistics based on an average ±3 standard deviations
4. Construction of the control chart, including average, control limits, and plotting historical data
5. Continue plotting future data, looking for signs of a process change

The concepts of the median-individual/range chart will be illustrated using the following example:

A filling line has five filling heads being supplied by a master pump. Each of the individual filling heads is independently adjustable as well as the master pump. The characteristic being monitored is the weight of the material in the containers in ounces.

First we will briefly discuss how to calculate the median. The median is the middle value of a set of numbers and, as a statistic, is abbreviated $\tilde{X}$. The median of an odd set of values is determined simply by eliminating the largest followed by the smallest and continuing until only one value remains. For the case where the number of values in the sample is even, we eliminate largest and smallest until two values remain then calculate the average to obtain the median.

Example:
Determine the median of the following five values by dropping the largest, smallest, largest, smallest, and so on until only the median remains.

$$
\begin{array}{ccccccccc}
8 & & & & & & & & \\
4 & & 8 & & 8 & & & & \\
11 & \rightarrow & 4 & \rightarrow & 11 & \rightarrow & 8 & \rightarrow & 8 = \text{median} \\
7 & & 11 & & 7 & & 11 & & \\
13 & & 7 & & & & & &
\end{array}
$$

Step 1. Collect historical data.

Samples are collected twice per day for five days, giving a total of 10 samples for our example. In a real application, one should obtain a minimum of 25 sets of samples to accurately characterize the process. Individuals within a sample should be identified as to the specific head from which they were obtained. The median and range for each sample will be determined and recorded.

Sample #1 3/6/95 AM		Sample #2 3/6/95 PM		Sample #3 3/7/95 AM		Sample #4 3/7/95 PM	
Head 1	42.0	Head 1	46.0	Head 1	45.8	Head 1	43.9
Head 2	44.7	Head 2	45.4	Head 2	44.9	Head 2	46.2
Head 3	46.1	Head 3	40.1	Head 3	43.7	Head 3	45.2
Head 4	43.8	Head 4	44.5	Head 4	41.5	Head 4	43.2
Head 5	43.6	Head 5	45.6	Head 5	42.4	Head 5	40.9
Median	43.8	Median	45.4	Median	43.7	Median	43.9
Range	4.1	Range	5.9	Range	4.3	Range	5.3

Sample #5 3/8/95 AM		Sample #6 3/8/95 PM		Sample #7 3/9/95 AM		Sample #8 3/9/95 PM	
Head 1	40.0	Head 1	45.7	Head 1	43.0	Head 1	46.9
Head 2	42.7	Head 2	43.0	Head 2	42.8	Head 2	44.8
Head 3	44.5	Head 3	45.4	Head 3	45.0	Head 3	43.4
Head 4	43.3	Head 4	45.5	Head 4	42.0	Head 4	43.2
Head 5	46.2	Head 5	42.2	Head 5	42.9	Head 5	41.5
Median	43.3	Median	45.4	Median	42.9	Median	43.4
Range	6.2	Range	3.5	Range	3.0	Range	5.4

Sample #9 3/10/95 AM		Sample #10 3/10/95 PM	
Head 1	45.2	Head 1	43.1
Head 2	45.7	Head 2	46.5
Head 3	46.0	Head 3	44.4
Head 4	43.0	Head 4	43.8
Head 5	42.5	Head 5	45.6
Median	45.2	Median	44.4
Range	3.5	Range	3.4

Step 2. Calculate the location and variation statistic.

The location statistic will be the grand average median $\bar{\bar{X}}$ and is determined by averaging the 10 median values:

$$\bar{\bar{X}} = \frac{43.8 + 45.4 + 43.7 + 43.9 + 43.3 + 45.4 + 42.9 + 43.4 + 45.2 + 44.4}{10}$$

$$\bar{\bar{X}} = 44.14$$

The variation statistic will be the average range $\bar{R}$ and is determined by averaging the 10 ranges:

$$\bar{R} = \frac{4.1 + 5.9 + 4.3 + 5.3 + 6.2 + 3.5 + 3.0 + 5.4 + 3.5 + 3.4}{10}$$

$$\bar{R} = 4.46$$

Step 3. Determine the statistical control limits for the location and variation statistic.

Control limits for the median:

Upper and lower control limits for the median are calculated by adding three standard deviations for the median distribution for the upper control limit (UCL) and subtracting three standard deviations for the lower control limit (LCL).

$$\text{UCL}_{\tilde{x}}/\text{LCL}_{\tilde{x}} = \bar{\bar{X}} \pm \tilde{A}_2 \bar{R}$$

where: $\bar{\bar{X}}$ = average median
$\tilde{A}_2$ = factor dependent on the subgroup sample size n
$\bar{R}$ = average range

The value for the factor $\tilde{A}_2$ for this example where $n = 5$ is 0.691 and is found in Table 1.

The upper and lower control limits for this example given that $\bar{\bar{X}} = 44.14$ and $\bar{R} = 4.46$ are

$$\text{UCL}_{\tilde{x}}/\text{LCL}_{\tilde{x}} = 44.14 \pm (0.691)(4.46)$$

$$\text{UCL}_{\tilde{x}} = 44.14 + 3.08 = 47.22$$

$$\text{UCL}_{\tilde{x}} = 44.14 - 3.08 = 41.06$$

Control limits for the individuals:

The control limits for the individuals are based on the average of the individuals, ± 3 standard deviations, and are calculated as follows:

$$\text{UCL}_{Xi}/\text{LCL}_{Xi} = \bar{\bar{X}} \pm E_2 \bar{R}$$

where: $\bar{\bar{X}}$ = average median
E_2 = factor dependent on the subgroup sample size
$\bar{R}$ = average range

For our example where $\bar{\bar{X}} = 44.14$, $\bar{R} = 4.46$, and $E_2 = 1.290$, the upper and lower control limits for individuals are

$\text{UCL}_{Xi}/\text{LCL}_{Xi} = \bar{\bar{X}} \pm E_2 \bar{R}$ $\rightarrow$ $\text{UCL}_{Xi}/\text{LCL}_{Xi} = 44.14 \pm (1.290)(4.46)$

$\text{UCL}_{Xi} = 44.14 + 5.75 = 49.89$ $\text{LCL}_{Xi} = 44.14 - 5.75 = 38.39$

Control limits for the ranges:

The upper and lower control limits are determined respectively by multiplying the average range $\bar{R}$ by factors D_4 and D_3.

$$\text{UCL}_R = D_4 \bar{R} \quad \text{and} \quad \text{LCL}_R = D_3 \bar{R}$$

where: D_4 and D_3 = factors dependent on the subgroup sample size n
$\bar{R}$ = the average range

Table 1. Factors Used for Median-Individual/Range Charts.				
Subgroup n	$\bar{\bar{A}}_2$ (medians)	E_2 (individuals)	D_3 (LCL range)	D_4 (UCL range)
2	1.880	2.660	–	3.267
3	1.187	1.772	–	2.574
4	0.796	1.457	–	2.282
5	0.691	1.290	–	2.114
6	0.548	1.184	–	2.004
7	0.508	1.109	0.076	1.924
8	0.433	1.054	0.136	1.864
9	0.412	1.010	0.184	1.816
10	0.362	0.975	0.223	1.777
11	0.350	0.945	0.256	1.744
12	0.316	0.921	0.283	1.717
13	0.308	0.899	0.307	1.693
14	0.282	0.811	0.328	1.672
15	0.276	0.864	0.347	1.653
16	0.255	0.849	0.363	1.637
17	0.251	0.836	0.378	1.622
18	0.234	0.824	0.391	1.608
19	0.231	0.813	0.403	1.597
20	0.218	0.803	0.415	1.585
21	0.215	0.794	0.425	1.575
22	0.203	0.786	0.434	1.566
23	0.201	0.778	0.443	1.557
24	0.191	0.770	0.451	1.548
25	0.189	0.763	0.459	1.541

For our example, $UCL_R = (2.114)(4.46) = 9.43$

and LCL_R = not defined (no values for D_3 are defined until the sample size reaches seven)

$LCL_R = 0$ as a default

Summary of control limits:

Upper control limit for medians, $UCL_{\tilde{x}} = 47.22$

Lower control limit for medians, $UCL_{\tilde{x}} = 41.06$

Upper control limit for individuals, $UCL_{Xi} = 49.89$

Lower control limit for individuals, $LCL_{Xi} = 38.39$

Upper control limit for ranges, $UCL_R = 9.43$

Lower control limit for ranges, $LCL_R = 0$

Average median, $\bar{\bar{X}} = 44.14$

Average range, $\bar{R} = 4.46$

Step 4. Construct the control chart, sketch in control limits and averages, and plot historical data.

Control limits are traditionally drawn with a broken line (----------------------), and averages are drawn with a solid line (————————). For subgroup sample sizes less than seven, the average line for the range is not drawn.

Since we are plotting two location statistics (median and individuals) on one chart, we will use broken lines for the control limits for the individuals and a short-long dash for the median control limits.

Notes on plotting data:

All individual points should be identified as a single dot (.), and the particular individual that is identified as the median will be represented by an "x." The median values will be connected with a line. Any median values falling outside of the median limits represent a rule violation, and any individual values (including the median) that fall outside of the individual control limits also represent a rule violation. If an individual value falls outside its control limit, it should be identified relative to its origin (such as head #1, head #2, and so on) by placing a number next to the dot.

The completed chart showing the first 10 sets of data follows:

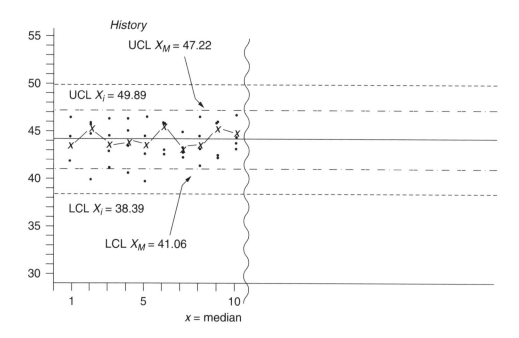

x = median

Note that all of the medians and individuals are within their respective control limits. This process is statistically stable and well behaved.

Step 5. Continue to monitor the process, looking for evidence of a process change.

Signs of a process change will be indicated when one of the SPC rules is violated. These rules are numerous, and most have the following two common characteristics:

1. They are patterns that are statistically rare to occur with a normal distribution.
2. They indicate a direction in which the process has changed, either increased or decreased.

The three more frequently used SPC detection rules are:

1. A single observation outside of the upper or lower control limit
2. Seven consecutive points below the average or above the average
3. A trend of seven consecutive points all steadily increasing or decreasing in value

These rules will be applied to both the movement of the individuals with respect to their control limits and to the median with respect to its control limit. The only rule applicable to monitoring the range will be rule #1, when the sample size is less than seven.

Data for the next seven samples are provided below. Plot the data, and determine if there is any evidence of a process change.

Sample #11		Sample #12		Sample #13		Sample #14	
3/11/95		**3/11/95**		**3/12/95**		**3/12/95**	
AM		**PM**		**AM**		**PM**	
Head 1	42.1	Head 1	43.4	Head 1	46.5	Head 1	45.1
Head 2	44.8	Head 2	40.4	Head 2	44.0	Head 2	40.0
Head 3	44.7	Head 3	43.3	Head 3	42.4	Head 3	38.0
Head 4	43.3	Head 4	45.1	Head 4	43.5	Head 4	42.5
Head 5	42.5	Head 5	40.7	Head 5	41.2	Head 5	43.3
Median	43.3	Median	43.3	Median	43.5	Median	42.5
Range	2.7	Range	4.7	Range	5.3	Range	7.1

Sample #15		Sample #16		Sample #17	
3/13/95		**3/13/95**		**3/14/95**	
AM		**PM**		**AM**	
Head 1	46.0	Head 1	43.0	Head 1	42.0
Head 2	40.6	Head 2	44.4	Head 2	43.0
Head 3	45.0	Head 3	40.6	Head 3	37.6
Head 4	41.8	Head 4	43.6	Head 4	45.0
Head 5	44.4	Head 5	42.4	Head 5	40.0
Median	44.4	Median	43.0	Median	42.0
Range	5.4	Range	3.8	Range	7.4

The completed control chart follows. Note that there are two incidences where head #3 had individual values that were below the lower control limit for individuals (38.39). This indicates that this individual fill head is delivering less than would be expected based on the historical characterization.

Control limits for the range are calculated in the traditional manner for Shewhart charts. The UCL for the range is $UCL = D_4 \bar{R}$, and the LCL for the range is $LCL = D_3 \bar{R}$. For this example, the LCL is undefined as $n = 5$, and the UCL is calculated to be $UCL = 9.43$.

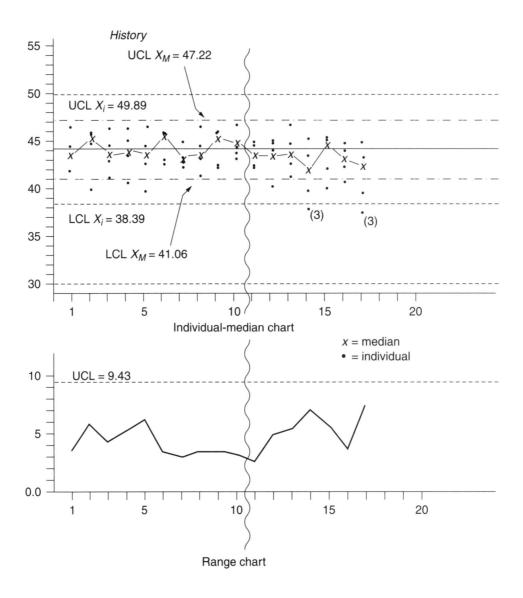

Individual-median chart

Range chart

Sample #14 and sample #17 have individuals from filling head #3 falling below the LCL for individuals, while the medians for these samples remain in control. This supports the conclusion that filling head #3 is not performing as was established for the historical data. Perhaps the filling nozzle is blocked slightly, causing a restriction in the filling rate. In any case, the output has been reduced enough to cause a process change in this position.

BIBLIOGRAPHY

van der Veen, J., and T. Holst, personal correspondence, Portland, OR: Northwest Analytical.

Individual/Moving Range Chart

Control charts are used to detect changes in processes. When processes are being monitored using variables data and the opportunity for obtaining data is relatively infrequent, we may elect to use an individual/moving range control chart. An example of such a case would be measurements of a characteristic of a batch process where individual batches are made once every four hours.

The ability of a control chart to detect a change in the process is related to:

1. The subgroup sample size
2. The amount of change that has occurred

The individual/moving range control chart has a subgroup sample size of $n = 1$ and is the weakest of all Shewhart control charts with respect to its ability to detect a process change.

The general principles and use of control charts are universal and are based on the following elements:

1. Collection of "historical data"
2. Calculation of *a location and variation statistic*
3. Determination of statistical control limits around the location and variation *statistic*
4. Graphical presentation of the data and control limits
5. Continuation of data collection and plotting, looking for changes in the future

The following example will illustrate how to develop an individual/moving range control chart.

Step 1. Collect historical data.

Typically, 40 to 100 individual data points should be taken over the period of time that we wish to serve as a baseline period. If we want to detect a process change relative to the last month of performance, then we will take the 40–100 measurements over a one-month period.

For demonstration purposes, we will use a total of 15 samples for our historical database. Our samples were taken three times per day over a five-day period. The sample size is $n = 1$, and the total number of samples taken to characterize the process is $k = 15$. The process that our control chart will monitor is the operation of filling out patient information during the admission process at a hospital. The times will be recorded in units of seconds. Samples are taken at approximately 9:00 AM, 1:00 PM, and again at 4:00 PM each day for five days.

215

The best single value to express the *location* of the performance of this process at a specific time is the data point taken at that time. The best indication of the *variation* of the process at a given point in time is the range between two successive values around that point in time. The range between successive points is defined as the *moving range* and is calculated by taking the difference between the current value and the previous value.

Moving ranges are always positive in value. The first data point collected will not have a moving range, because it does not have a previous data point associated with it. There will always be one less moving range than there are data points.

At the time we collect our initial data, we will also calculate and record the moving ranges. The first moving range is 260 – 196 = 64, the second moving range is 260 – 187 = 73, and the last moving range is 230 – 210 = 20.

Date	Time	Admission time	Moving range
11/2/94	8:45A	196	—
11/2/94	1:12P	260	64
11/2/94	3:48P	187	73
11/3/94	9:04A	170	17
11/3/94	1:00P	296	126
11/3/94	3:55P	171	125
11/4/94	9:00A	320	149
11/4/94	12:55P	139	181
11/4/94	4:10P	196	57
11/5/94	8:53A	228	32
11/5/94	1:07P	176	52
11/5/94	3:58P	125	51
11/6/94	9:11A	226	101
11/6/94	12:55P	210	16
11/6/94	3:57P	230	20

Step 2. Calculate a location and variation statistic.

All processes can be described or characterized by two types of descriptive statistics:

Location—The central tendency or where a process is tending to perform. Examples of location statistics are average, median, and mode.

Variation—The amount of change in the values obtained from a process. Examples of variation statistics are standard deviation, range, and moving range.

For the individual/moving control charts, the location statistic will be the average of all of the individual data points collected for the historical period. To calculate the average, add all of the values and divide by the number of data points used to obtain the sum.

$$\text{Average, } \overline{X} = \frac{196 + 260 + 187 + 170 + \ldots + 230}{15} = \frac{3130}{15} = 208.6$$

The variation statistic we will use is the average moving range. It will be calculated by averaging the moving ranges.

$$\text{Average moving range, } \overline{MR} = \frac{64 + 73 + 17 + 126 + \ldots 20}{14} = \frac{1064}{14} = 76.0$$

Step 3. Determine control limits for the location and variation statistics.

The average of the individuals represents the best overall location statistic for the process. There is variation of the individual observations relative to the average. Control limits will be determined for the individual values that make up the average. These limits will define the upper and lower values we expect to encounter approximately 99.7 percent of the time and will be based on the average ±3 standard deviations.

Our original variation statistic was the moving range, which was determined for each successive pair of data points. Using these moving ranges, we calculated an average moving range $\overline{MR}$. An estimate of three standard deviations can be made from the relationship

$$3S = 2.66\overline{MR}$$

$$3S = 2.66 \times 76.0 = 202.2$$

The upper and lower control limits for the individual values are determined by adding and subtracting three standard deviations to the average.

$$\text{Upper control limit, UCL} = \overline{X} + 2.66\overline{MR}$$

$$\text{Lower control limit, LCL} = \overline{X} - 2.66\overline{MR}$$

$$\text{UCL} = 208.6 + 202.2 = 410.8$$

$$\text{LCL} = 208.6 - 202.2 = 6.4$$

Based on the historical data we have collected, there is a 99.7 percent probability that we will find an individual value between 6.4 and 410.8 seconds. We used the moving range to calculate the average moving range, and from the average moving range we estimated three standard deviations. There was variation among the moving ranges just as there was variation among the individual values. A single UCL for the moving ranges is calculated using the relationship:

There is no LCL for the moving range.

$$\text{UCL}_{MR} = 3.267 \times 76.0 = 248.3$$

$$\text{Upper control limit } MR = 3.267\overline{MR}$$

Step 4. Construct the control chart, and plot the data.

Note that the first individual observation does not have an associated moving range. Locate the position of the control limits for the individuals and the average of the individuals, and label appropriately $\overline{X} = 208.6$, LCL = 6.4, and UCL = 410.8. Draw the average line using a solid line, and draw the control limits using a broken line. Repeat this for the UCL for the moving range. It is not necessary to show the average of the moving range. Label the UCL for the moving range as UCL = 248.3.

Plot the individual and moving range values, and connect the points as they are located. Draw a vertical wavy line between sample #15 and sample #16. This line will serve to separate the *historical* data used to characterize the process and establish control limits from values plotted from the future. In the area above the UCL for the individuals in the vicinity of sample #6 through sample #9, record the word *History* to identify those data points used for historical characterization.

An individual/moving range chart (historical data only) follows:

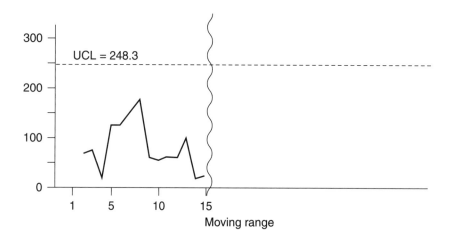

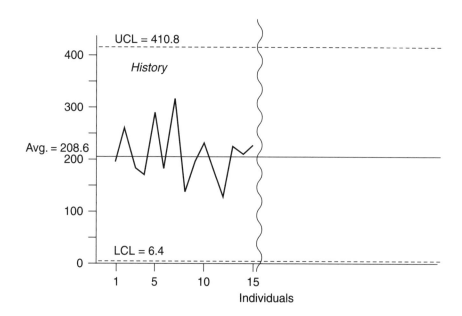

Step 5. Continue plotting future data, looking for process changes.

Continue to collect data from the process, and plot the results on the control chart. If the process has not changed, then the distribution of data in the future will support the statistical characterization of a process as described by the *historical data*. For our example, the distribution would conform to one having an average of 208.6 with a standard deviation for the distribution of individuals of 67.4 (one third of 202.2, which is an estimate of three standard deviations), and we would expect the maximum moving range to be 248.3. If either the process average or the normal expected variation were to change, we would expect to see this change manifest as a change in the relationship of the individual values to the control limits. There are several *rules* that, if violated, assume that a change in the process has occurred or that a lack of control is indicated. It is not impossible for these rules to be violated even when there has been no real process change, but the likelihood is so small that when we do see a *rule* violation, we assume that a change has occurred. All of these rules have the following common features:

1. They are statistically rare to occur.
2. They indicate a direction in which the process parameter is changing.

Rule One: A lack of control is indicated whenever a single point falls outside of the three standard deviation limits (shift detection).

Rule Two: A lack of control is indicated when seven consecutive points are on the same side of the average line with none outside of the three standard deviation limits (shift detection).

Rule Three: A lack of control is indicated when seven consecutive points are steadily increasing or decreasing in value (trend detection).

Rule Four: Two out of three consecutive points are greater than the two standard deviation limits with the third falling anywhere. No points fall outside of the three standard deviation limits.

Rule Five: Four out of five points are greater than one standard deviation, and the fifth point falls anywhere. No points fall outside of the three standard deviation limits.

There are other patterns representing nonstable situations, some of which may not indicate a direction of change. The five patterns described by the aforementioned rules are the most frequent used to detect process changes. Only rule one through rule three utilize the ±3 standard deviation limits and are the rules traditionally used in an operational environment.

Only rule one should be applied to the moving range portion of the control chart for the individual/moving range. This is due to the fact that the distribution of the moving range is not normal and is not symmetrical with respect to the average moving range. For a more detailed discussion of control chart patterns, see *Statistical Quality Control Handbook* (Western Electric Company: "Statistical Quality Control Handbook," 2d ed. Western Electric Company, Inc., New York, 1958. 149–183).

Based on rule one, rule two, and rule three, plot the following additional data points, and determine if there is evidence of a process change. Has the process average changed? Has the process variation changed?

Date	Time	Admission time
11/7/94	8:56A	115
11/7/94	12:40P	175
11/7/94	3:50P	250
11/8/94	9:00A	215
11/8/94	1:10P	330
11/8/94	4:05P	220
11/9/94	9:05P	280
11/9/94	12:58P	215
11/9/94	3:50P	305
11/10/94	8:30A	275
11/10/94	1:10P	125
11/10/94	4:00P	150
11/11/94	8:50A	250
11/11/94	1:11A	285
11/11/94	3:51P	190
11/12/94	9:11A	105
11/12/94	1:00A	330
11/12/94	4:00P	250

After completing your chart, compare to the following finished chart.

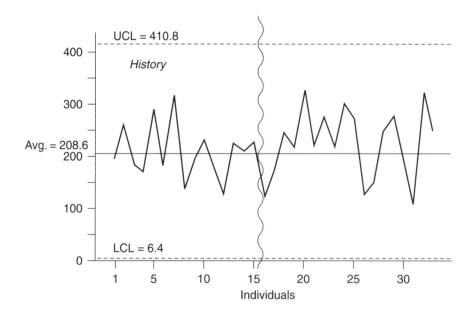

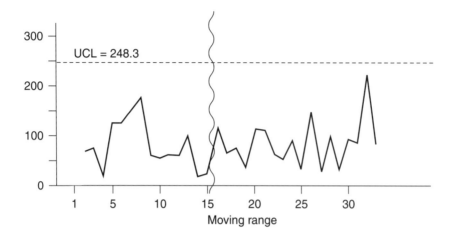

The chart indicates the possibility of an increase in the average with no change in the variation. Eight consecutive points are seen, starting with sample #18 (11/7/94 3:50 PM).

BIBLIOGRAPHY

Besterfield, D. H. 1994. *Quality Control.* 4th edition. Englewood Cliffs, NJ: Prentice Hall.

Grant, E. L., and R. S. Leavenworth. 1996. *Statistical Quality Control.* 7th edition. New York: McGraw-Hill.

Hayes, G. E., and H. G. Romig. 1982. *Modern Quality Control.* 3rd edition. Encino, CA: Glencoe.

Montgomery, D. C. 1996. *Introduction to Statistical Quality Control.* 3rd edition. New York: John Wiley and Sons.

Western Electric Company. 1958. *Statistical Quality Control Handbook.* 2nd edition. New York: Western Electric Company, inc.

Wheeler, D. J., and D. S. Chambers. 1992. *Understanding Statistical Process Control.* 2nd edition. Knoxville, TN: SPC Press.

MEASUREMENT ERROR ASSESSMENT

All of the conclusions reached when using data obtained from observations depend on the accuracy and validity of the data. We assume that the information fairly represents the truth. Measurement data can be derived from either pure attribute inputs such as visual standards or from direct measured variables such as gages. The latter type of measurement error will be discussed first.

VARIABLES MEASUREMENT ERROR

Terms and Concepts

Accuracy: The degree to which the observed value agrees with the true value.

Precision: The degree to which repeated observations vary from the true value or from each other; a statement of variation.

Repeatability: A measure of how well one can obtain the same observed value when measuring the same part or sample over and over using the same measuring device. Repeatability is sometimes referred to as equipment variation, or *EV.*

Reproducibility: How well others agree with our measurement when measuring the same sample using the same measuring device. Reproducibility is sometimes referred to as appraiser variation, or *AV.*

Example of repeatability and reproducibility:
Bob has been asked to measure the diameter of a steel shaft. His results for five measurements on the same part are

0.50012

0.50001

0.50013

0.50011

0.50009

The fact that there is variation can be attributed to the following three causes:

1. Measurement variation due to the measuring device
2. Measurement variation due to positional variation of the part
3. Measurement variation due to inconsistency of the measuring technique

We will *assume* that while Bob's technique may or may not be correct, he is using the same technique each time he makes a measurement. We will exclude technique as a contributing factor. Also exclude the possibility of positional variation, as this is generally controlled by specifying where the part is to be measured. This leaves only the measuring device as a source of error. For this reason, error or variation from repeated measures is called repeatability error or equipment variation (*EV*).

If we ask others in the department to measure the same part using the same measuring device, the results might be

Bob	0.50011
Mary	0.49995
Jim	0.50008
Mark	0.50000

We will *assume* the lack of agreement is associated not with the measuring device (this has already been acknowledged via the repeatability, or *EV*) but rather with the different techniques in using the measuring device. The lack of agreement is due to the reproducibility error or appraiser variation (*AV*).

The amount of variation for each of these contributing factors could have been determined by simply calculating the standard deviation of each set of data for the repeatability and reproducibility:

Standard deviation for Bob's five measurements: 0.0000482

Standard deviation for four measurements from all operators: 0.0000733

In most formal studies to assess measurement, since there is a relatively small number of operators that typically measure a given part, the range of measurements is used to estimate the standard deviation. From statistical process control (SPC) the relationship

$$\hat{\sigma} = \frac{\overline{R}}{d_2^{\,*}}$$

May be used to estimate the standard deviation of measurement error. In this relationship, $\overline{R}$ is the average range of repeated measures, and d_2 is a constant that is dependent on the number of observations to calculate the range. In pure SPC calculations, the actual number of samples from which the average range is determined is assumed to be very large ($n > 25$) for most control chart applications. For this reason, the d_2 factors must be modified to compensate for the small number of samples. These modified factors are abbreviated $d_2^{\,*}$. These modified values depend on both the size of the sample from which the range is determined and on the number of samples used to determine the average.

The values for $d_2^{\,*}$ can be found in Table 1.

Table 1. d_2^* Values for the Distribution of the Average Range.

	\multicolumn{14}{c}{Size of sample m}
	2	3	4	5	6	7	8	9	10	11	12	13	14	15
1	1.41	1.91	2.24	2.48	2.67	2.83	2.96	3.08	3.18	3.27	3.35	3.42	3.49	3.55
2	1.28	1.81	2.15	2.40	2.60	2.77	2.91	3.02	3.13	3.22	3.30	3.38	3.45	3.51
3	1.23	1.77	2.12	2.38	2.58	2.75	2.89	3.01	3.11	3.21	3.29	3.37	3.43	3.50
4	1.21	1.75	2.11	2.37	2.57	2.74	2.88	3.00	3.10	3.20	3.28	3.36	3.43	3.49
5	1.19	1.74	2.10	2.36	2.56	2.73	2.87	2.99	3.10	3.19	3.28	3.35	3.42	3.49
6	1.18	1.73	2.09	2.35	2.56	2.73	2.87	2.99	3.10	3.19	3.27	3.35	3.42	3.49
7	1.17	1.73	2.09	2.35	2.55	2.72	2.87	2.99	3.10	3.19	3.27	3.35	3.42	3.48
8	1.17	1.72	2.08	2.35	2.55	2.72	2.87	2.98	3.09	3.19	3.27	3.35	3.42	3.48
9	1.16	1.72	2.08	2.34	2.55	2.72	2.86	2.98	3.09	3.18	3.27	3.35	3.42	3.48
10	1.16	1.72	2.08	2.34	2.55	2.72	2.86	2.98	3.09	3.18	3.27	3.34	3.42	3.48
11	1.16	1.71	2.08	2.34	2.55	2.72	2.86	2.98	3.09	3.18	3.27	3.34	3.41	3.48
12	1.15	1.71	2.07	2.34	2.55	2.72	2.85	2.98	3.09	3.18	3.27	3.34	3.41	3.48
13	1.15	1.71	2.07	2.34	2.55	2.71	2.85	2.98	3.09	3.18	3.27	3.34	3.41	3.48
14	1.15	1.71	2.07	2.34	2.54	2.71	2.85	2.98	3.08	3.18	3.27	3.34	3.41	3.48
15	1.15	1.71	2.07	2.34	2.54	2.71	2.85	2.98	3.08	3.18	3.26	3.34	3.41	3.48
>15	1.128	1.693	2.059	2.326	2.534	2.704	2.847	2.970	3.078	3.173	3.258	3.336	3.407	3.472

(left axis label: Number of samples g)

Example of repeatability (equipment variation *EV*) error determination:
Five samples are chosen for measurement. The samples are selected with typical measurement values over the entire range of values expected to be found. Two operators are chosen to make the measurements. Each of the five parts will be measured two times. The results are tallied as follows:

	Operator A			Operator B		
Part #	Trial 1	Trial 2	Range	Trial 1	Trial 2	Range
1	0.889	0.886	0.003	0.888	0.886	0.002
2	0.855	0.859	0.004	0.852	0.851	0.001
3	0.868	0.870	0.002	0.866	0.866	0.000
4	0.888	0.890	0.002	0.885	0.884	0.001
5	0.867	0.870	0.003	0.862	0.863	0.001

Estimation of one standard deviation of EV_0:

Use the relationship of

$$\hat{\sigma} = \frac{\bar{R}}{d_2^*}$$

where the average range $\bar{R}$ is the grand average range of the two operators' ranges for the duplicate measurements for each of the five sample parts.

The average range for operator A is

$$\bar{R}_a = \frac{0.003 + 0.004 + 0.002 + 0.002 + 0.003}{5} = 0.0028$$

The average range for operator B is

$$\bar{R}_b = \frac{0.002 + 0.001 + 0.000 + 0.001 + 0.001}{5} = 0.0010$$

The grand average range for the two operators is

$$\bar{R}_1 = \frac{0.0028 + 0.0010}{2} = 0.0019$$

This average range is the same as if we had averaged all 10 sets of ranges. Each operator measured the samples two times. In terms of SPC, the subgroup sample size is two in terms of measurement error, $m = 2$ (m = the number of trials). The total number of samples measured g is determined by multiplying the number of parts (five) by the number of operators participating in the study (two). In this case, $g = 2 \times 5 = 10$.

Looking up the appropriate d_2^* in Table 1, where $m = 2$ and $g = 10$, we find $d_2^* = 1.16$.

$$\bar{\sigma} = \frac{\bar{R}_1}{d_2^*} \qquad \sigma = \frac{0.0019}{1.16} = 0.0016$$

This estimate is for *one* standard deviation for the repeatability error and should be written as σ_{EV}. Traditionally, the appraiser variation or repeatability is reported as 5.15σ to reflect a 99 percent level of confidence.

$$EV = 5.15 \times 0.0016$$

$$EV = 0.0082$$

By dividing the EV by two and rewriting, we may also express the AV as ± 0.0041.

Any measurement we make with the system under evaluation is subject to a measurement of ± 0.0041 99 percent of the time due to equipment error.

Example of reproducibility (appraiser variation, AV) error determination:
In the example illustrating the concept of repeatability, each operator measured the same five parts twice. If there were no measurement error due to the actual measuring device ($EV = 0$), then the difference between the grand average for all ten observations (5 parts $\times$ 2 measurements each = 10 measurements) would be equal. Any differences in the average of operator A and operator B would be attributed to differences in technique. This error factor is referred to as reproducibility error or appraiser variation (AV).

The AV standard deviation is calculated similar to that of the EV. In this case, we are using a range of one sample of two observations. The appropriate d_2^* is obtained from

Table 1, and it is dependent on the number of operators ($m = 2$) and the number of samples ($g = 1$), since there is only one range calculation.

$$d_2^* = 1.41$$

Each operator's average is determined by averaging all 10 samples. The range is determined by subtracting the lowest from the highest.

Average operator $A = 0.8742$

Average operator $B = \underline{0.8704}$

Difference $R_2 = 0.0038$

Estimate of one standard deviation for appraiser variation:

$$\hat{\sigma} = \frac{\overline{R}_2}{d_2^*} \qquad \sigma = \frac{0.0038}{1.41} = 0.0027 \qquad \sigma_{AV} = 0.0027$$

Unadjusted $AV = 5.15 \times \sigma_{AV}$

Unadjusted $AV = 0.014$

This unadjusted AV is contaminated by the variation due to the measuring device (EV) and must be adjusted by subtracting a proportion of the EV from it. Since standard deviations cannot be added or subtracted, we convert to the variance by squaring the standard deviation, performing operations, and then taking the square root to convert back to standard deviations.

Corrected AV compensated for EV:

$$AV = \sqrt{\left[\frac{5.15R_2}{d_2^*}\right]^2 - \left[\frac{(5.15\sigma_{EV})^2}{nt}\right]}$$

where: R_2 = range for average values of operator measurements
        $R_2 = 0.0038$
        n = number of parts used in evaluation
        $n = 5$
        t = number of times (trials) each operator measures the part
        $t = 2$
    $\sigma_{EV} = 0.0016$

$$AV = \sqrt{\left[\frac{(5.15)(0.0038)}{1.41}\right]^2 - \left[\frac{[(5.15)(0.0016)]^2}{(5)(2)}\right]}$$

$$AV = \sqrt{0.000193 - 0.0000068}$$

$$AV = 0.0136 \text{ or } \pm 0.0068$$

Note: It is technically possible that a negative AV could be encountered. If this happens, default to zero.

Combined (Gage) Repeatability and Reproducibility Error (GR&R)

Since each of these contributing factors are forms of standard deviation, they cannot be added together directly. Variances (square of the standard deviation) can be added. To combine the two error factors, we square each, add them together, and take the square root.

$$GR \& R = \sqrt{(AV)^2 + (EV)^2} \qquad GR \& R = \sqrt{(0.0136)^2 + (0.0082)^2}$$

$$GR \& R = 0.016 \, (\text{or } GR \& R = \pm 0.008)$$

Any measurement made with this system is subject to an error of ±0.008 units of measurement 99 percent of the time.

In order to make a more quantitative judgement as to the significance of this amount of error, we compare the total GR&R to the specification tolerance. The GR&R is expressed as a percentage of the total tolerance (upper specification – lower specification).

Assume that for this example the specification is 0.5000 ± 0.030. The %GR&R would be

$$\%GR \& R = \left(\frac{R \& R}{T.T} \right) \times 100$$

$$\%GR \& R = \left(\frac{0.016}{0.060} \right) \times 100$$

$$\%GR \& R = 26.7\%$$

26.7 percent of the total tolerance is being consumed as a result of measurement error. As a rule of thumb, the following guidelines may be used:

%GR&R   Less than 10 percent acceptable

%GR&R   10 percent to 30 percent may be acceptable depending on application

%GR&R   Greater than 30 percent not acceptable

Consequences of Measurement Error

The following two erroneous inspection conclusions can be made, each of which is amplified when measurement error is excessive:

1. The rejection of materials and parts that truly meet specification requirements
2. The acceptance of materials or parts that truly do not meet specification requirements

In the previous example, the total repeatability and reproducibility error was determined to be 0.016. Dividing by 5.15, the standard deviation of measurement error $\sigma_{R\&R}$ is 0.0032. If a part were measured and found to be 0.529 and the upper specification were 0.530, then the conclusion would be to accept the part as meeting the specification requirement. Based on this information, one can calculate the probability that the part, in fact, exceeds the upper specification and that measurement error has led to a false acceptance when we should, in fact, reject the part. This probability can be determined using the normal distribution, the standard deviation of measurement, and a point of interest as the upper specification limit.

Given an upper specification of 0.530, a standard deviation of measurement error of 0.0032, and an observed measurement of 0.529, what is the probability that the true measurement is greater than 0.530?

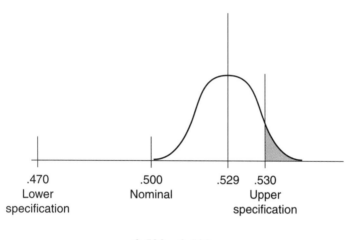

$$Z_U = \frac{0.530 - 0.529}{0.0032} = 0.31$$

A Z-score of 0.31 corresponds to a probability of exceeding 0.530, the upper specification of 0.3783. There is a 37.8 percent chance that the part actually exceeds the upper specification limit of 0.530; this is approximately a one-in-three chance.

Supplemental problem:

An R&R of 0.355 is reported for a measuring system. The lower specification is 12.0. An inspected unit is deemed out of specification with an observed measurement of 11.92. What is the probability that the part actually meets the specification requirement?

Confidence Interval for Repeatability, Reproducibility and Combined R&R

Confidence intervals for the *AV*, *EV*, and R&R can be calculated similarly to those of the standard deviation. Recall that the confidence interval for the standard deviation is determined by

$$\sqrt{\frac{(n-1)s^2}{X^2_{\alpha/2}}} < \sigma < \sqrt{\frac{(n-1)s^2}{X^2_{1-\frac{\alpha}{2}}}}$$

where: n = sample size
s = standard deviation
χ^2 = chi square value for appropriate degrees of freedom and confidence level

The equation will be modified such that the sample size and degrees of freedom will be taken from Table 2. The degrees of freedom will depend on the d_2^* value.

Confidence Interval for Repeatability, *EV*

Lower confidence limit *Upper confidence limit*

$$\text{LCL} = \sqrt{\frac{(EV)^2}{\dfrac{\chi^2_{\alpha/2}}{v_1}}} \qquad \text{UCL} = \sqrt{\frac{(EV)^2}{\dfrac{\chi^2_{1-\alpha/2}}{v_1}}}$$

where: EV = calculated EV from measurement error study
v_1 = degrees of freedom for appropriate m (sample size) and g (number of samples) from Table 2
χ^2 = chi square for specified confidence and v_1 degrees of freedom

Table 2. Degrees of Freedom v as a Function of d_2^*.

Number of samples g	Size of sample m: 2	3	4	5	6	7	8	9	10	11	12	13	14	15
1	1.0	2.0	2.9	3.8	4.7	5.5	6.3	7.0	7.7	8.4	9.0	9.6	10.2	10.8
2	1.9	3.8	5.7	7.5	9.2	10.8	12.3	13.8	15.1	16.5	17.8	19.0	20.2	21.3
3	2.8	5.7	8.4	11.1	13.6	16.0	18.3	20.5	22.6	24.6	26.5	28.4	30.2	31.9
4	3.7	7.5	11.2	14.7	18.1	21.3	24.4	27.3	30.1	32.7	35.3	37.8	40.1	42.4
5	4.6	9.3	13.9	18.4	22.6	26.6	30.4	34.0	37.5	40.9	44.1	47.1	50.1	52.9
6	5.5	11.4	16.6	22.0	27.1	31.8	36.4	40.8	45.0	49.0	52.8	56.5	60.1	63.5
7	6.4	12.9	19.4	25.6	31.5	37.1	42.5	47.5	52.4	57.1	61.6	65.9	70.0	74.0
8	7.2	14.8	22.1	29.3	36.0	42.4	48.5	54.3	59.9	65.2	70.3	75.3	80.0	84.6
9	8.1	16.6	24.8	32.9	40.5	47.7	54.5	61.0	67.3	73.3	79.1	84.6	90.0	95.1
10	9.0	18.4	27.6	36.5	44.9	52.9	60.6	67.8	74.8	81.5	87.9	94.0	99.9	105.6
11	9.9	20.2	30.3	40.1	49.4	58.2	66.6	74.6	82.3	89.6	96.6	103.4	109.9	116.2
12	10.8	22.0	33.0	43.7	53.9	63.5	72.7	81.3	89.7	97.7	105.4	112.8	119.9	126.7
13	11.6	23.9	35.7	47.4	58.4	68.8	78.7	88.1	97.2	105.8	114.1	122.2	129.9	137.3
14	12.5	25.7	38.5	51.0	62.8	74.0	84.7	94.8	104.6	113.9	122.9	131.5	139.8	147.8
15	13.4	27.5	41.2	54.6	67.3	79.3	90.8	101.6	112.1	122.1	131.7	140.9	149.8	158.3
>15														
c.d	0.88	1.82	2.74	3.62	4.47	5.27	6.03	6.76	7.45	8.12	8.76	9.38	9.97	10.54

Example calculation:

Calculate the 90 percent confidence interval for *EV* given the following (from the previous example):

Part I. Lower confidence interval limit:

$EV = 0.0082$

Number of samples = 5 parts

Number of operators = 2

Total measurements taken, $g = 10$ (Parts $\times$ Operators = $5 \times 2 = 10$)

Number of trials, $m = 2$

Level of confidence = 90%

We get v_1 from Table 2. For the *EV* calculation, $m = 2$ and $g = 10$.

$v_1 = 9.0$

$\chi^2_{\alpha/2}$ for 9.0 degrees of freedom and $\alpha/2 = 0.05$

$\chi^2_{\alpha/2} = 16.92$

$$\text{LCL} = \sqrt{\frac{(EV)^2}{\frac{\chi^2_{\alpha/2}}{v_1}}} \qquad \text{LCL} = \sqrt{\frac{(0.0082)^2}{\frac{16.92}{9}}} \qquad \text{LCL} = 0.0059$$

Part II: Upper confidence interval limit:

$EV = 0.0082$

Number of samples = 5 parts

Number of operators = 2

Total measurements taken, $g = 10$ (Parts $\times$ Operators = $5 \times 2 = 10$)

Number of trials, $m = 2$

Level of confidence = 90%

We get v_1 from Table 2. For the *EV* calculation, $m = 2$ and $g = 10$.

$v_1 = 9.0$

$\chi^2_{1-\alpha/2}$ for 9.0 degrees of freedom and $1 - \alpha/2 = 0.95$

$\chi^2_{1-\alpha/2} = 3.33$

$$UCL = \sqrt{\frac{(EV)^2}{\frac{\chi^2_{1-\alpha/2}}{v_1}}} \quad UCL = \sqrt{\frac{(0.0082)^2}{\frac{3.33}{9}}} \quad UCL = 0.0135$$

Summary:

The 90 percent confidence interval for the EV is 0.0059 to 0.0135.

An appropriate statement would be: "We do not know the true repeatability error (EV), but we are 90 percent confident it is between 0.0059 and 0.0135."

Confidence Interval for Reproducibility, *AV*

Lower confidence limit   *Upper confidence limit*

$$LCL = \sqrt{\frac{(AV)^2}{\frac{\chi^2_{\alpha/2}}{v_2}}} \qquad UCL = \sqrt{\frac{(AV)^2}{\frac{\chi^2_{1-\alpha/2}}{v_2}}}$$

Calculate the 90 percent confidence interval for EV given the following (from the previous example):

The calculations for the AV are done exactly as for the EV except that the degrees of freedom are changed. With this example, the number of trials is two since we are dealing with a range of two averages, operator A and operator B, and we only have one such sample range. For this case, $m = 2$ and $g = 1$.

Part I. Lower confidence interval limit:

$AV = 0.0136$

Number of samples, $m = 2$ (each operator has an average, and there are two operators)

Total measurements taken, $g = 1$

Level of confidence = 90%

We get v_2 from Table 2. For the AV calculation, $m = 2$ and $g = 1$.

$v_2 = 1.00$

$\chi^2_{\alpha/2}$ for 1.00 degrees of freedom and $\alpha/2 = 0.05$

$\chi^2_{\alpha/2} = 3.84$

$$\text{LCL} = \sqrt{\frac{(AV)^2}{\dfrac{X^2_{\alpha/2}}{v_2}}} \qquad \text{LCL} = \sqrt{\frac{(0.0136)^2}{\dfrac{3.84}{1}}} \qquad \text{LCL} = 0.0069$$

Part II. Upper confidence interval limit:

$v_2 = 1.00$ (from Table 2, $m = 2$ and $g = 1$)

$X^2_{1-\alpha/2}$ for $\alpha = 0.10$ and one degree of freedom = 0.0039

$$\text{UCL} = \sqrt{\frac{(AV)^2}{\dfrac{X^2_{1-\alpha/2}}{v_2}}} \qquad \text{UCL} = \sqrt{\frac{(0.0136)^2}{\dfrac{0.0039}{1}}} \qquad \text{UCL} = 0.2178$$

Summary:

The 90 percent confidence interval for the *AV* is 0.0069 to 0.2178.

An appropriate statement would be: "We do not know the true repeatability error (*EV*), but we are 90 percent confident it is between 0.0069 and 0.2178."

Confidence Interval for Combined Repeatability and Reproducibility, R&R

Lower confidence limit   *Upper confidence limit*

$$\text{LCL} = \sqrt{\frac{(R\,\&\,R)^2}{\dfrac{X^2_{\alpha/2}}{v_1 + v_2}}} \qquad\qquad \text{UCL} = \sqrt{\frac{(R\,\&\,R)^2}{\dfrac{X^2_{1-\alpha/2}}{v_1 + v_2}}}$$

Part I. Lower confidence interval limit:

Degrees of freedom

$v_1 = 9.0$

$v_2 = 1.0$

R&R = 0.016

$X^2_{\alpha/2}$ for $\alpha = 0.10$ and degrees of freedom = 10.0

$X^2_{0.05,10.0} = 18.3$

$$\text{LCL} = \sqrt{\frac{(R\,\&\,R)^2}{\dfrac{X^2_{\alpha/2}}{v_1 + v_2}}} \qquad \text{LCL} = \sqrt{\frac{(0.016)^2}{\dfrac{18.3}{10.0}}} \qquad \text{LCL} = 0.0118$$

Part II. Upper confidence interval limit:

$v_1 = 9.0$

$v_2 = 1.0$

R&R = 0.016

$\chi^2_{1-\alpha/2}$ for $\alpha = 0.10$ and degrees of freedom = 10.0.

$\chi^2_{0.95,10} = 3.94$

$$\text{UCL} = \sqrt{\dfrac{(R\&R)^2}{\dfrac{X^2_{1-\alpha/2}}{v_1 + v_2}}} \qquad \text{UCL} = \sqrt{\dfrac{(0.016)^2}{\dfrac{3.94}{10}}} \qquad \text{UCL} = 0.0254$$

Summary:

The 90 percent confidence interval for the repeatability and reproducibility (R&R) is 0.0118 to 0.0254.

An appropriate statement would be: "We do not know the true R&R, but we are 90 percent confident that it is between 0.0118 and 0.0254."

In the event that the measurement R&R is excessive, a short-term fix could be realized by reporting an average of several observations. The amount of error would decrease proportionally to the square root of the sample size. This is a direct result of the central limit theorem. Reporting an average of $n = 4$ will cut the measurement error in half relative to the error expected by reporting a single observation.

ATTRIBUTE AND VISUAL INSPECTION ERROR

When working with visual standard, it is appropriate to maintain a set of reference standards that represent acceptable and nonacceptable quality. Examples of such standards that have been developed internally are soldering standards, textile workmanship, paint quality, and print quality, to name a few.

Periodically, inspectors should be judged or evaluated in their ability to inspect those quality visual standards that have been established.

The following method is suitable for determining the inspection efficiency accuracy and developed in 1928 by J. M. Juran (1935, 643–644) and C. A. Melsheimer.

It essentially involves the determination of the percentage accuracy of inspection or the percentage of materials or items correctly classified as conforming or nonconforming to an attribute requirement as judged by a reference or check inspector.

A collection of items to be inspected is assembled. This "test" group of items contains a mixture of both defective and nondefective items. The specific items are tagged with an identification number, and a consensus is reached by several individuals as to the classification of the item as being defective or nondefective. The description of the exact location and type of defect(s) found are also noted. It may be required that items be synthetically rendered defective. A master list is made of the item number, defect, and location. The items are then renumbered by the person conducting the attribute inspection efficiency study.

The lot of items subject to inspection is given to the inspector participating in the evaluation. The inspector is requested to inspect the lot and identify the nonconforming items, the location of the defect, and the nature of the defect. The units found to be conforming are screened by the check inspector. The defective units missed by the participating inspector but found by the check inspector are tallied as B. The items found defective by the participating inspector are tallied as D. These units are then reinspected by the check inspector. Any items in the defective group D that are judged acceptable by the check inspector are tallied as K. The inspection efficiency or accuracy is determined by

$$\% \text{ inspection accuracy or } \% \text{ of defects correctly identified} = \frac{D - K}{D - K + B}$$

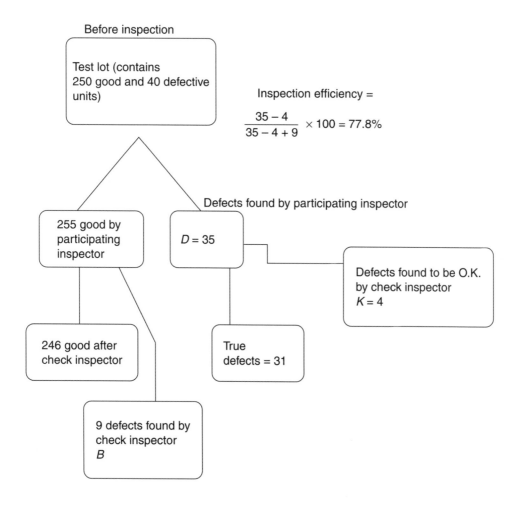

where: D = defects reported by the participating inspector

K = number of "good" units rejected by participating inspector, as determined by the check inspector

B = defects missed by the participating inspector as determined by the check inspector

BIBLIOGRAPHY

Automotive Industry Action Group (AIAG). 1995. *Measurement Systems Analysis.* Southfield, MI: AIAG.

Duncan, A. J. 1974. *Quality Control and Industrial Statistics.* 4th edition. Homewood, IL: Richard D. Irwin.

Juran, J. M. 1935. Inspector's Errors in Quality Control. *Mechanical Engineering* 59, no. 10 (October): 643–644.

Montgomery, D. C. 1996. *Introduction to Statistical Quality Control.* 3rd edition. New York: John Wiley and Sons.

Wheeler, D. J., and R. W. Lyday. 1989. *Evaluating the Measurement Process.* Knoxville, TN: SPC Press.

MULTIVARIATE CHARTS

Often processes will have variables that are related to each other such as the percentage of calcium carbonate in limestone and the pH of a 25 percent slurry of calcium carbonate. Rather than maintain a control chart for both of these characteristics, a single control chart can respond to a process change in either the concentration of the slurry and/or the pH of the slurry. A control chart that will allow the monitoring of two or more related variables is referred to as a *multivariate control chart*. This chart will detect a change in the relationship of two or more variables.

The two methods of constructing multivariate control charts that will be presented are:

1. Hotelling's T^2 statistic (T^2 chart)
2. Standardized Euclidean distance (D_E chart)

T^2 CHART

The following example will be used to illustrate the T^2 control chart.

Two process characteristics are monitored in the production of breaded shrimp:

X_1 = weight of the batter pick-up in grams

X_2 = volume of the shrimp in cm^3

These two characteristics are correlated, and both affect the ratio of shrimp meat to the total breaded weight.

Samples are taken from the breading line every hour. The weight of the breading is determined by the difference between the before-washing weight and the after-washing weight for five randomly selected shrimp. The volume of the shrimp is also determined by water displacement. Data for the first 14 hours of production are taken.

For each sample, the average, standard deviation (s), and range will be determined.

	Sample #1		Sample #2		Sample #3		Sample #4	
	X_1	X_2	X_1	X_2	X_1	X_2	X_1	X_2
	6.08	11.97	6.09	11.97	6.05	11.91	6.40	12.30
	5.70	11.65	5.92	11.82	5.86	11.74	6.14	12.22
	5.54	11.58	5.52	11.48	6.06	11.94	6.19	12.25
	6.53	12.53	5.98	11.85	5.21	11.27	6.02	11.89
	6.45	12.36	5.10	10.96	6.13	12.17	6.38	12.27
Average:	6.06	12.02	5.72	11.62	5.86	11.81	6.23	12.19
S:	0.44	0.42	0.41	0.41	0.38	0.34	0.16	0.17
Range:	0.92	0.95	0.99	1.01	0.92	0.90	0.38	0.41

	Sample #5		Sample #6		Sample #7		Sample #8	
	X_1	X_2	X_1	X_2	X_1	X_2	X_1	X_2
	5.96	11.83	5.92	11.80	6.11	12.15	6.53	12.55
	6.68	12.72	6.49	12.37	6.13	12.18	6.53	12.53
	5.81	11.72	5.79	11.69	6.09	12.08	5.72	11.65
	5.53	11.54	5.59	11.62	5.99	11.87	6.05	11.91
	5.83	11.74	6.19	12.23	5.80	11.71	5.80	11.70
Average:	5.96	11.91	5.99	11.94	6.02	12.00	6.13	12.07
S:	0.43	0.46	0.35	0.34	0.14	0.20	0.39	0.44
Range:	1.15	1.18	0.90	0.75	0.33	0.47	0.81	0.90

	Sample #9		Sample #10		Sample #11		Sample #12	
	X_1	X_2	X_1	X_2	X_1	X_2	X_1	X_2
	5.45	11.42	6.02	11.89	6.05	11.92	6.39	12.29
	6.52	12.43	6.01	11.87	5.83	11.73	6.56	12.63
	6.14	12.19	6.02	11.88	5.77	11.69	6.15	12.23
	6.15	12.22	5.74	11.68	5.73	11.65	5.94	11.83
	6.39	12.29	6.12	12.16	5.86	11.75	6.11	12.14
Average:	6.13	12.11	5.98	11.90	5.85	11.75	6.23	12.22
S:	0.41	0.40	0.14	0.17	0.12	0.10	0.24	0.29
Range:	1.07	1.01	0.38	0.48	0.32	0.27	0.62	0.80

	Sample #13		Sample #14	
	X_1	X_2	X_1	X_2
	5.81	11.71	5.87	11.79
	5.52	11.49	6.39	12.28
	5.26	11.39	6.84	12.74
	5.94	11.83	5.67	11.64
	6.45	12.36	5.59	11.63
Average:	5.80	11.76	6.07	12.02
S:	0.45	0.38	0.53	0.48
Range:	1.19	0.97	1.25	1.11

Each of the process variables could be monitored via an average/range control chart that would, in effect, provide two areas of joint process control. The following square represents the control region using an $\overline{X}$ chart for each parameter. Any point outside of the square indicates that the process has changed—such as a subgroup outside of the limits for either or both of the variables.

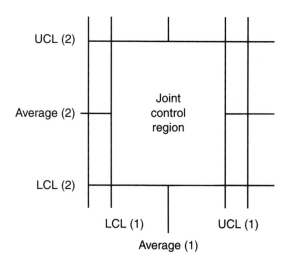

However, if we consider the correlation, we can identify a better region of the square. If a particular unit has a low value for X_1, it is likely not to yield a high X_2, due to the positive correlation. So, given that the process is under control, a unit produced is unlikely to yield a low value for X_1 and a high value for X_2. This means that points are unlikely to occur in the upper-left or lower-right corner of the square.

A better control region would be one that excludes the two unlikely corners and enlarges the other remaining corners where the likelihood of occurrence increases. This is accomplished using the T^2 chart.

Individual charts cannot address the correlation between parameters. However, this does not discount the value in maintaining individual control charts for specific characteristics. If the characteristics are not correlated, then the maintenance of separate control charts is required.

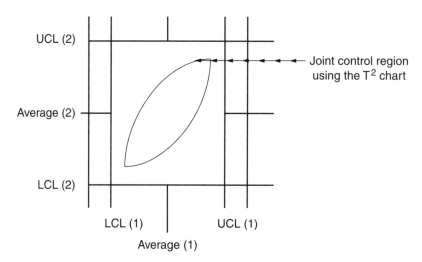

The data used in this example are highly correlated with $r = 0.9770$. The joint control region using the T^2 chart is shown in the following figure:

The T^2 Equation

T^2 control chart for two parameters:

The value for T^2 for each subgroup is calculated as

$$T_i^2 = \frac{n}{\overline{S_1^2}\,\overline{S_2^2} - \overline{S}_{(12)}^2} \left[\overline{S_2^2}(\overline{X}_{1_i} - \overline{\overline{X}}_1)^2 + \overline{S_1^2}(\overline{X}_{2_i} - \overline{\overline{X}}_2)^2 - 2\overline{S_{(12)}}(\overline{X}_{1_i} - \overline{\overline{X}}_1)(\overline{X}_{2_i} - \overline{\overline{X}}_2) \right]$$

where: $\overline{\overline{X}}_{1_i}$ and $\overline{\overline{X}}_{2_i}$ are subgroup averages

$\overline{S_1^2}$ and $\overline{S_2^2}$ are overall averages of the variance for the within-subgroup variance

$\overline{\overline{X}}_1$ and $\overline{\overline{X}}_2$ are overall averages for all subgroups

$\overline{S_{(12)}}$ is the overall average of the covariance of X_1 and X_2

n is the subgroup sample size

Calculation of Individual and Average Covariance

$S_{(12)}$ is the overall average of the covariance of X_1 and X_2. Each of the subgroups will yield an independent covariance. The average of the covariance will be used in calculating the T^2 statistic used in the control chart. The individual subgroup covariance is calculated by

$$S_{(12)i} = \frac{\Sigma X_{11i} X_{21i} - \dfrac{(\Sigma X_{11i})(\Sigma X_{21i})}{n}}{n-1}$$

Example:
For subgroup sample #1:

	Sample #1	
X_1	X_2	$X_1 X_2$
6.08	11.97	72.78
5.70	11.65	66.41
5.54	11.58	64.15
6.53	12.53	81.82
6.45	12.36	79.72
Average: 6.06	12.02	
S: 0.44	0.42	
Range: 0.92	0.95	
Sum, Σ: 30.30	60.09	364.88

$$S_{(12)i} = \frac{\Sigma X_{11i} X_{21i} - \dfrac{(\Sigma X_{11i})(\Sigma X_{21i})}{n}}{n-1}$$

$$S_{(12)i} = \frac{364.88 - \dfrac{(30.30)(60.09)}{5}}{4} = 0.1837$$

The remaining covariances are calculated in a similar manner for subgroups 2–14. A summary of the $S_{(12)}$, subgroup averages, and variances are found in Table 1.

Table 1. Statistics for the T^2 chart.

Subgroup (I)	$\bar{X}_{1i}$	$\bar{X}_{2i}$	S_{1i}^2	S_{2i}^2	$S_{(12)i}$
1	6.06	12.02	0.1936	0.1764	0.1837
2	5.72	11.62	0.1681	0.1681	0.1666
3	5.86	11.81	0.1444	0.1156	0.1233
4	6.23	12.19	0.0256	0.0289	0.0221
5	5.96	11.91	0.1849	0.2116	0.1982
6	5.99	11.94	0.1225	0.1156	0.1152
7	6.02	12.00	0.0196	0.0400	0.0267
8	6.13	12.07	0.1521	0.1936	0.1708
9	6.13	12.11	0.1681	0.1600	0.1610
10	5.98	11.90	0.0196	0.0289	0.0218
11	5.85	11.75	0.0144	0.0100	0.0128
12	6.23	12.22	0.0576	0.0841	0.0671
13	5.80	11.76	0.2025	0.1444	0.1687
14	6.07	12.02	0.2809	0.2304	0.2557
Averages:	$\bar{\bar{X}}_1 = 6.00$	$\bar{\bar{X}}_2 = 11.95$	$\overline{S_1^2} = 0.1252$	$\overline{S_2^2} = 0.1220$	$\overline{S_{(12)i}} = 0.1210$

Calculation of Individual T^2 Values

$$T_i^2 = \frac{n}{\overline{S_1^2}\,\overline{S_2^2} - \overline{S_{(12)}^2}}\left[\overline{S_2^2}(\overline{X}_{1i} - \overline{\overline{X}}_1)^2 + \overline{S_1^2}(\overline{X}_{2i} - \overline{\overline{X}}_2)^2 - 2\overline{S_{(12)}}(\overline{X}_{1i} - \overline{\overline{X}}_1)(\overline{X}_{2i} - \overline{\overline{X}}_2)\right]$$

$$\overline{S_1^2} = 0.1252 \quad \overline{S_2^2} = 0.1220 \quad \left(\overline{S_{(12)i}}\right)^2 = (0.1210)^2 \quad n = 5$$

$$\frac{n}{\overline{S_1^2}\,\overline{S_2^2} - \overline{S_{(12)}^2}} = \frac{5}{(0.1252)(0.1220) - (0.1210)^2} = 7893.9$$

The term 7893.9 will be utilized for all T^2 calculations.

$$T_i^2 = 7893.9\left(\begin{array}{c}0.1220(\overline{X}_{1i} - 6.00)^2 + 0.1252(\overline{X}_{2i} - 11.95)^2 \\ -(2)(0.121)(\overline{X}_{1i} - 6.00)(\overline{X}_{2i} - 11.95)\end{array}\right)$$

$$T_i^2 = 7893.9\left(\begin{array}{c}0.1220(6.06 - 6.00)^2 + 0.1252(12.02 - 11.95)^2 \\ -(2)(0.121)(6.06 - 6.00)(12.02 - 11.95)\end{array}\right)$$

$$T_1^2 = 7893.9(0.0004392 + 0.0006135 - 0.0010164)$$

$$T_1^2 = 0.2866$$

The remaining T^2 values are calculated using

$$T_i^2 = 7893.9\left(\begin{array}{c}0.1220(\overline{X}_{1i} - 6.00)^2 + 0.1252(\overline{X}_{2i} - 11.95)^2 \\ -(2)(0.121)(\overline{X}_{1i} - 6.00)(\overline{X}_{2i} - 11.95)\end{array}\right)$$

$$T_2^2 = 7893.9\left(\begin{array}{c}0.1220(5.72 - 6.00)^2 + 0.1252(11.62 - 11.95)^2 \\ -(2)(0.121)(5.72 - 6.00)(11.62 - 11.95)\end{array}\right)$$

$$T_2^2 = 7893.9565 + 0.013634 - 0.0223608$$

$$T_2^2 = 6.6167$$

$$T_i^2 = 7893.98\left(\begin{array}{c}0.1220(\overline{X}_{1i} - 6.00)^2 + 0.1252(\overline{X}_{2i} - 11.95)^2 \\ -(2)(0.121)(\overline{X}_{1i} - 6.00)(\overline{X}_{2i} - 11.95)\end{array}\right)$$

$$T_3^2 = 7893.98\left(\begin{array}{c}0.1220(5.86 - 6.00)^2 + 0.1252(11.81 - 11.95)^2 \\ -(2)(0.121)(5.86 - 6.00)(11.81 - 11.95)\end{array}\right)$$

$$T_3^2 = 7893.9(0.0023391 + 0.002454 - 0.0047432)$$

$$T_3^2 = 0.8036$$

A summary of all of the individual T^2 follows.

Sample, i	T^2
1	0.2866
2	6.6167
3	0.8036
4	2.4226
5	0.0657
6	0.0041
7	0.9457
8	0.7063
9	1.8418
10	0.9457
11	3.8192
12	4.3628
13	1.6081
14	0.2011

Calculation of the Control Limits

There is no lower control limit. The upper control limit is calculated using the following relationship:

$$\text{Upper Control Limit, } T^2{}_{\alpha,p,n-1} = \frac{p(n-1)}{n-p} F_{\alpha,p,n-p}$$

where: α = probability of a point above the control limit when there has been no process change

p = number of variables (2 for this example)

n = subgroup sample size (5 for this example)

$F_{\alpha,p,n-p}$ = value from F distribution table for level of significance α with p and $n - p$ degrees of freedom

For this example, $F_{.05,2,3} = 9.55$.

When looking up the F value, use p degrees of freedom for the numerator and $n - p$ degrees of freedom for the denominator.

The upper control limit is

$$T^2{}_{\alpha,p,n-1} = \frac{p(n-1)}{n-p} F_{\alpha,p,n-p} = \frac{2(5-1)}{3}(9.55) = 25.47$$

The individual T^2 values are plotted, and the upper control limit of 25.47 is drawn using a broken line.

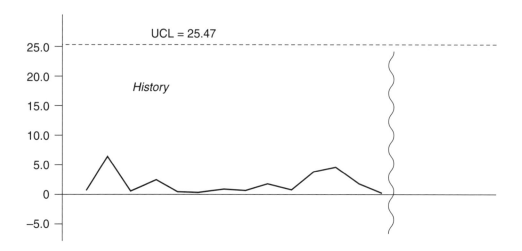

The process appears to be very well behaved, indicating that the correlation of the data is stable and well defined.

The control chart of averages for each of the two variables, plotted independently follows:

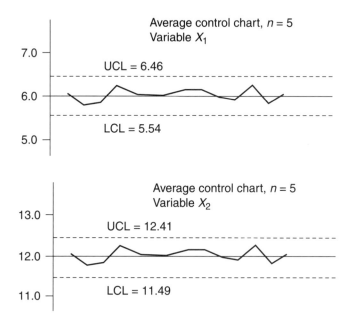

Both control charts are in a state of good control (statistically stable).

The first 14 points were used to provide a historical period from which the process was characterized. Calculate the T^2 for the next six data sets, and determine if the process has changed.

	Sample #15		Sample #16		Sample #17		Sample #18	
	X_1	X_2	X_1	X_2	X_1	X_2	X_1	X_2
	5.60	11.91	6.22	12.17	6.25	11.30	5.94	12.16
	5.91	12.05	6.35	12.05	6.55	12.15	5.53	12.01
	5.44	11.75	5.84	11.73	5.95	12.05	5.61	11.62
	5.80	11.53	6.01	11.55	6.12	12.55	6.05	11.51
	6.20	11.61	6.08	12.10	6.38	12.55	5.52	11.95
Average:	5.79	11.77	6.10	11.92	6.25	12.10	5.73	11.85
S:	0.29	0.21	0.20	0.27	0.23	0.49	0.25	0.27
Range:	0.76	0.52	0.51	0.62	0.60	1.25	0.53	0.65

	Sample #19		Sample #20	
	X_1	X_2	X_1	X_2
	6.44	12.21	6.22	11.75
	6.35	12.46	5.87	12.10
	6.18	12.10	6.05	11.88
	6.17	12.27	6.28	12.00
	5.91	12.36	6.08	11.92
Average:	6.21	12.28	6.10	11.93
S:	0.20	0.14	0.16	0.13
Range:	0.53	0.36	0.41	0.35

The new *future* data values are used to calculate the T^2 points. Use the process averages, variances, and covariance from the historical data. That is

$$T_i^2 = 7893.9$$

$$\left[0.1220(\overline{X}_{1i} - 6.00)^2 + 0.1252(\overline{X}_{2i} - 11.95)^2 - (2)(0.121)(\overline{X}_{1i} - 6.00)(\overline{X}_{2i} - 11.95) \right]$$

A summary of the T^2 for subgroups 15–20 follows:

Sample, *I*	T^2
15	2.282
16	16.250
17	10.790
18	28.511
19	17.713
20	13.847

The completed control chart with both the historical data and the new *future* data follows.

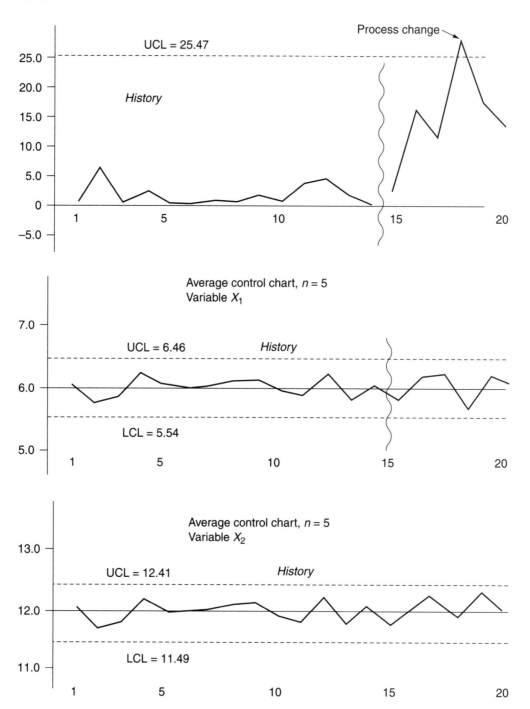

Note that the average control chart for both variables X_1 and X_2 remain in control, failing to detect the change in process correlation.

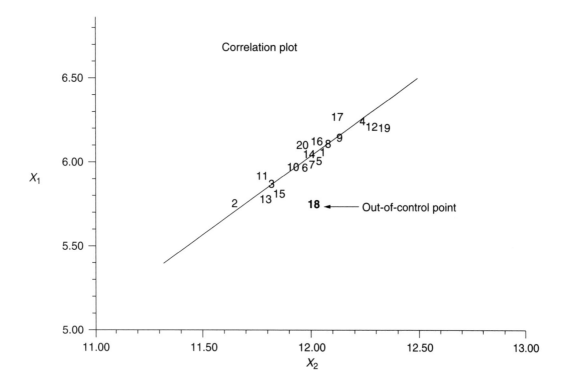

In the above plot, the numbered values represent the subgroup sample average. The bold values are for the *future* data, and the normal text represents the *historical* data from which the T^2 chart parameters were derived. Notice that the subgroup sample #18 falls significantly off the best linear line, indicating a departure from the expected value per the historical data. This departure from the expected correlation is detected using the T^2 but not using the traditional $\overline{X}/R$ chart.

STANDARD EUCLIDEAN DISTANCE D_E

An alternative to the more complex T^2 is the D_E control chart. This chart is based on calculating a standardized value of the monitored characteristics, combining them vectorially in the same manner that one would combine variances, and reporting an overall single value to plot.

Plotting Statistic

The characteristic plotted is a combined standardized Z-score from two or more characteristics:

$$Z_i = \frac{X_i - T_i}{\text{Sigma}\,(T_i)}$$

where: X_i = the location statistic for the characteristic such as an individual observation or an average of a subgroup as appropriate

T_i = the target value for the location statistic, may be the average of the individuals or the average of averages for subgroup data $n \geq$ 2 as appropriate

Sigma (T_i) = the standard deviation of individuals or standard deviation of averages as appropriate

The D_E may be used for individuals or averages.
The Z-score calculations are based on:
For individuals:

$$Z_i = \frac{X_i - \overline{X}_i}{S_i}$$

where: X_i = individual observation
       $\overline{X}_i$ = average of the individuals
       S_i = sample standard deviation used as an estimate of the ι standard deviation, σ

For averages:

$$Z_i = \frac{\overline{X}_i - \overline{\overline{X}}_i}{S_{\overline{X}}}$$

where $\overline{X}_i$ = individual subgroup average
      $\overline{\overline{X}}_i$ = average of subgroup averages

$S_{\overline{X}}$ = standard deviation of the averages can be estimated from $\left(\dfrac{A_2 \overline{R}}{3} \right)$

Values for A_2 are dependent on the subgroup sample size n

Several Z-scores are combined to yield a single D_E value to plot:

$$D_E = \sqrt{(Z_1)^2 + (Z_2)^2 + \ldots + (Z_k)^2}$$

where: Z_1 = Z-score for 1st variable
       Z_2 = Z-score for 2nd variable

Control Limits and Center Line

There is no lower control limit, and the upper control limit for the individual D_E values is

$$\mathrm{UCL}_{D_E} = \sqrt{\chi^2_{0.995}(p)}$$

where: $\chi^2_{0.995}$ = the 99.5 percentile of the chi-square distribution using p degrees of freedom (p = number of variables considered)

Note: Some chi-square tables will require looking up $1 - C$, or in this case $\alpha = 0.005$. The center line CL for the D_E chart is determined by

$$CL = \sqrt{p}$$

The following table gives the center line values CL and upper control limits $UCL_{D_E} = \sqrt{\chi^2_{0.995}(p)}$ as a function of the number of variables p:

p	CL	UCL
2	1.41	3.26
3	1.73	3.58
4	2.00	3.85
5	2.24	4.09
6	2.45	4.31
7	2.65	4.50
8	2.83	4.69
9	3.00	4.86
10	3.16	5.02

The following example will be used to demonstrate the construction and use of the $UCL_{D_E} = \sqrt{\chi^2_{0.995}(p)}$

Two characteristics that are monitored in the production of a water-based nondrying cement are the batch viscosity and the peel strength of a test strip contact bonded to a substrate. Twenty batches have been tested for these two parameters. A traditional individual/moving range control chart will be developed for each of the independent parameters, and a D_E chart combining the two parameters.

Data:

Batch	X_1 Peel, lb/in	X_2 Viscosity, cps	Batch	X_1 Peel, lb/in	X_2 Viscosity, cps
1	17.5	1200	13	15.5	1550
2	18.3	1700	14	17.8	1500
3	14.2	825	15	15.2	950
4	16.4	1025	16	17.5	1300
5	19.0	1600	17	21.0	1850
6	16.3	1320	18	17.3	1375
7	19.8	1725	19	18.2	1400
8	19.8	1600	20	20.0	1925
9	16.1	1150			
10	17.5	1300		$\bar{X}_1 = 17.76$	$\bar{X}_2 = 1433.5$
11	19.4	1825			
12	18.3	1550		$S_1 = 1.78$	$S_2 = 305.2$

For each of the individual pairs of data points (X_1 and X_2), a standard Z-score and a combined Z-score is calculated.

Batch 1:

$$\text{Standard peel strength, } Z_{1,X1} = \frac{X_{11} - \overline{X}_1}{S_1} = \frac{17.5 - 17.76}{1.78} = -0.03$$

$$\text{Standard viscosity, } Z_{1,X2} = \frac{X_{12} - \overline{X}_2}{S_2} = \frac{1200 - 1443.5}{305.2} = -0.77$$

$$\text{Combined Z-score, } D_{1E} = \sqrt{(-0.03)^2 + (-0.77)^2} = 0.77$$

Batch 2:

$$\text{Standard peel strength, } Z_{2,X1} = \frac{X_{21} - \overline{X}_1}{S_1} = \frac{18.3 - 17.76}{1.78} = 0.30$$

$$\text{Standard viscosity, } Z_{2,X2} = \frac{X_{22} - \overline{X}_2}{S_2} = \frac{1700 - 1433.5}{305.2} = 0.87$$

$$\text{Combined Z-score, } D_{2E} = \sqrt{(0.30)^2 + (-0.87)^2} = 0.92$$

Batch 3:

$$\text{Standard peel strength, } Z_{3,X1} = \frac{X_{31} - \overline{X}_1}{S_1} = \frac{14.2 - 17.76}{1.78} = -2.00$$

$$\text{Standard viscosity, } Z_{3,X2} = \frac{X_{32} - \overline{X}_2}{S_2} = \frac{825 - 1433.5}{305.2} = -1.99$$

$$\text{Combined Z-score, } D_{3E} = \sqrt{(-2.00)^2 + (-1.99)^2} = 2.82$$

The remaining batches are calculated, and a summary of the results follows.

Batch	X_1 Peel, lb/in	X_2 Viscosity, cps	Standardized peel, Z_1	Standardized viscosity, Z_2	Combined D_E
1	17.5	1200	−0.03	−0.77	0.77
2	18.3	1700	0.30	0.87	0.92
3	14.2	825	−2.00	−1.99	2.82
4	16.4	1025	−0.76	−1.34	1.54
5	19.0	1600	0.70	0.55	0.39
6	16.3	1320	−0.82	−0.37	0.90
7	19.8	1725	1.15	0.96	1.50
8	19.8	1600	1.15	0.55	1.27
9	16.1	1150	−0.93	−0.93	1.32
10	17.5	1300	−0.03	−0.44	0.44
11	19.4	1825	0.92	1.28	1.58
12	18.3	1550	0.30	0.38	0.48
13	15.5	1550	−1.27	0.38	1.32
14	17.8	1500	0.02	0.22	0.22
15	15.2	950	−1.44	−1.58	2.14
16	17.5	1300	−0.03	−0.44	0.44
17	21.0	1850	1.82	1.36	2.27
18	17.3	1375	−0.26	−0.19	0.32
19	18.2	1400	0.25	−0.11	0.27
20	20.0	1925	1.26	1.61	2.04

$\overline{X}_1 = 17.76$ $\overline{X}_2 = 1433.5$

$S_1 = 1.78$ $S_2 = 305.2$

Using an upper control limit of 3.26 and a center line of 1.41, the D_E control chart using the historical data follows:

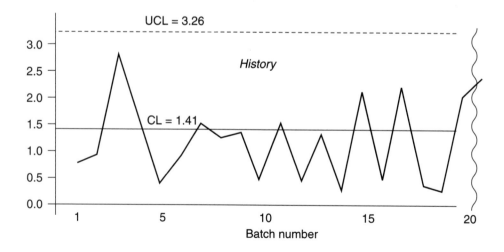

Continue the D_E chart using the following information:

Batch	Peel, lb/in	Viscosity, cps	Peel, Z_1	Viscosity, Z_2	D_E
21	13.5	1250	-2.39	-0.60	2.46
22	14.8	1450	-1.66	0.05	1.66
23	15.3	1975	-1.38	1.77	2.24
24	21.3	2235	1.99	2.63	3.30
25	14.1	1000	-2.06	-1.42	2.50

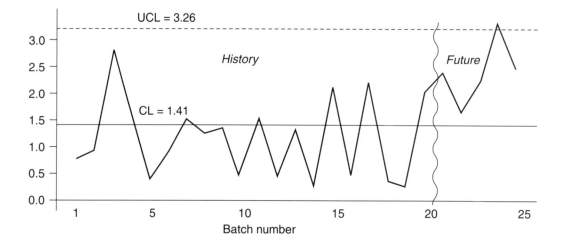

Control charts for the individual characteristics are presented in the following table. The control limits are based on $\overline{X} \pm 3S$. S is from the sample standard deviation using samples 1–20.

Characteristic	Average	LCL	UCL
Peel, X_1	17.76	12.42	23.10
Viscosity, X_2	1433.50	517.90	2349.10

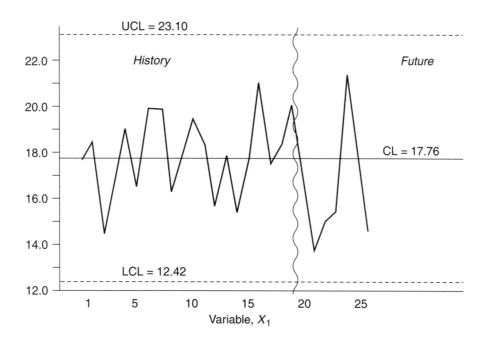

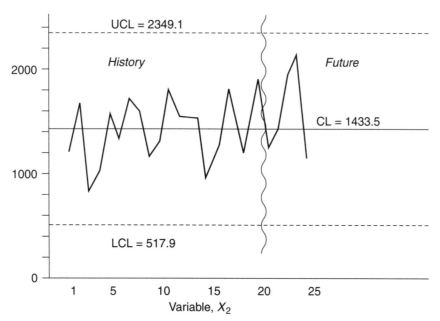

BIBLIOGRAPHY

Montgomery, D. C. 1996. *Introduction to Statistical Quality Control.* 3rd edition. New York: John Wiley and Sons.

Wheeler, D. J. 1995. *Advanced Topics in Statistical Process Control.* Knoxville, TN: SPC Press.

Nonnormal Distribution C_{PK}

The traditional calculations for C_{pk} (and other process capability indices) are based on the assumption of a normal distribution. Some of the distinguishing characteristics of a normal distribution are:

1. The distribution is symmetrical about the average.
2. The mean, median, and mode are equal.
3. The kurtosis $Ku = 0$ and the skewness $Sk = 0$

The calculation of C_{pk} and C_r can be adjusted to compensate for nonnormalcy by using the median rather than the average and applying corrections based on Pearson frequency curves.

Knowing the skewness and kurtosis for a given distribution of data will allow the determination of the specific values that will result with a probability of 0.00135 and a probability of 0.99865. These two probability values represent the probability of getting a value less than the average minus three standard deviations and the probability of getting a value less than the average plus three standard deviations. These specific values are used for the calculation of C_{pk} for all distributions, including the normal distribution.

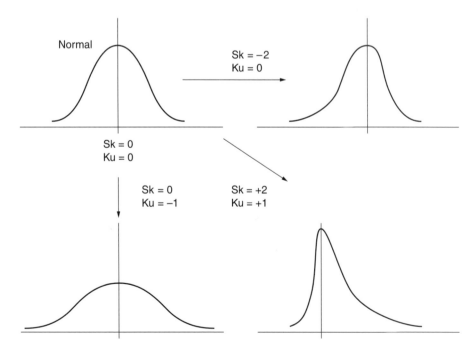

The Pearson distributions can be used for many distributions, including the normal where Sk = 0 and Ku = 0.

As skewness and kurtosis change, so does the shape of the distribution. The following illustrations demonstrate the effect of these parameter changes on a normal distribution.

Skewness measures how unsymmetrical the distribution is with respect to the average. The greater the positive skewness, the more data will be greater than the average. A negative skewness means that more data will be less than the average.

Kurtosis is a measure of the peakness of the distribution. A negative kurtosis is indicative of a relative flatness compared to a normal distribution, and a positive kurtosis reflects a distribution with more peak than a normal distribution.

There are several statistics used to measure skewness and kurtosis. The calculation of these statistics as applied in this discussion will be defined in terms of various moments about the arithmetic mean, defined by the formula

$$\mu_k = \frac{\Sigma(X_i - \bar{X})^k}{n}$$

where: $\bar{X}$ = average
X_i = an individual observation

The first moment about the mean is always zero, and the second moment about the mean adjusted for bias is the variance.

$$\mu_1 = 0$$

$$\mu_2\left(\frac{n}{n-1}\right) = s^2$$

A measure of skewness often used is

$$Sk = \sqrt{\frac{\mu_3^2}{\mu_2^3}}$$

and a measure of kurtosis is

$$Ku = \frac{\mu_4}{\mu_2^2} - 3$$

Note: Occasionally Ku will be defined as

$$Ku = \frac{\mu_4}{\mu_2^2}$$

in which case a normal distribution will have a skewness of Sk = 0 and a kurtosis of Ku = 3.

The following example will illustrate a technique used to calculate a C_{pk} compensating for skewness and kurtosis.

Step 1. Collect data and determine the Sk and Ku values.

X_i	Frequency f	$(X_i - \bar{X})$	$(X_i - \bar{X})^2$	$(X_i - \bar{X})^3$	$(X_i - \bar{X})^4$
4	1	−2.046	4.186	−8.565	17.524
5	8	−1.046	1.094	−1.144	1.197
6	6	−0.046	0.002	−0.000	0.000
7	4	0.954	0.910	0.868	0.828
8	2	1.954	3.818	7.461	14.578
9	1	2.954	8.726	25.777	76.145

$\bar{X} = 6.046$

$S = 1.253$

$n = 22$

$$\text{Sk} = \sqrt{\frac{\mu_3^2}{\mu_2^3}} \quad \text{Sk} = \sqrt{\frac{(1.206)^2}{(1.498)^3}} = 0.66$$

$$\text{Ku} = \frac{\mu_4}{\mu_2^2} - 3 \quad \text{Ku} = \left(\frac{6.169}{(1.498)^2}\right) - 3 = -0.25$$

Step 2. Summarize the process characterization statistics.

Average, $\bar{X} = 6.046$

Standard deviation, $S = 1.253$

Upper specification, Uspec = 14.0

Lower specification, Lspec = 2.0

Step 3. Determine the standardized value for getting a probability of less than 0.00135 with specified Ku and Sk, $P_{.00135}$.

For +Sk, use Table 1.

For −Sk, use Table 2.

More exact values are obtained by linear interpolation:

	Sk:		
	0.60	**0.66**	0.70
Ku:			
−0.40	1.496		1.299
−0.25		1.486	
−0.20	1.655		1.434

Table 1.

Ku	0.0	0.1	0.2	0.3	0.4	0.5	0.6	0.7	0.8	0.9	1.0	1.1	1.2	1.3	1.4	1.5	1.6	1.7	1.8	1.9
-1.4	1.512	1.421	1.317	1.206	1.092	0.979	0.868	0.762												
-1.2	1.727	1.619	1.496	1.364	1.230	1.100	0.975	0.858	0.747											
-1.0	1.966	1.840	1.696	1.541	1.384	1.232	1.089	0.957	0.836											
-0.8	2.210	2.072	1.912	1.736	1.5551.	1.212	1.062	0.927	0.804	0.692										
-0.6	2.442	2.298	2.129	1.941	1.740	1.539	1.348	1.175	1.023	0.887	0.766	0.656								
-0.4	2.653	2.506	2.335	2.141	1.930	1.711	1.496	1.299	1.125	0.974	0.841	0.723	0.616							
-0.2	2.839	2.692	2.522	2.329	2.116	1.887	1.655	1.434	1.235	1.065	0.919	0.791	0.677	0.574						
0.0	3.000	2.856	2.689	2.500	2.289	2.059	1.817	1.578	1.356	1.163	1.000	0.861	0.739	0.630	0.531					
0.2	3.140	2.986	2.834	2.653	2.447	2.220	1.976	1.726	1.485	1.269	1.086	0.933	0.801	0.686	0.583					
0.4	3.261	3.088	2.952	2.785	2.589	2.368	2.127	1.873	1.619	1.382	1.178	1.008	0.865	0.742	0.634	0.536				
0.6	3.366	3.164	3.045	2.896	2.714	2.502	2.267	2.015	1.754	1.502	1.277	1.087	0.931	0.799	0.685	0.583	0.489			
0.8	3.458	3.222	3.118	2.986	2.821	2.622	2.396	2.148	1.887	1.625	1.381	1.172	1.000	0.857	0.736	0.629	0.533			
1.0	3.539	3.266	3.174	3.058	2.910	2.727	2.512	2.271	2.013	1.748	1.491	1.262	1.072	0.917	0.787	0.675	0.575	0.484		
1.2	3.611	3.300	3.218	3.115	2.983	2.817	2.616	2.385	2.132	1.876	1.602	1.357	1.149	0.979	0.840	0.721	0.617	0.524		
1.4	3.674	3.327	3.254	3.161	3.043	2.893	2.708	2.488	2.243	1.981	1.713	1.456	1.230	1.045	0.894	0.768	0.659	0.562	0.475	
1.6	3.731	3.349	3.282	3.199	3.092	2.957	2.787	2.581	2.345	2.089	1.821	1.556	1.316	1.113	0.950	0.815	0.701	0.600	0.510	
1.8	3.782	3.367	3.306	3.229	3.133	3.011	2.855	2.664	2.438	2.189	1.925	1.664	1.404	1.185	1.008	0.863	0.743	0.638	0.546	0.461
2.0	3.828	3.382	3.325	3.255	3.167	3.055	2.914	2.736	2.524	2.283	2.023	1.755	1.494	1.261	1.068	0.913	0.785	0.676	0.580	0.494
2.2	3.870	3.395	3.342	3.277	3.196	3.093	2.964	2.800	2.600	2.369	2.116	1.850	1.584	1.339	1.132	0.964	0.828	0.714	0.615	0.526
2.4	3.908	3.405	3.356	3.295	3.220	3.126	3.006	2.855	2.669	2.448	2.202	1.940	1.673	1.420	1.198	1.018	0.873	0.752	0.649	0.557
2.6	3.943	3.415	3.367	3.311	3.241	3.153	3.043	2.904	2.730	2.521	2.283	2.026	1.760	1.501	1.267	1.073	0.918	0.791	0.683	0.589
2.8	3.975	3.423	3.378	3.324	3.259	3.177	3.075	2.946	2.784	2.586	2.358	2.107	1.844	1.581	1.338	1.131	0.965	0.830	0.717	0.620
3.0	4.004	3.430	3.387	3.326	3.274	3.198	3.103	2.983	2.831	2.646	2.427	2.183	1.924	1.661	1.410	1.191	1.013	0.870	0.752	0.651
3.2	4.031	3.436	3.395	3.346	3.288	3.216	3.127	3.015	2.874	2.699	2.491	2.254	2.000	1.738	1.483	1.253	1.063	0.911	0.787	0.681
3.4	4.056	3.441	3.402	3.356	3.300	3.233	3.149	3.043	2.911	2.747	2.549	2.321	2.072	1.813	1.555	1.317	1.115	0.953	0.822	0.712
3.6	4.079	3.446	3.408	3.364	3.311	3.247	3.168	3.069	2.945	2.790	2.602	2.383	2.140	1.888	1.626	1.381	1.169	0.996	0.858	0.744
3.8	4.101	3.450	3.414	3.371	3.321	3.259	3.184	3.091	2.974	2.829	2.651	2.440	2.205	1.953	1.695	1.446	1.224	1.041	0.895	0.775
4.0	4.121	3.454	3.419	3.378	3.329	3.271	3.200	3.111	3.001	2.864	2.695	2.494	2.265	2.018	1.762	1.510	1.281	1.088	0.932	0.807
4.2	4.140	3.458	3.423	3.384	3.337	3.281	3.213	3.129	3.025	2.895	2.735	2.543	2.321	2.080	1.827	1.574	1.338	1.135	0.971	0.839
4.4	4.157	3.461	3.428	3.389	3.344	3.290	3.225	3.145	3.047	2.923	2.771	2.588	2.374	2.138	1.889	1.636	1.396	1.184	1.011	0.872
4.6	4.174	3.464	3.431	3.394	3.350	3.299	3.236	3.160	3.066	2.949	2.805	2.629	2.424	2.194	1.948	1.697	1.453	1.234	1.052	0.905
4.8	4.189	3.466	3.435	3.399	3.356	3.306	3.246	3.173	3.084	2.972	2.835	2.668	2.470	2.246	2.005	1.756	1.510	1.285	1.094	0.939
5.0	4.204	3.469	3.438	3.403	3.362	3.313	3.256	3.186	3.100	2.994	2.863	2.703	2.513	2.296	2.059	1.813	1.566	1.336	1.137	0.975
5.2	4.218	3.471	3.441	3.406	3.367	3.320	3.264	3.197	3.114	3.013	2.888	2.735	2.562	2.342	2.111	1.867	1.621	1.387	1.181	1.010
5.4	4.231	3.473	3.444	3.410	3.371	3.326	3.272	3.207	3.128	3.031	2.911	2.765	2.589	2.386	2.160	1.920	1.675	1.438	1.225	1.047

Table 1. Continued.

Ku	0.0	0.1	0.2	0.3	0.4	0.5	0.6	0.7	0.8	0.9	1.0	1.1	1.2	1.3	1.4	1.5	1.6	1.7	1.8	1.9
5.6	4.243	3.475	3.446	3.413	3.375	3.331	3.279	3.216	3.140	3.047	2.933	2.793	2.624	2.427	2.206	1.970	1.727	1.489	1.270	1.085
5.8	4.255	3.477	3.448	3.416	3.379	3.336	3.286	3.225	3.152	3.062	2.952	2.818	2.656	2.465	2.250	2.019	1.778	1.539	1.316	1.123
6.0	4.266	3.478	3.451	3.419	3.383	3.341	3.292	3.233	3.162	3.076	2.970	2.841	2.685	2.501	2.292	2.065	1.827	1.588	1.361	1.162
6.2	4.276	3.480	3.453	3.422	3.386	3.345	3.297	3.240	3.172	3.089	2.987	2.863	2.713	2.535	2.332	2.109	1.874	1.635	1.407	1.202
6.4	4.286	3.481	3.454	3.424	3.389	3.349	3.303	3.247	3.181	3.100	3.003	2.883	2.739	2.567	2.369	2.151	1.919	1.682	1.452	1.242
6.6	4.296	3.483	3.456	3.426	3.392	3.353	3.308	3.254	3.189	3.111	3.017	2.902	2.763	2.597	2.405	2.191	1.962	1.727	1.496	1.282
6.8	4.305	3.484	3.458	3.429	3.395	3.357	3.312	3.260	3.197	3.122	3.030	2.919	2.785	2.624	2.438	2.229	2.004	1.771	1.540	1.323
7.0	4.313	3.485	3.459	3.431	3.398	3.360	3.316	3.265	3.204	3.131	3.043	2.936	2.806	2.651	2.469	2.265	2.044	1.814	1.583	1.363
7.2	4.322	3.486	3.461	3.432	3.400	3.363	3.321	3.270	3.211	3.140	3.054	2.951	2.825	2.675	2.499	2.300	2.083	1.855	1.625	1.403
7.4	4.330	3.487	3.462	3.434	3.403	3.366	3.324	3.275	3.218	3.148	3.065	2.965	2.843	2.698	2.527	2.333	2.120	1.895	1.666	1.443
7.6	4.337	3.488	3.464	3.436	3.405	3.369	3.328	3.280	3.224	3.156	3.075	2.978	2.860	2.720	2.554	2.364	2.155	1.933	1.706	1.482
7.8	4.344	3.489	3.465	3.437	3.407	3.372	3.331	3.284	3.229	3.164	3.085	2.990	2.876	2.740	2.579	2.394	2.189	1.970	1.744	1.521
8.0	4.351	3.490	3.466	3.439	3.409	3.374	3.335	3.289	3.235	3.171	3.094	3.002	2.891	2.759	2.603	2.422	2.221	2.005	1.782	1.559
8.2	4.358	3.491	3.467	3.440	3.411	3.377	3.338	3.292	3.240	3.177	3.103	3.013	2.906	2.777	2.625	2.449	2.252	2.040	1.818	1.596
8.4	4.365	3.492	3.468	3.442	3.412	3.379	3.340	3.296	3.244	3.183	3.111	3.023	2.919	2.794	2.646	2.475	2.282	2.073	1.854	1.632
8.6	4.371	3.492	3.469	3.443	3.414	3.381	3.343	3.300	3.249	3.189	3.118	3.033	2.932	2.810	2.666	2.499	2.310	2.104	1.888	1.667
8.8	4.377	3.493	3.470	3.444	3.416	3.383	3.346	3.303	3.253	3.195	3.125	3.042	2.943	2.825	2.685	2.522	2.337	2.135	1.921	1.702
9.0	4.382	3.494	3.471	3.445	3.417	3.385	3.348	3.306	3.257	3.200	3.132	3.051	2.955	2.839	2.703	2.544	2.363	2.164	1.953	1.736
9.2	4.388	3.495	3.472	3.447	3.418	3.387	3.351	3.309	3.261	3.205	3.138	3.059	2.965	2.853	2.720	2.565	2.388	2.192	1.984	1.768
9.4	4.393	3.495	3.473	3.448	3.420	3.388	3.353	3.312	3.265	3.209	3.144	3.067	2.975	2.866	2.736	2.585	2.411	2.219	2.014	1.800
9.6	4.398	3.496	3.473	3.449	3.421	3.390	3.355	3.315	3.268	3.214	3.150	3.075	2.985	2.878	2.752	2.604	2.434	2.245	2.042	1.831
9.8	4.403	3.496	3.474	3.450	3.422	3.392	3.357	3.317	3.272	3.218	3.158	3.082	2.994	2.890	2.766	2.622	2.456	2.271	2.070	1.861
10.0	4.408	3.497	3.475	3.451	3.424	3.393	3.359	3.320	3.275	3.222	3.161	3.088	3.003	2.901	2.780	2.639	2.476	2.295	2.097	1.890
10.2					3.425	3.395	3.361	3.322	3.278	3.226	3.166	3.095	3.011	2.911	2.793	2.655	2.496	2.318	2.123	1.918
10.4						3.396	3.363	3.325	3.281	3.230	3.171	3.101	3.019	2.921	2.806	2.671	2.515	2.340	2.148	1.945
10.6							3.364	3.327	3.283	3.233	3.175	3.107	3.026	2.930	2.818	2.686	2.533	2.361	2.172	1.972
10.8								3.329	3.286	3.237	3.179	3.112	3.033	2.940	2.829	2.700	2.551	2.382	2.196	1.998
11.0									3.289	3.240	3.184	3.118	3.040	2.948	2.840	2.714	2.567	2.401	2.218	2.023
11.2										3.243	3.188	3.123	3.046	2.956	2.851	2.727	2.583	2.420	2.240	2.047
11.4											3.191	3.128	3.053	2.964	2.861	2.739	2.598	2.438	2.261	2.070
11.6											3.195	3.132	3.058	2.972	2.870	2.751	2.613	2.456	2.281	2.093
11.8												3.137	3.064	2.979	2.879	2.762	2.627	2.473	2.301	2.115
12.0												3.141	3.070	2.986	2.888	2.773	2.641	2.489	2.320	2.136
12.2													3.075	2.993	2.896	2.784	2.653	2.505	2.338	2.157

Table 2.

Ku	0.0	0.1	0.2	0.3	0.4	0.5	0.6	0.7	0.8	0.9	1.0	1.1	1.2	1.3	1.4	1.5	1.6	1.7	1.8	1.9
-1.4	1.512	1.584	1.632	1.655	1.653	1.626	1.579	1.516												
-1.2	1.727	1.813	1.871	1.899	1.895	1.861	1.803	1.726	1.636											
-1.0	1.966	2.065	2.134	2.170	2.169	2.131	2.061	1.966	1.856											
-0.8	2.210	2.320	2.400	2.446	2.454	2.422	2.349	2.241	2.108	1.965	1.822									
-0.6	2.442	2.560	2.648	2.704	2.726	2.708	2.646	2.540	2.395	2.225	2.052	1.885								
-0.4	2.653	2.774	2.869	2.934	2.969	2.968	2.926	2.837	2.699	2.518	2.314	2.114	1.928							
-0.2	2.839	2.961	3.060	3.133	3.179	3.194	3.173	3.109	2.993	2.824	2.608	2.373	2.152	1.952						
0.0	3.000	3.123	3.224	3.303	3.358	3.387	3.385	3.345	3.259	3.116	2.914	2.665	2.405	2.169	1.960					
0.2	3.140	3.261	3.364	3.447	3.510	3.550	3.564	3.546	3.488	3.378	3.206	2.970	2.690	2.412	2.167					
0.4	3.261	3.381	3.484	3.570	3.639	3.688	3.715	3.715	3.681	3.603	3.468	3.264	2.993	2.687	2.398	2.149				
0.6	3.366	3.485	3.588	3.676	3.749	3.805	3.843	3.858	3.844	3.793	3.693	3.529	3.290	2.984	2.658	2.366	2.119			
0.8	3.458	3.576	3.678	3.768	3.844	3.905	3.951	3.978	3.981	3.900	3.883	3.758	3.561	3.283	2.945	2.609	2.322			
1.0	3.539	3.654	3.757	3.847	3.926	3.991	4.044	4.080	4.096	4.087	4.043	3.952	3.797	3.561	3.243	2.881	2.547	2.269		
1.2	3.611	3.724	3.826	3.917	3.997	4.066	4.124	4.167	4.194	4.208	4.177	4.115	3.998	3.808	3.529	3.172	2.798	2.476		
1.4	3.674	3.786	3.887	3.978	4.060	4.131	4.193	4.243	4.278	4.296	4.290	4.252	4.168	4.020	3.789	3.463	3.075	2.705	2.399	
1.6	3.731	3.842	3.942	4.033	4.115	4.189	4.253	4.308	4.351	4.378	4.386	4.367	4.311	4.200	4.015	3.736	3.364	2.961	2.609	
1.8	3.782	3.891	3.990	4.081	4.164	4.239	4.307	4.365	4.414	4.449	4.468	4.472	4.431	4.352	4.209	3.979	3.646	3.238	2.840	2.511
2.0	3.828	3.936	4.034	4.125	4.208	4.285	4.354	4.416	4.468	4.511	4.539	4.549	4.532	4.479	4.372	4.189	3.907	3.522	3.095	2.719
2.2	3.870	3.976	4.073	4.164	4.248	4.325	4.396	4.460	4.517	4.564	4.600	4.620	4.619	4.587	4.510	4.369	4.137	3.796	3.370	2.949
2.4	3.908	4.013	4.109	4.199	4.283	4.361	4.433	4.500	4.559	4.611	4.653	4.682	4.693	4.678	4.627	4.521	4.336	4.047	3.648	3.201
2.6	3.943	4.046	4.142	4.231	4.315	4.394	4.467	4.535	4.597	4.653	4.700	4.736	4.757	4.756	4.725	4.649	4.506	4.269	3.916	3.471
2.8	3.975	4.077	4.172	4.261	4.344	4.423	4.498	4.567	4.631	4.690	4.741	4.783	4.812	4.824	4.809	4.758	4.650	4.460	4.160	3.745
3.0	4.004	4.105	4.199	4.287	4.371	4.450	4.525	4.596	4.662	4.723	4.777	4.824	4.860	4.882	4.881	4.850	4.771	4.623	4.376	4.007
3.2	4.031	4.131	4.224	4.312	4.396	4.475	4.550	4.622	4.689	4.752	4.810	4.861	4.903	4.932	4.944	4.929	4.875	4.762	4.563	4.247
3.4	4.056	4.155	4.247	4.335	4.418	4.498	4.573	4.645	4.714	4.779	4.839	4.893	4.940	4.976	4.997	4.996	4.963	4.880	4.723	4.461
3.6	4.079	4.177	4.269	4.356	4.439	4.518	4.594	4.667	4.737	4.803	4.865	4.922	4.973	5.015	5.044	5.055	5.038	4.980	4.859	4.647
3.8	4.101	4.197	4.288	4.375	4.458	4.537	4.614	4.687	4.757	4.825	4.888	4.948	5.002	5.049	5.085	5.106	5.103	5.066	4.976	4.806
4.0	4.121	4.217	4.307	4.393	4.476	4.555	4.631	4.705	4.776	4.845	4.910	4.972	5.029	5.080	5.122	5.150	5.159	5.139	5.075	4.943
4.2	4.140	4.234	4.324	4.410	4.492	4.571	4.648	4.722	4.794	4.863	4.929	4.993	5.052	5.107	5.153	5.189	5.208	5.202	5.159	5.059
4.4	4.157	4.251	4.340	4.425	4.508	4.587	4.663	4.737	4.809	4.879	4.947	5.012	5.074	5.131	5.181	5.223	5.250	5.257	5.232	5.159
4.6	4.174	4.267	4.355	4.440	4.522	4.601	4.677	4.752	4.824	4.895	4.963	5.029	5.093	5.152	5.207	5.253	5.288	5.305	5.295	5.244
4.8	4.189	4.281	4.369	4.454	4.535	4.614	4.691	4.765	4.838	4.909	4.978	5.045	5.110	5.172	5.229	5.280	5.321	5.346	5.349	5.318
5.0	4.204	4.295	4.383	4.467	4.548	4.627	4.703	4.778	4.851	4.922	4.992	5.060	5.126	5.190	5.249	5.303	5.350	5.383	5.396	5.381
5.2	4.218	4.308	4.395	4.479	4.560	4.638	4.715	4.789	4.862	4.934	5.004	5.073	5.141	5.206	5.267	5.325	5.376	5.415	5.437	5.436
5.4	4.231	4.321	4.407	4.490	4.571	4.649	4.725	4.800	4.873	4.945	5.016	5.086	5.154	5.220	5.284	5.344	5.399	5.443	5.474	5.483
5.6	4.243	4.332	4.418	4.501	4.581	4.659	4.736	4.810	4.884	4.956	5.027	5.097	5.166	5.233	5.299	5.361	5.418	5.468	5.505	5.525

Table 2. *Continued.*

Ku	0.0	0.1	0.2	0.3	0.4	0.5	0.6	0.7	0.8	0.9	1.0	1.1	1.2	1.3	1.4	1.5	1.6	1.7	1.8	1.9
5.8	4.255	4.343	4.429	4.511	4.591	4.669	4.745	4.820	4.893	4.966	5.037	5.108	5.177	5.246	5.312	5.376	5.436	5.491	5.533	5.561
6.0	4.266	4.354	4.439	4.521	4.600	4.678	4.754	4.829	4.902	4.975	5.046	5.117	5.188	5.257	5.325	5.390	5.452	5.511	5.558	5.593
6.2	4.276	4.364	4.448	4.530	4.609	4.695	4.763	4.837	4.911	4.983	5.055	5.126	5.197	5.267	5.336	5.403	5.467	5.529	5.581	5.621
6.4	4.286	4.373	4.457	4.538	4.618	4.703	4.771	4.845	4.919	4.991	5.063	5.135	5.206	5.276	5.346	5.414	5.480	5.542	5.600	5.646
6.6	4.296	4.382	4.466	4.547	4.626	4.710	4.778	4.853	4.926	4.999	5.071	5.143	5.214	5.285	5.356	5.425	5.492	5.557	5.618	5.669
6.8	4.305	4.391	4.474	4.554	4.633	4.717	4.785	4.860	4.933	5.006	5.078	5.150	5.222	5.293	5.364	5.434	5.503	5.569	5.634	5.688
7.0	4.313	4.399	4.481	4.562	4.640	4.724	4.792	4.867	4.940	5.013	5.085	5.157	5.229	5.301	5.372	5.443	5.513	5.581	5.648	5.706
7.2	4.322	4.406	4.489	4.569	4.647	4.730	4.799	4.873	4.946	5.019	5.091	5.164	5.236	5.308	5.380	5.451	5.522	5.591	5.658	5.722
7.4	4.330	4.414	4.496	4.576	4.654	4.736	4.805	4.879	4.952	5.025	5.097	5.170	5.242	5.314	5.387	5.459	5.530	5.601	5.669	5.736
7.6	4.337	4.421	4.503	4.582	4.660	4.742	4.811	4.885	4.958	5.031	5.103	5.175	5.248	5.320	5.393	5.466	5.538	5.609	5.679	5.749
7.8	4.344	4.428	4.509	4.588	4.666	4.747	4.817	4.890	4.963	5.036	5.109	5.181	5.253	5.326	5.399	5.472	5.545	5.617	5.688	5.760
8.0	4.351	4.434	4.515	4.594	4.672	4.753	4.822	4.896	4.969	5.041	5.114	5.186	5.259	5.331	5.404	5.478	5.551	5.624	5.696	5.771
8.2	4.358	4.441	4.521	4.600	4.677	4.758	4.827	4.901	4.974	5.046	5.118	5.191	5.263	5.336	5.410	5.483	5.557	5.631	5.704	5.775
8.4	4.365	4.447	4.527	4.605	4.682	4.762	4.832	4.905	4.978	5.051	5.123	5.195	5.268	5.341	5.414	5.488	5.562	5.637	5.710	5.783
8.6	4.371	4.452	4.532	4.611	4.687	4.767	4.837	4.910	4.983	5.055	5.127	5.200	5.272	5.345	5.419	5.493	5.567	5.642	5.717	5.790
8.8	4.377	4.458	4.538	4.616	4.692	4.772	4.841	4.914	4.987	5.059	5.132	5.204	5.276	5.349	5.423	5.497	5.572	5.647	5.722	5.797
9.0	4.382	4.463	4.543	4.621	4.697	4.776	4.845	4.918	4.991	5.063	5.135	5.208	5.280	5.353	5.427	5.501	5.576	5.652	5.727	5.803
9.2	4.388	4.468	4.548	4.625	4.701	4.780	4.850	4.923	4.995	5.067	5.139	5.211	5.284	5.357	5.431	5.505	5.580	5.656	5.732	5.808
9.4	4.393	4.473	4.552	4.630	4.705	4.784	4.854	4.926	4.999	5.071	5.143	5.215	5.287	5.361	5.434	5.509	5.584	5.660	5.736	5.813
9.6	4.398	4.478	4.557	4.634	4.710	4.788	4.857	4.930	5.002	5.074	5.146	5.218	5.291	5.364	5.437	5.512	5.587	5.663	5.740	5.817
9.8	4.403	4.483	4.561	4.638	4.714	4.791	4.861	4.934	5.006	5.078	5.149	5.222	5.294	5.367	5.440	5.515	5.590	5.667	5.744	5.821
10.0	4.408	4.487	4.565	4.642	4.717	4.795	4.865	4.937	5.009	5.081	5.153	5.225	5.297	5.370	5.443	5.518	5.593	5.670	5.747	5.825
10.2					4.721	4.798	4.868	4.940	5.012	5.084	5.156	5.228	5.300	5.373	5.446	5.521	5.596	5.673	5.750	5.828
10.4							4.871	4.943	5.015	5.087	5.158	5.230	5.303	5.375	5.449	5.523	5.599	5.675	5.753	5.831
10.6							4.874	4.947	5.018	5.090	5.161	5.233	5.305	5.378	5.451	5.526	5.601	5.678	5.755	5.834
10.8								4.949	5.021	5.092	5.164	5.236	5.308	5.380	5.454	5.528	5.603	5.680	5.757	5.836
11.0									5.024	5.095	5.166	5.238	5.310	5.383	5.456	5.530	5.605	5.682	5.760	5.838
11.2										5.098	5.169	5.240	5.312	5.385	5.458	5.532	5.607	5.684	5.762	5.840
11.4											5.171	5.243	5.314	5.387	5.460	5.534	5.609	5.686	5.763	5.842
11.6											5.173	5.245	5.316	5.389	5.462	5.536	5.611	5.687	5.765	5.844
11.8												5.247	5.318	5.391	5.464	5.538	5.613	5.689	5.767	5.845
12.0												5.249	5.320	5.393	5.465	5.539	5.614	5.690	5.768	5.847
12.2													5.322	5.394	5.467	5.541	5.616	5.692	5.769	5.848

$$P_{.00135} = 1.486$$

Step 4. Determine the standardized value that will represent a probability of 0.99865 for obtaining a value less than $P_{.99865}$. Interpolation will yield a more precise result.

For +Sk, use Table 2.

For −Sk, use Table 1.

		Sk:	
	0.60	**0.66**	0.70
Ku:			
−0.40	2.926		2.837
−0.25		**3.069**	
−0.20	3.173		3.109

$$P_{.99865} = 3.069$$

Step 5. Look up the standardized median M' using Table 3.

		Sk:	
	0.60	**0.66**	0.70
Ku:			
−0.40	0.161		0.212
−0.25		**0.175**	
−0.20	0.141		0.183

Standardized median $M' = 0.175$

Step 6. From the $P_{.00135}$, calculate the percentile point in units that the process is being measured. This is the Lp value.

$$Lp = \bar{X} - SP_{.00135} \qquad Lp = 6.046 - 1.253(1.486) \qquad Lp = 4.18$$

Step 7. From the $P_{.99865}$, calculate the percentile point in units that the process is being measured. This is the Up value.

$$Up = \bar{X} + SP_{.99865} \qquad Up = 6.046 + 1.253(3.069) \qquad Up = 9.891$$

Step 8. Calculate the estimated median $\hat{M}$.

If Sk is positive (as in this case), subtract SM' from $\bar{X}$.

$$\hat{M} = \bar{X} - SM' \qquad \hat{M} = 6.046 - 1.253(0.175) \qquad \hat{M} = 5.82$$

If Sk is negative, add SM' to $\bar{X}$.

Table 3.

Ku	0.0	0.1	0.2	0.3	0.4	0.5	0.6	0.7	0.8	0.9	1.0	1.1	1.2	1.3	1.4	1.5	1.6	1.7	1.8	1.9
-1.4	0.000	0.053	0.111	0.184	0.282	0.424	0.627	0.754												
-1.2	0.000	0.039	0.082	0.132	0.196	0.284	0.412	0.591	0.727											
-1.0	0.000	0.031	0.065	0.103	0.151	0.212	0.297	0.419	0.586											
-0.8	0.000	0.026	0.054	0.085	0.123	0.169	0.231	0.317	0.439	0.598										
-0.6	0.000	0.023	0.047	0.073	0.104	0.142	0.190	0.254	0.343	0.468	0.681	0.633								
-0.4	0.000	0.020	0.041	0.064	0.091	0.122	0.161	0.212	0.280	0.375	0.504	0.653	0.616							
-0.2	0.000	0.018	0.037	0.058	0.081	0.108	0.141	0.183	0.237	0.311	0.413	0.542	0.638	0.574						
0.0	0.000	0.017	0.034	0.053	0.073	0.097	0.126	0.161	0.206	0.266	0.347	0.456	0.579	0.621						
0.2	0.000	0.016	0.032	0.049	0.068	0.089	0.114	0.145	0.183	0.233	0.299	0.388	0.501	0.605	0.531					
0.4	0.000	0.015	0.029	0.045	0.063	0.082	0.105	0.132	0.165	0.208	0.263	0.336	0.433	0.545	0.582	0.536				
0.6	0.000	0.014	0.028	0.043	0.059	0.077	0.097	0.122	0.151	0.188	0.235	0.297	0.379	0.481	0.579	0.579	0.489			
0.8	0.000	0.013	0.026	0.040	0.055	0.072	0.091	0.113	0.140	0.172	0.213	0.266	0.336	0.425	0.527	0.590	0.533			
1.0	0.000	0.012	0.025	0.038	0.053	0.068	0.082	0.106	0.130	0.159	0.196	0.242	0.301	0.379	0.474	0.563	0.569	0.484		
1.2	0.000	0.011	0.024	0.036	0.050	0.065	0.080	0.100	0.122	0.148	0.181	0.222	0.274	0.341	0.426	0.520	0.576	0.524		
1.4	0.000	0.011	0.023	0.035	0.048	0.062	0.078	0.095	0.116	0.140	0.169	0.206	0.252	0.310	0.385	0.474	0.554	0.555	0.475	
1.6	0.000	0.010	0.022	0.034	0.046	0.060	0.074	0.091	0.110	0.132	0.159	0.192	0.233	0.285	0.351	0.432	0.518	0.564	0.510	
1.8	0.000	0.010	0.021	0.032	0.044	0.057	0.072	0.087	0.105	0.126	0.151	0.180	0.217	0.264	0.323	0.396	0.480	0.549	0.540	0.461
2.0	0.000	0.009	0.020	0.031	0.043	0.055	0.069	0.084	0.101	0.120	0.143	0.171	0.204	0.246	0.299	0.365	0.443	0.521	0.552	0.494
2.2	0.000	0.009	0.020	0.030	0.042	0.054	0.067	0.081	0.097	0.115	0.137	0.162	0.193	0.231	0.279	0.338	0.410	0.488	0.544	0.522
2.4	0.000	0.009	0.019	0.030	0.040	0.052	0.065	0.078	0.094	0.111	0.131	0.155	0.183	0.218	0.261	0.315	0.381	0.456	0.524	0.538
2.6	0.000	0.008	0.018	0.029	0.039	0.051	0.063	0.076	0.091	0.107	0.126	0.148	0.175	0.207	0.246	0.295	0.355	0.426	0.498	0.539
2.8	0.000	0.008	0.018	0.028	0.038	0.049	0.061	0.074	0.088	0.104	0.122	0.143	0.167	0.197	0.233	0.278	0.333	0.398	0.470	0.526
3.0	0.000	0.008	0.017	0.027	0.037	0.048	0.059	0.072	0.085	0.101	0.118	0.138	0.161	0.189	0.222	0.263	0.313	0.374	0.443	0.508
3.2	0.000	0.008	0.017	0.027	0.037	0.047	0.058	0.070	0.083	0.098	0.114	0.133	0.155	0.181	0.212	0.250	0.296	0.352	0.417	0.483
3.4	0.000	0.008	0.017	0.026	0.036	0.046	0.057	0.068	0.081	0.095	0.111	0.129	0.150	0.174	0.203	0.239	0.281	0.333	0.394	0.460
3.6	0.000	0.007	0.016	0.025	0.035	0.045	0.056	0.067	0.079	0.093	0.108	0.125	0.145	0.168	0.196	0.228	0.268	0.316	0.373	0.437
3.8	0.000	0.007	0.016	0.025	0.034	0.044	0.054	0.066	0.078	0.091	0.105	0.122	0.141	0.163	0.188	0.219	0.256	0.301	0.354	0.415
4.0	0.000	0.007	0.015	0.025	0.034	0.043	0.053	0.064	0.076	0.089	0.103	0.119	0.137	0.158	0.182	0.211	0.246	0.288	0.337	0.395
4.2	0.000	0.007	0.015	0.024	0.033	0.043	0.053	0.063	0.075	0.087	0.101	0.116	0.133	0.153	0.176	0.204	0.236	0.276	0.322	0.376
4.4	0.000	0.007	0.015	0.024	0.033	0.042	0.052	0.062	0.073	0.085	0.099	0.113	0.130	0.149	0.171	0.197	0.228	0.265	0.308	0.359
4.6	0.000	0.007	0.015	0.023	0.032	0.041	0.051	0.061	0.072	0.084	0.097	0.111	0.127	0.145	0.167	0.191	0.220	0.255	0.296	0.344
4.8	0.000	0.006	0.015	0.023	0.032	0.041	0.050	0.060	0.071	0.082	0.095	0.109	0.124	0.142	0.162	0.186	0.213	0.246	0.285	0.330
5.0	0.000	0.006	0.014	0.023	0.031	0.040	0.049	0.059	0.070	0.081	0.093	0.107	0.122	0.139	0.158	0.181	0.207	0.238	0.274	0.317
5.2	0.000	0.006	0.014	0.023	0.031	0.040	0.049	0.058	0.069	0.080	0.092	0.106	0.119	0.136	0.155	0.176	0.201	0.231	0.265	0.306
5.4	0.000	0.006	0.014	0.022	0.030	0.039	0.048	0.057	0.068	0.078	0.090	0.103	0.117	0.133	0.151	0.172	0.196	0.224	0.257	0.295
5.6	0.000	0.006	0.014	0.022	0.030	0.039	0.047	0.057	0.067	0.077	0.089	0.101	0.115	0.131	0.148	0.168	0.191	0.218	0.249	0.285

Table 3. Continued.

Ku	0.0	0.1	0.2	0.3	0.4	0.5	0.6	0.7	0.8	0.9	1.0	1.1	1.2	1.3	1.4	1.5	1.6	1.7	1.8	1.9
5.8	0.000	0.006	0.014	0.022	0.030	0.038	0.047	0.056	0.066	0.076	0.087	0.100	0.113	0.128	0.145	0.164	0.186	0.212	0.242	0.277
6.0	0.000	0.006	0.014	0.021	0.029	0.038	0.046	0.055	0.065	0.075	0.086	0.098	0.111	0.126	0.142	0.161	0.182	0.207	0.235	0.268
6.2	0.000	0.006	0.013	0.021	0.029	0.037	0.046	0.055	0.064	0.074	0.085	0.097	0.110	0.124	0.140	0.158	0.178	0.202	0.229	0.261
6.4	0.000	0.006	0.013	0.021	0.029	0.037	0.045	0.054	0.063	0.073	0.084	0.096	0.108	0.122	0.137	0.155	0.175	0.197	0.223	0.254
6.6	0.000	0.006	0.013	0.021	0.028	0.037	0.045	0.054	0.063	0.073	0.083	0.094	0.107	0.120	0.135	0.152	0.171	0.193	0.218	0.247
6.8	0.000	0.006	0.013	0.020	0.028	0.036	0.044	0.53	0.062	0.072	0.082	0.093	0.105	0.118	0.133	0.150	0.168	0.189	0.213	0.241
7.0	0.000	0.005	0.013	0.020	0.028	0.036	0.044	0.053	0.061	0.071	0.081	0.092	0.104	0.117	0.131	0.147	0.165	0.185	0.209	0.236
7.2	0.000	0.005	0.013	0.020	0.028	0.036	0.044	0.052	0.061	0.070	0.080	0.091	0.103	0.115	0.129	0.145	0.162	0.182	0.205	0.230
7.4	0.000	0.005	0.013	0.020	0.027	0.035	0.043	0.052	0.060	0.070	0.079	0.090	0.101	0.114	0.128	0.143	0.160	0.179	0.201	0.226
7.6	0.000	0.005	0.012	0.020	0.027	0.035	0.043	0.051	0.060	0.069	0.079	0.089	0.100	0.113	0.126	0.141	0.157	0.176	0.197	0.221
7.8	0.000	0.005	0.012	0.020	0.027	0.035	0.043	0.051	0.059	0.068	0.078	0.088	0.099	0.111	0.124	0.139	0.155	0.173	0.193	0.217
8.0	0.000	0.005	0.012	0.019	0.027	0.034	0.042	0.050	0.059	0.068	0.077	0.087	0.098	0.110	0.123	0.137	0.153	0.170	0.190	0.213
8.2	0.000	0.005	0.012	0.019	0.027	0.034	0.042	0.050	0.058	0.067	0.076	0.086	0.097	0.109	0.121	0.135	0.151	0.168	0.187	0.209
8.4	0.000	0.005	0.012	0.019	0.026	0.034	0.042	0.050	0.058	0.067	0.076	0.086	0.096	0.108	0.120	0.134	0.149	0.165	0.184	0.205
8.6	0.000	0.005	0.012	0.019	0.026	0.034	0.041	0.049	0.057	0.066	0.075	0.085	0.095	0.107	0.119	0.132	0.147	0.163	0.181	0.202
8.8	0.000	0.005	0.012	0.019	0.026	0.033	0.041	0.049	0.057	0.066	0.075	0.084	0.094	0.106	0.118	0.131	0.145	0.161	0.179	0.199
9.0	0.000	0.005	0.012	0.019	0.026	0.033	0.041	0.049	0.057	0.065	0.074	0.084	0.094	0.105	0.116	0.129	0.143	0.159	0.176	0.196
9.2	0.000	0.005	0.012	0.019	0.026	0.033	0.041	0.049	0.056	0.065	0.073	0.083	0.093	0.104	0.115	0.128	0.142	0.157	0.174	0.193
9.4	0.000	0.005	0.012	0.019	0.026	0.033	0.040	0.048	0.056	0.064	0.073	0.082	0.092	0.103	0.114	0.127	0.140	0.155	0.172	0.190
9.6	0.000	0.005	0.012	0.019	0.025	0.033	0.040	0.048	0.055	0.064	0.072	0.082	0.091	0.102	0.113	0.125	0.139	0.153	0.170	0.188
9.8	0.000	0.005	0.012	0.018	0.025	0.032	0.040	0.048	0.055	0.063	0.072	0.081	0.091	0.101	0.112	0.124	0.137	0.152	0.168	0.185
10.0	0.000	0.005	0.011	0.018	0.025	0.032	0.040	0.047	0.055	0.063	0.071	0.080	0.090	0.100	0.111	0.123	0.136	0.150	0.166	0.183
10.2	0.000					0.032	0.040	0.047	0.054	0.063	0.071	0.080	0.089	0.099	0.110	0.122	0.135	0.149	0.164	0.181
10.4	0.000					0.032	0.039	0.047	0.054	0.062	0.071	0.079	0.089	0.099	0.109	0.121	0.133	0.147	0.162	0.179
10.6	0.000					0.032	0.039	0.047	0.054	0.062	0.070	0.079	0.088	0.098	0.109	0.120	0.132	0.146	0.160	0.177
10.8	0.000						0.039	0.046	0.054	0.061	0.070	0.078	0.088	0.097	0.108	0.119	0.131	0.144	0.159	0.176
11.0	0.000							0.046	0.053	0.061	0.070	0.078	0.087	0.097	0.107	0.118	0.130	0.143	0.157	0.173
11.2	0.000									0.061	0.069	0.078	0.087	0.096	0.106	0.117	0.129	0.142	0.156	0.171
11.4	0.000										0.069	0.077	0.086	0.095	0.105	0.116	0.128	0.141	0.154	0.169
11.6	0.000										0.069	0.077	0.086	0.095	0.104	0.116	0.127	0.139	0.153	0.168
11.8	0.000										0.068	0.076	0.085	0.094	0.104	0.115	0.126	0.138	0.152	0.166
12.0	0.000											0.076	0.085	0.094	0.104	0.114	0.125	0.137	0.150	0.165
12.2	0.000											0.076	0.084	0.093	0.103	0.113	0.124	0.136	0.149	0.163

Step 9. Calculate the process capability indices based on normal distribution compensated for skewness and kurtosis.

A. Capability ratio, C_r and C_p

$$C_p = \frac{USpec - LSpec}{Up - Lp} \qquad C_p = \frac{14.0 - 2.0}{9.891 - 4.180}$$

$$C_r = \frac{1}{C_p} = 0.48$$

B. Process capability index, C_{pk}

As with the traditional C_{pk}, C_{pk} = minimum of C_{pl} and C_{pu}.

$$C_{pl} = \frac{\hat{M} - LSpec}{\hat{M} - Lp} \qquad C_{pl} = \frac{5.82 - 2.00}{5.82 - 4.18} = 2.33$$

$$C_{pu} = \frac{USpec - \hat{M}}{Up - \hat{M}} \qquad C_{pu} = \frac{14.0 - 5.82}{9.891 - 5.82} = 2.01$$

$$C_{pk} = 2.01$$

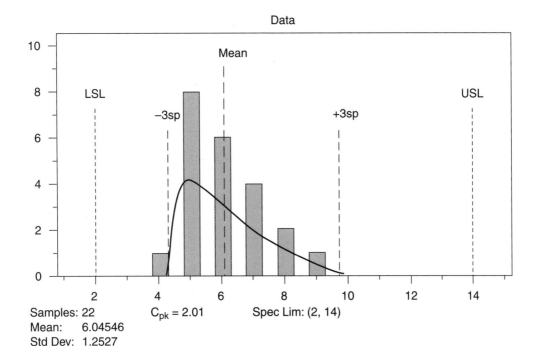

Samples: 22 $C_{pk} = 2.01$ Spec Lim: (2, 14)
Mean: 6.04546
Std Dev: 1.2527

Process capability where Sk = 0.66 and Ku = –0.25.

Had the C_{pk} and C_r been determined without adjusting for skewness and kurtosis, the results would have been:

$$C_{pl} = \frac{\overline{X} - LSpec}{3s} \quad C_{pl} = \frac{6.046 - 2.000}{3(1.253)} \quad C_{pl} = 1.08$$

$$C_{pu} = \frac{USpec - \overline{X}}{3s} \quad C_{pu} = \frac{14.000 - 6.046}{3(1.253)} \quad C_{pl} = 2.12$$

$$C_{pk} = C_{pl} = 1.08$$

and

$$C_r = \frac{6s}{USpec - LSpec} \quad \frac{7.518}{12.00} = 0.63$$

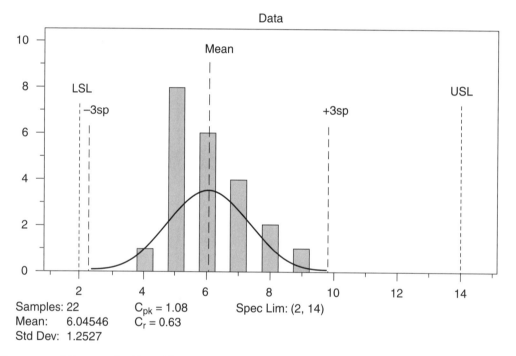

Samples: 22 $C_{pk} = 1.08$ Spec Lim: (2, 14)
Mean: 6.04546 $C_r = 0.63$
Std Dev: 1.2527

Process capability assuming Normal Distribution.

C_{PK} ADJUSTED FOR SK AND KU USING NORMAL PROBABILITY PLOTS

(For a more in depth discussion on normal probability plots [NOPP], see *Testing for a Normal distribution.*)

The calculations for C_{pk} using a normal distribution are based on the fact that the proportion of the data below the average minus three standard deviations is 0.000135, and the portion greater than the average plus three standard deviations is 0.99865. The median for the nonnormal distribution can be used in place of the average. The median can be determined by locating the point at which 50 percent of the data is below (or above).

The points at which specified proportions of the data are below the p*th* percentile are designated as X_p, where p = the proportion less than the value X.

$X_{0.00135}$ is the data value that 0.135% of the data is below and $X_{0.99865}$ is the data value for which 99.865 percent of the data is below. The point located at $X_{0.50000}$ always represents the median and for a normal distribution will also represent the average.

Calculations for the adjusted C_{pk} will be made using the following substitutions:

1. In the equation for the traditional $C_{pl} = \dfrac{(\overline{\overline{X}} - \text{LSpec})}{3S}$

For: *Use:*

$\overline{\overline{X}}$ $X_{0.5000}$

$3S$ $X_{0.5000} - X_{0.00135}$

2. In the equation for the traditional $C_{pu} = \dfrac{(\text{USpec} - \overline{\overline{X}})}{3S}$

For: *Use:*

$\overline{\overline{X}}$ $X_{0.5000}$

$3S$ $X_{0.99865} - X_{0.5000}$

C_{pk} is now determined as the minimum of C_{pl} and C_{pu}.

BIBLIOGRAPHY

Bothe, D. R. 1997. *Measuring Process Capability.* New York: McGraw-Hill.

Clements, J. A. "Process Capability Calculations for Non-Normal Distributions". ASQC *Quality Progress* (September 1989): 95–101.

Kotz, S., and N. L. Johnson. 1993. *Process Capability Indices.* New York: Chapman and Hall.

NONPARAMETRIC STATISTICS

Many of the traditional statistical techniques used in quality are based on specific distributions such as the normal distribution. Statistical techniques that are **not** dependent on a particular distribution (distribution free) but rather on a relative ranking are referred to as nonparametric methods. There are several such methods available. The methods discussed here are the Wilcoxon-Mann-Whitney test and the Kruskall-Wallis rank sum test.

COMPARING TWO AVERAGES: WILCOXON-MANN-WHITNEY TEST

Does the average of product *A* exceed that of product *B?*
Stated in the form of a hypothesis test:

Ho: A = B

Ha: A > B

A manufacturer of an exterior protective coating is evaluating two test formulations, *A* and *B*. Five test panels made with formula *A* and eight test panels using formula *B* are placed on test. Each panel is continuously monitored, and the time at which 50 percent of the coating thickness has worn is recorded in days. The objective is to maximize the length of time when 50 percent removal has been reached.

Step 1. Choose α, the significance level, or risk, for the test. Risks are traditionally 0.025, 0.05 and 0.10 for this one-sided test. For this example, let $\alpha = 0.05$.

Step 2. Collect data, then combine the two population data samples and rank in order of increasing response from smallest to largest. The failure times are ranked from the lowest to the highest with identification for the formula type. Combine the responses and rank from 1 (lowest) to 13 (highest). In the case of ties, assign to each the average rank had ties not been observed.

Formula	Time, days	Rank	Formula A	Formula B
A	26	1	26 (1)	30 (3)
A	28	2	28 (2)	41 (4)
B	30	3	50 (6.5)	46 (5)
B	41	4	68 (8)	50 (6.5)
B	46	5	72 (10)	70 (9)
A	50	6.5		73 (11)
B	50	6.5		78 (12)
A	68	8		80 (13)
B	70	9		
A	72	10		
B	73	11		
B	78	12		
B	80	13		

If more than 20 percent of the observations are involved in ties, this procedure should not be used.

Step 3. Let n_1 = the smaller sample = 5 (n_a)

n_2 = the larger sample = 8 (n_b)

$n = n_1 + n_2 = 13$

Step 4. Look up $R\alpha$ (n_1, n_2) in Table 1. This table gives critical values for n_1 and n_2 to 20. For larger values of n_1 and n_2, critical values can be estimated by

$$\frac{n_1}{2}(n_1 + n_2 + 1) - Z_\alpha \sqrt{\frac{n_1 n_2 (n_1 + n_2 + 1)}{12}}$$

where: $Z = 1.28$ for $\alpha = 0.10$
$Z = 1.65$ for $\alpha = 0.05$
$Z = 1.96$ for $\alpha = 0.025$

Step 5a. If the two sample sizes are equal or if n_b is smaller, compute R_B (the sum of ranks for sample B). If $R_B \leq R\alpha$ (n_1, n_2), conclude that the average of formula A exceeds formula B; otherwise, there is no reason to believe that the average for formula A is greater than the average for formula B.

Step 5b. If n_a is the smaller (as in this case), compute R_A (the sum of ranks for formula A), and compute $R'_A = n_A (n + 1) - R_A$. If $R'_A \leq R\alpha(n_1, n_2)$, conclude that formula A exceeds formula B; otherwise, there is no reason to believe that the two formulas are different.

Since the smaller sample is associated with formula A (n_a is the smaller sample), R_A (the sum of ranks for formula A) is calculated as follows

Table 1. Critical Values of Smaller Rank Sum for Wilcoxon-Mann-Whitney Test.

n_2	α	3	4	5	6	7	8	9	10	11	12	13	14	15
									n_1 (Smaller sample)					
3	0.10	7												
	0.05	6												
	0.025	—												
4	0.10	7	13											
	0.05	6	11											
	0.025	—	10											
5	0.10	8	14	20										
	0.05	7	12	19										
	0.025	6	11	17										
6	0.10	9	15	22	30									
	0.05	8	13	20	28									
	0.025	7	12	18	26									
7	0.10	10	16	23	32	41								
	0.05	8	14	21	29	39								
	0.025	7	13	20	27	36								
8	0.10	11	17	25	34	44	55							
	0.05	9	15	23	31	41	51							
	0.025	8	14	21	29	38	49							
9	0.10	11	19	27	36	46	58	70						
	0.05	10	16	24	33	43	54	66						
	0.025	8	14	22	31	40	51	62						
10	0.10	12	20	28	38	49	60	73	87					
	0.05	10	17	26	35	45	56	69	82					
	0.025	9	15	23	32	42	53	65	78					
11	0.10	13	21	30	40	51	63	76	91	106				
	0.05	11	18	27	37	47	59	72	86	100				
	0.025	9	16	24	34	44	55	68	81	96				
12	0.10	14	22	32	42	54	66	80	94	110	127			
	0.05	11	19	28	38	49	62	75	89	104	120			
	0.025	10	17	26	35	46	58	71	84	99	115			
13	0.10	15	23	33	44	56	69	83	98	114	131	149		
	0.05	12	20	30	40	52	64	78	92	108	125	142		
	0.025	10	18	27	37	48	60	73	88	103	119	136		
14	0.10	16	25	35	46	59	72	86	102	118	136	154	174	
	0.05	13	21	31	42	54	67	81	96	112	129	147	166	
	0.025	11	19	28	38	50	62	76	91	106	123	141	160	
15	0.10	16	26	37	48	61	75	90	106	123	141	159	179	200
	0.05	13	22	33	44	56	69	84	99	116	133	152	171	192
	0.025	11	20	29	40	52	65	79	94	110	127	145	164	184

$$R_A = 1 + 2 + 6.5 + 8 + 10 = 27.5$$

$$R_A' = n_A(n+1) - R_A = 5(13+1) - 27.5 = 42.5$$

$$R\alpha(n_1, n_2)$$

where: $\alpha = 0.05$
$n_1 = 5$
$n_2 = 8$
$R_{0.05(5,8)} = 23$

If $R'_A \leq R\alpha$ (n_1, n_2), conclude that formula A exceeds formula B.
$R'_A = 42.5$ and $R_{0.05,5,8} = 23$ R'_A is not $\leq R_{0.05,5,8}$; therefore, we cannot conclude that formula A exceeds the performance of formula B.

COMPARING MORE THAN TWO AVERAGES: THE KRUSKALL-WALLIS RANK SUM TEST

Do the averages of K products differ?
The objective of this nonparametric test is to determine if several samples could have all come from the same population; that is, the true averages of all of the samples are equal.

Data:

Design 1	Design 2	Design 3
1.7 (1)*	13.6 (6)	13.4 (5)
1.9 (2)	19.8 (8)	20.9 (9)
6.1 (3)	25.2 (12)	25.1 (10.5)
12.5 (4)	46.2 (16.5)	29.7 (13)
16.5 (7)	46.2 (16.5)	46.9 (18)
25.1 (10.5)	46.2 (16.5)	
30.5 (14)		
42.1 (15)		
82.5 (20)		

*The numbers shown in parentheses are the ranks, from lowest to highest, for all observations combined, as required in step 3 of the following procedure.

Procedure:

Step 1. Choose a level of significance, or alpha risk α. For this example, let $\alpha = 0.10$.

Step 2. Look up chi-square for $\chi^2_{1-\alpha}$ for $K - 1$ degrees of freedom (K = 3).

$$X^2_{0.90,2} = 4.62$$

Step 3. We have $n_1, n_2, n_3, \ldots, n$ observations on each product design 1, 2, 3, $\ldots, K$.

$$N = n_1 + n_2 + n_3 + \ldots + n_K$$

Assign ranks to each observation according to its size in relationship to all n observations. That is, assign rank 1 to the smallest, rank 2 to the next largest, and so on. In cases of ties, assign to each the rank that would have been assigned had the observations differed slightly. If more than 20 percent of the observations are involved in ties, this procedure should not be used.

Step 4. Calculate R_i (the sum of the ranks of the observations on the ith product) for each of the products.

$$R_1 = 76.5 \qquad R_2 = 78.0 \qquad R_3 = 55.5$$

Step 5. Calculate the following:

$$H = \frac{12}{N(N+1)} \sum \frac{R_i^2}{n_i} - 3(N+1)$$

$$H = \frac{12}{420}(2280.30) - 63$$

$$H = 2.15$$

Step 6. If $H > \chi^2_{1-\alpha}$, conclude that the averages of the K designs differ; otherwise, there is no reason to believe that the averages differ.

When using this procedure, there should be a minimum of five observations for each sample.

BIBLIOGRAPHY

Kohler, H. 1988. *Statistics for Business and Economics.* 2nd edition. Glenview, IL: Scott, Foresman and Company.

Siegel, S. 1956. *Nonparametric Statistics.* New York: McGraw-Hill.

NORMAL DISTRIBUTION

Distributions are mathematical functions that describe data. Graphically they can describe unique shapes. One of the most frequently used distributions is the normal distribution.

The normal distribution is used exclusively with the application of statistical process control (SPC) of variables data and is assumed when working with process capability indices.

The pattern or histogram of normally distributed data is one shaped like a bell and, hence, is sometimes referred to as the bell curve. One of the subjective tests for a normal distribution is to construct a frequency histogram and look for a bell-shaped curve. There are other tests for a normal distribution such as performing a normal probability plot or performing a chi-square goodness of fit test (see the module, *Testing for a Normal Distribution*).

All distributions have the following two defining characteristics:

1. A location statistic that defines the central tendency of the data (examples are the average, median, and mode)
2. A variation statistic that defines the amount of dispersion or variation of the data (examples are standard deviation and range)

The central tendency of data can be described by one of the following statistics:

1. Average $\bar{X}$: The sum of the data divided by the total observations
2. Mode M: The most frequently occurring data value
3. Median $\tilde{X}$: The middle value

The variation of a distribution can be measured by

1. Range R: The difference between the largest and smallest data values
2. Standard deviation S:

$$S = \sqrt{\frac{\Sigma(X - \bar{X})^2}{n-1}}$$

where: X = individual value
$\bar{X}$ = average of individuals
n = sample size
Σ = sum

The standard deviation is a standard way of measuring the deviation (variation) of the individuals from their average. The standard deviation is a universal index that can communicate the degree of variation of data. The greater the standard deviation, the greater the variation.

The normal distribution has several important characteristics:

1. The distribution is divided in half by the average.
2. The average, mode, and median are equal.
3. Approximately 68 percent of the data are contained within the limits of $\bar{X} \pm 1S$.
4. Approximately 95 percent of the data are contained within the limits of $\bar{X} \pm 2S$.
5. Approximately 99.7 percent of the data are contained within the limits of $\bar{X} \pm 3S$.

While the limits of $\pm 1S$, $\pm 2S$, and $\pm 3S$ are important from a practical perspective such as in the application of SPC, there are other times we might want to apply other multiples. Using the probability density function for the normal distribution, we can determine the probability of events occurring given an average and standard deviation. The following examples will illustrate this use of the normal distribution.

Example 1:
Given an average of $\bar{X} = 23.0$ and a standard deviation of $S = 3.1$, what is the probability of getting a value greater than 28.58?

Step 1. Sketch out the problem.

Draw a normal distribution. The vertical line dividing the distribution in half represents the average. A vertical line drawn in the tail area will represent the point of interest; in this case it is 28.58 The standard deviation is not drawn in but can be recorded near the average. The exact location of the point of interest is not critical, but it should be on the appropriate side of the average line. It is drawn so that there will be some tail area remaining outside the limit. There is an assumption that there is a measured quantity greater than (for this case) the point of interest.

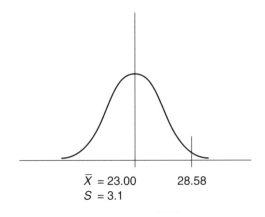

$\bar{X} = 23.00$ 28.58
$S = 3.1$

Step 2. Calculate a Z-score.

The Z-score is the difference between the average and the point of interest in units of standard deviation. Z-scores can be labeled as Z upper or Z lower depending on the location of the point of interest relative to the average. For our example, the Z-score is a Z upper score:

$$Z_U = \frac{X' - \overline{X}}{S} \quad Z_U = \frac{28.58 - 23.00}{3.1} = Z_U = 1.80$$

Calculate Z-scores to the nearest second decimal place.

Step 3. Look up the Z-score using the standard normal distribution table.

Go down the left column to 1.8, then proceed across this row until intercepting the *x.x0* column.

x.x0	*x.x1*	*x.x2*	*x.x3 . . . x.x9*
$|z|$			
4.0			
3.9			
3.8			
3.7			
.			
.			
.			
.			
1.8——— 0.03593			

0.03593 is the proportion of the distribution of data that is greater than 28.58. An alternative way of expressing this is 3.593 percent. This proportion may also be expressed as a rate greater than 28.58 by taking the reciprocal and rounding to the nearest integer.

$$\frac{1}{0.03593} = 27.83 \Rightarrow 28$$

This would imply that 1 out of 28 will exceed 27.83.

Example 2:

The specification for the amount of current drawn for a small electric motor under test is 3.0 ± 0.50. Thirty-five motors were tested with an average draw of 2.82 amps and a standard deviation of 0.45. What percentage of the motors fail to meet the specification requirement? What is the combined rate of nonconformance?

Step 1. Sketch out the problem:

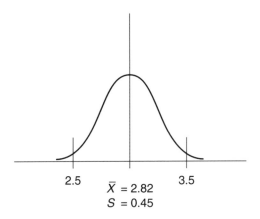

2.5 $\overline{X} = 2.82$ 3.5
         $S = 0.45$

Step 2. Calculate the Z-scores:

$$Z_L = \frac{\overline{X} - X'}{S} \qquad Z_L = \frac{2.82 - 2.50}{0.45} \qquad Z_L = 0.71$$

$$Z_U = \frac{X' - X}{S} \qquad Z_U = \frac{3.50 - 2.82}{0.45} \qquad Z_L = 1.51$$

Step 3. Look up the Z-scores.

	x.x0	**x.x1**	**x.x2**	**x.x3 . . . x.x9**		**x.x0**	**x.x1**	**x.x2**	**x.x3 . . . x.x9**
$|z|$					$|z|$				
4.0					4.0				
3.9					3.9				
3.8					3.8				
3.7					3.7				
3.6					3.6				
.					.				
.					.				
.					.				
.					.				
.					.				
0.7—— 0.23885					1.5—— 0.06552				

A Z_L of 0.71 indicates that a proportion of 0.23885 (23.885 percent) will be below the lower specification of 2.50. The rate of nonconformance below the lower specification is

$$\frac{1}{0.23885} = 4.186 = 4 \quad \text{1 out of 4 will fall below the lower specification}$$

A Z_U of 1.51 indicates that a proportion of 0.06552 (6.552 percent) will be greater than the upper specification of 3.50. The rate of nonconformance above the upper specification is

$$\frac{1}{0.06552} = 15.263 = 15 \quad \text{1 out of 15 will exceed the upper specification}$$

The total exceeding the specification is

$$\frac{1}{0.23885 + 0.06552} = \frac{1}{0.30437} = 3 \quad \text{1 out of 3 will not meet the specification}$$

Example 3:
Samples of cat food are analyzed, and the average sodium chloride content is 110.5 parts per 1000 with a standard deviation of 15.0. What proportion of the product is expected to have between 80 and 90 parts per thousand sodium chloride?

Step 1. Sketch out the problem.

In this example, we have two points of interest. Both are below the average; therefore, we will designate the Z-score for the 80.0 point of interest as Z_{LL} and the Z-score for the 90.0 point of interest as Z_L.

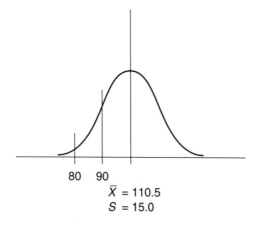

80   90
$\bar{X}$ = 110.5
S = 15.0

Step 2. Calculate the two Z-scores.

A. Calculate Z_{LL}.

$$Z_{LL} = \frac{\overline{X} - X'_{LL}}{S} \qquad Z_{LL} = \frac{110.5 - 80.0}{15.0} \qquad Z_{LL} = 2.03$$

B. Calculate Z_L.

$$Z_L = \frac{\overline{X} - X'_L}{S} \qquad Z_L = \frac{110.5 - 90.0}{15.0} \qquad Z_L = 1.37$$

Step 3. Look up the Z-scores.

	For Z = 2.03				For Z = 1.37		
x.x0	**x.x1**	**x.x2**	**x.x3**	**x.x0**	**x.x1**	**x.x2 . . . x.x7**
$\|z\|$				$\|z\|$			
4.0				4.0			
3.9				3.9			
3.8				3.8			
3.7				3.7			
3.6				3.6			
3.5				3.5			
.				.			
.				.			
.				.			
.				.			
2.0			0.02118	1.3			0.08534

The proportion of the data less than 80.0 is 0.02118, and the proportion of the data that is less than 90.0 is 0.08534. Therefore, the proportion between these two limits is

$$\begin{array}{r} 0.08534 \\ -0.02118 \\ \hline 0.06416 \end{array}$$

6.416 percent of the data is between 90.0 and 80.0.

CENTRAL LIMIT THEOREM

The standard deviation σ and the corresponding statistic S represent the variation of individuals. The distribution and variation of averages chosen from a population of individuals can be related to the distribution of the individuals from where they came. This relationship is defined by the central limit theorem.

$$S_{\overline{X}} = \frac{\sigma}{\sqrt{n}}$$

Note: The best estimate for σ will be the statistic S.

The standard deviation of averages will equal the true standard deviation of the individuals (from which the averages were determined) divided by the square root of the sample size.

Two things result from the central limit theorem:

1. The distribution of averages will tend to be normally distributed, regardless of the distribution of the individuals from where they are chosen.
2. The standard deviation of averages $S_{\bar{x}}$ will be less than the standard deviation of the individuals σ.

Example:

A population is defined with a sample average of $\bar{X} = 40$ and a sample standard deviation of $S = 8$.

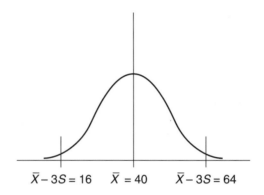

$$\bar{X} - 3S = 16 \qquad \bar{X} = 40 \qquad \bar{X} - 3S = 64$$

If samples of $n = 5$ are chosen from this distribution and a frequency distribution drawn of the averages, the following will result.

The standard deviation of the averages $S_{\bar{X}}$ will be

$$S_{\bar{X}} = \frac{\sigma}{\sqrt{n}} \qquad S_{\bar{X}} = \frac{8}{\sqrt{5}} \qquad S_{\bar{X}} = 3.8$$

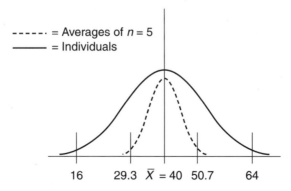

----- = Averages of $n = 5$
——— = Individuals

$$16 \qquad 29.3 \quad \bar{X} = 40 \quad 50.7 \qquad 64$$

Comparative Distribution of Individual vs. Averages of $n = 5$.

PARETO ANALYSIS

Pareto analysis is used to assist in prioritizing or focusing activities. In 1897, Vilfredo Pareto (1848–1923), an Italian economist, showed that the distribution of income was not evenly distributed, but that it was concentrated to a relatively small percentage of the population. Pareto's "Law" of income distribution was published in *Manuale d'Economia Politica* (1906). The relationship was based on the following formula:

$$\log N = \log A - a \log x$$

where: N = number of individuals
           x = income
       A and a = constants

Since 1951, various questions have been raised to the effect that Pareto was the wrong man to identify with the concentration of defects. Another candidate who might be associated with the application of defect concentration would be M. O. Lorenz, who depicted the concentration of wealth in the now familiar curves, using cumulative percent of population on one axis and cumulative percent of wealth on the other.

The following illustrates the basic procedure for performing a Pareto analysis:

1. Decide the objective of the Pareto analysis: (examples include defects by type, dollars of sales by location, defects by location, or numbers of physicians by specialty). Develop a list of the responses to be classified. A list of defects for a paint finish might include smears, voids, scratches, and runs. A list of geographical locations might include Seattle, Chicago, New York, and so on.
2. Design a tally sheet to collect the data.
3. Collect data.
4. Arrange the collected data in order of the smallest frequency occurrence first and the largest frequency last.
5. Determine the percent of each classification as a percent of the total.
6. Accumulate the percent distribution in a cumulative manner until a total of 100 percent has been obtained.

An example of a Pareto analysis follows:

Step 1–3: objective, talley, and data collection.

Customer-satisfaction survey cards have been collected during the first quarter of 1998 from Dr. Wise's dental practice. A score on a Likert-type format of three or less is considered not acceptable from a quality-improvement perspective. Dr. Wise wants to find out what areas he needs to focus his attention on in order to improve the quality of his service. The initial collected data yields the following:

Area of concern	Number of responses
Courteousness	10
Cost	35
Procedure explanation	11
Attractiveness of office	3
Mouthwash taste	1
Pain	60
Magazine selection	8
Uniforms	2
Promptness of schedule	2
Professionalism	6
Cleanliness	12
Total	150

Steps 4–6: Rearrange the data in descending order, and determine the percent each area is of the total and the cumulative percent total.

Area	Number of responses	Percentage of total responses	Cumulative percentage of total
Pain	60	40.0	40.0
Cost	35	23.3	63.3
Cleanliness	12	8.0	71.3
Procedure explanation	11	7.3	78.6
Courteousness	10	6.7	85.3
Magazine selection	8	5.3	90.6
Professionalism	6	4.0	94.6
Attractiveness of office	3	2.0	96.6
Promptness of schedule	2	1.3	97.9
Uniforms	2	1.3	99.2
Mouthwash taste	1	0.8	100.0
Totals:	150	100.0	

From this *tabular Pareto analysis,* it can be readily seen that 63 percent of Dr. Wise's problems are caused by 18 percent of the problem list.

This information can be presented in a graphical manner as follows:

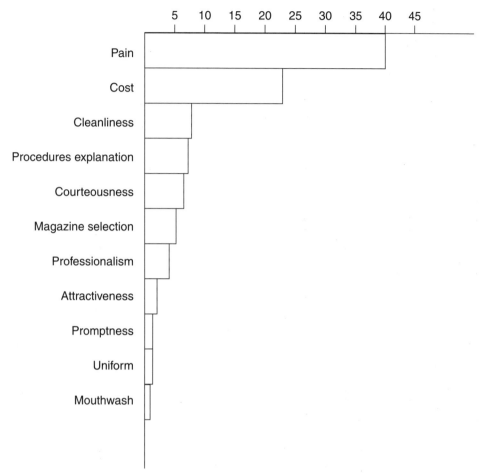

It is clear that Dr. Wise should focus his attention on reducing patient discomfort and lowering his fees.

In this example, an incident of *Magazine* is no more important than an incident of *Cleanliness* when, in fact, an incident of *Cleanliness* could result in an infection and possible litigation and a potential malpractice suit.

By assigning weighting factors to different levels of severity such as critical, major, and minor, the Pareto analysis can be augmented to reflect the level of impact for the respective assignments.

This assignment of weighting factors will result in a *weighted pareto analysis.*

Assume that Dr. Wise has categorized the various areas of concern into three groups and has assigned a weighting factor of 30, 5, and 1 in order of decreasing severity.

Category	Weighting factor
Critical	30
Major	5
Minor	1

With this scale, it will require 30 minor incidences to impact the Pareto analysis as much as one critical incident. Likewise, it will require six major incidences to impact the analysis as much as one critical incident.

Each one of the incidences will be multiplied by the weighting factor, and the resulting product will be used to develop the Pareto analysis.

Area of concern	Number of responses	Category	Factor	Pareto response
Courteousness	10	Major	5	50
Cost	35	Major	5	175
Procedure explanation	11	Major	5	55
Attractiveness of office	3	Major	5	15
Mouthwash taste	1	Minor	1	1
Pain	60	Major	5	300
Magazine selection	8	Minor	1	8
Uniforms	2	Minor	1	2
Promptness of schedule	2	Major	5	10
Professionalism	6	Critical	30	180
Cleanliness	12	Critical	30	360
Total	150			1156

The Pareto responses are reordered in a descending manner, and the percentage of total and cumulative percentage of total are determined.

Area of concern	Weighted Pareto response	Percentage of Total	Cumulative percent of total age
Cleanliness	360	31.1	31.1
Pain	300	26.0	57.1
Professionalism	180	15.6	72.7
Cost	175	15.1	87.8
Procedure explanation	55	4.8	92.6
Courteousness	50	4.3	96.9
Attractiveness of office	15	1.2	98.1
Promptness of schedule	10	0.9	99.0
Magazine selection	8	0.6	99.6
Uniforms	2	0.3	99.9
Mouthwash	1	0.1	100.0
Total	1156	100	

Note that the number one and number two rankings of concerns are now Cleanliness and Pain. 72.7 percent of the Pareto responses are described by 27.3 percent of the concerns list (3 out of 11).

Optionally, a Pareto diagram can be constructed.

Problem:

Construct a Pareto diagram for the distribution of letters in the following paragraph. What percentage of the letters are responsible for 80 percent of the total letters in the passage?

> "**Black Huckleberry** has fair landscape value and is quite attractive in the fall. Provides fair cover but no nesting to speak of. Fruit is available for a short period of time during the summer only and is a favorite for more than forty-five species of birds such as grosbeaks, towhees, bluebirds, robins, chickadees, and catbirds."

BIBLIOGRAPHY

Besterfield, D. H. 1994. *Quality Control.* 4th edition. Englewood Cliffs, NJ: Prentice Hall.

Grant, E. L., and R. S. Leavenworth. 1996. *Statistical Quality Control.* 7th edition. New York: McGraw-Hill.

Hayes, G. E., and H. G. Romig. 1982. *Modern Quality Control,* 3rd edition. Encino, CA: Glencoe.

Juran, J. M. 1960. Pareto, Lorenz, Cournot, Bernoulli, Juran and Others. *Industrial Quality Control* 17, no. 4 (October 1960): 25.

Juran, J. M., and F. M. Gryna. 1980. *Quality Planning and Analysis,* 3rd edition. New York, NY, McGraw-Hill.

PRE-CONTROL

Pre-control is a technique for *validating* a process to perform at a minimal level of conformance to the specification. Satterthwaite, (1954) initiated the introduction of pre-control in 1954 and identified the development team as consisting of himself, C. W. Carter, W. R. Purcell, and Dorian Shainin. The latter team member has been responsible for the popularization of pre-control.

Pre-control attempts to control a process relative to the specification requirements, while statistical process control (SPC) monitors the process relative to natural variations as described by the normal distribution and its associated average and standard deviation. SPC is truly based on *historical* characterization of a process and is independent of any specification requirements.

DESCRIPTION

The specification (bilateral) is divided into four equal parts. The areas from the specification nominal to one-half of the upper specification and the area from the nominal to one-half of the lower specification are called the *pre-control limits*. The area between the two pre-control limits is called the *green zone*. The areas outside of the pre-control limits but not exceeding the upper or lower specification limits are called the *yellow zone*. Areas outside of the specification limits are called the *red zone*. Data used in the pre-control technique are frequently placed on a chart, and the zones are colored according to the aforementioned scheme, leading to what is referred to as a *rainbow chart*.

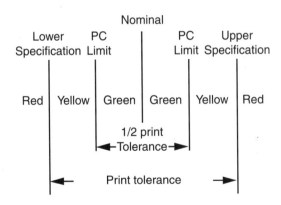

Example:
What are the pre-control limits if the specification is 4.50 ± 0.40?

1. Divide the tolerance by 4:

$$\frac{0.80}{4.0} = 0.20$$

2. Subtract the value from nominal for the lower PC limit:

$$4.50 - 0.20 = 4.30$$

3. Add the value to nominal for the upper PC limit:

$$4.50 + 0.20 = 4.70$$

Thus, the two pre-control limits are located at 4.30 and 4.70, one-half the normal tolerance.

Note: If the specification is unilateral (single sided), the sole pre-control limit is placed halfway between the target, or nominal, and the single specification limit.

Pre-Control Rules

1. Initial setup: five consecutive samples are measured. All measurements are required to fall within the pre-control limits before proceeding.
2. Following the initial setup approval, two consecutive samples are taken from the process and measured.
   a. If both units fall in the green zone, continue.
   b. If one unit falls inside the pre-control limits and the other falls in the yellow zone, continue.
   c. If both units fall inside the same yellow zone, the process is stopped and adjustments are made. After adjustments have been made, the setup approval is again required. Five consecutive units must fall within the pre-control limits.
   d. If both of the units measured fall in opposite yellow zones, then it is likely that variation has increased. The process is stopped, and efforts are made to reduce variation. Again, five consecutive measurements are required to fall within the pre-control limits.
   e. A single measurement outside of the specification limits (red zone) will result in stopping the process and adjusting. Following adjustment, setup approval is required.

After any adjustment or process stoppage, five consecutive measurements are required to fall inside the pre-control limits.

The statistical foundation of pre-control is based on the assumption that the process average is equal to the nominal or target value and that normal variation (average ± 3 standard deviations) is equal to the product or process tolerance. That is, $C_{pk} = 1.00$ and Cp

= 1.00. For a normal distribution, approximately 86 percent of the data (12 out of 14) will fall inside of the pre-control limits (the green zone), and approximately 7 percent (1 out of 14) will fall outside of the pre-control limits but inside the specification (the yellow zone). The probability of getting measurements outside of the specification is very low, approximately 3/1000.

The probabilities for various outcomes can be seen in the following table:

			Color zones			
Decision **Stop** **Adjust**	**Red** **A**	**Yellow**	**Green**	**Yellow** **A**	**Red**	**Probability** **Nil** **Nil**
Stop		A		B		1/14 × 1/14 = 1/196
Get help		B		A		1/14 × 1/14 = 1/196
Stop		A,B				1/14 × 1/14 = 1/196
Adjust				A,B		1/14 × 1/14 = 1/196
			A,B			12/14 × 12/14 = 144/196
		A	B			1/14 × 12/14 = 12/196
Continue		B	A			1/14 × 12/14 = 12/196
			A	B		12/14 × 1/14 = 12/196
			B	A		12/14 × 1/14 = 12/196

Lower	PC	PC	Upper	
Specification	limit	limit	Specification	Total = 196/196
	Nominal (target)			

Continue to monitor the process, sampling at a frequency equivalent to six pairs (A,B) chosen between each adjustment or stoppage.

Time between adjustments, hours	*Time between measurements, minutes*
1	10
2	20
3	30
4	40
5	50
etc.	etc.

A COMPARISON OF PRE-CONTROL AND SPC

Pre-control is specification oriented and detects changes in the capability of a process to meet the specification, whereas the objective of SPC is to detect changes in the location and variation statistic (average and standard deviation, for example). Both will force the operator to take measurements of the process. One argument for the use of pre-control over

SPC is that pre-control does not burden the operator with the task of plotting the data. Plotting the data using traditional SPC methods is vital in detecting process changes, identifying and removing (special) causes of variation, and distinguishing the differences in assignable and chance (common) causes.

If we set the specification limits at ±2, the pre-control limits become ±1. Pre-control assumes that the process average equals the nominal and that $Cp = 1.00$. This implies that $\sigma = 2/(3\ Cp)$. We may now determine the probability that a pre-control chart scheme will give a signal (both observations in a yellow zone or one in the red zone) for and degree of process change δ and the process Cp. This probability β^* will be the probability of getting a signal and is calculated as follows:

$$g = \Pr[\ X \text{ in green zone; } \delta\ ]$$

$$= \Pr[-1 \leq X \leq +1;\ \delta\ ]$$

$$= \Pr\left[\frac{(-1-\delta)}{\sigma} \leq Z \leq \frac{(1-\delta)}{\sigma}\right]$$

$$= \Pr\left[\frac{3(-1-\delta)Cp}{2} \leq Z \leq \frac{3(1-\delta)Cp}{2}\right]$$

$$= \phi\left[\frac{3(-1-\delta)Cp}{2}\right] - \phi\left[\frac{3(1-\delta)Cp}{2}\right]$$

$$y = \Pr[\ X \text{ in yellow zone, } \delta\ ]$$

$$= \Pr[-2 \leq X \leq +2;\ \delta\ ]$$

$$= \phi\left[\frac{3(2-\delta)Cp}{2}\right] - \phi\left[\frac{-3(2+\delta)Cp}{2}\right]$$

$$\beta^* = \Pr[\text{signal; } \delta\ ]$$

$$= 1 - \{g(g+2y)\}$$

$1/\beta^*$ = average run length (ARL) for a signal given Cp and amount of process shift δ

where: δ = amount of process shift from nominal
   ϕ = cumulative probability density function for the normal distribution

Similarly, the probability of detecting a process change (a single point outside of a control limit) using SPC can be determined.

$$\beta = \Pr[\text{sample average exceeds a control limit; } \delta\ ]$$

where: δ = process shift from nominal

Table 1. ARLs of Pre-control (PC) and Average Control Charts with Subgroup size $n = 2$ as a Function of Level Shift and Process Capability Index, Cp.

δ	Cp = 0.75 Percent defective	ARL CC	ARL PC	Cp = 1.00 Percent defective	ARL CC	ARL PC	Cp = 1.30 Percent defective	ARL CC	ARL PC
0.00	2.445	370.4	9.6	0.270	370.4	44.4	0.009	370.4	370.3
0.10	2.535	328.9	9.4	0.300	302.0	41.4	0.012	266.1	310.7
0.20	2.810	243.0	8.7	0.395	188.3	33.9	0.022	135.0	198.1
0.30	3.274	164.6	7.7	0.567	108.8	25.6	0.044	66.3	110.6
0.40	3.940	108.8	6.7	0.836	63.4	18.5	0.087	34.2	59.9
0.50	4.821	72.4	5.7	1.231	38.1	13.2	0.166	18.8	33.1
0.60	5.935	48.9	4.8	1.791	23.8	9.5	0.307	11.1	19.2
0.70	7.299	33.7	4.1	2.561	15.4	6.9	0.547	6.9	11.7
0.80	8.932	23.8	3.5	3.594	10.4	5.2	0.941	4.6	7.5
0.90	10.850	17.1	3.0	4.948	7.3	4.0	1.565	3.3	5.1
1.00	13.066	12.6	2.6	6.681	5.3	3.1	2.514	2.4	3.6

δ	Cp = 1.75 Percent defective	ARL CC	ARL PC
0.00	0.000	370.4	13273.1
0.10	0.000	214.6	8200.0
0.20	0.000	82.8	2847.0
0.30	0.000	33.7	890.7
0.40	0.001	15.4	297.6
0.50	0.004	7.9	110.5
0.60	0.012	4.5	45.9
0.70	0.032	2.9	21.3
0.80	0.082	2.0	11.0
0.90	0.194	1.6	6.3
1.00	0.433	1.3	3.9

REFERENCE

1. Ledolter, J., and A. Swersey. An Evaluation of Pre-Control. *Journal of Quality Technology* 29, no. 2 (April 1997).

BIBLIOGRAPHY

Besterfield, D. H. 1994. *Quality Control.* 4th edition. Englewood Cliffs, NJ: Prentice Hall.

Grant, E. L., and R. S. Leavenworth, 1996. *Statistical Quality Control.* 7th edition. New York: McGraw-Hill.

Montgomery, D. C., 1996. *Introduction to Statistical Quality Control.* 3rd edition. New York: John Wiley and Sons.

Satterthwaite, F. E. 1954. *A Simple, Effective Process Control Method,* Report 54-1. Boston, MA: Rath & Strong.

Wheeler, D. J., and D. S. Chambers. 1992. *Understanding Statistical Process Control,* 2nd edition. Knoxville, TN: SPC Press.

PROCESS CAPABILITY INDICES

The relationship of the process specification requirements and the actual performance of the process can be described by one of several index numbers: C_{pk}, Cr, and Cp. Each of these index numbers represents a single value that serves as a measure of how well the process can produce parts or services that comply with the specification requirement. All process capability indices are applicable only to variables measurements and assume that the data are normally distributed.

C_{PK}

This index uses both the average and standard deviation of the process to determine if the process is capable of meeting the specification. For processes where the average is closer to one specification than the other (process not centered), which is more common than being centered, or where there is only a single specification limit (unilateral), the following table gives the expected level of nonconformance in parts per million (ppm).

C_{pk} Value	nonconforming, ppm	Defective rate
1.60	1	1/1,000,000
1.50	3	1/333,333
1.40	13	1/76,923
1.30	48	1/20,833
1.20	159	1/6,289
1.10	483	1/2070
1.00	1350	1/741
0.90	3467	1/288
0.80	8198	1/122
0.70	17865	1/56

C_{pk} is defined as

$$C_{pk} = min\ (C_{pu}, C_{pl})$$

where:

$$C_{pu} = \frac{(USL - \overline{X})}{3\sigma}$$

$$C_{pl} = \frac{(\overline{X} - LSL)}{3\sigma}$$

USL = upper specification limit

LSL = lower specification limit

σ = population standard deviation

C_{pk} may be calculated for unilateral or bilateral specifications, with or without a target value.

σ may be estimated from:

1. The sample standard deviation S
2. Control chart data from average/range charts or average/standard deviation charts:

$$\sigma = \frac{\overline{R}}{d_2} \quad \text{or} \quad \sigma = \frac{\overline{S}}{c_4}$$

The determination of C_{pk} is accomplished in the following four basic steps:

1. Collect data to determine the average and standard deviation.
2. Sketch a linear graph indicating the upper and lower specification, average, and target if available.
3. Determine if the average is closer to the upper or lower specification.
4. Select the appropriate relationship for calculating the C_{pk}.

Example 1:

Step 1. Collect data.

Forty-five measurements have been made on the length of a part. The specification is bilateral with a requirement of 4.500 ± 0.010. The average, $\overline{X} = 4.505$, and the standard deviation $S = 0.0011$.

Step 2. Sketch a linear graph.

$$\overline{X} = 4.505$$

LSL = 4.49   Target = 4.500   USL = 4.510

Step 3 and 4. Determine the location of the average, and select the appropriate relationship for calculating the C_{pk}.

The average is closer to the upper specification; therefore, the defining relationship for C_{pk} is

$$C_{pk} = C_{pu} = \frac{(USL - \overline{X})}{3\sigma} = \frac{(4.510 - 4.505)}{(3)(.0011)} = 1.52$$

This process exhibits a high degree of capability, yielding approximately three ppm as a defective rate.

Example 2:
The minimum latching strength for a component is 7500 pounds. There is no maximum requirement or target. The average and standard deviation for 130 measurements are $\bar{X} = 6800$ and $S = 125$. What is the C_{pk}?

Sketch the linear graph as follows:

$\bar{X} = 6800$

LSL = 7500

$$C_{pk} = C_{pl} = \frac{(6800 - 7500)}{(3)(125)} = -1.87$$

Process is extremely poor; the process average is below the lower specification limit. Note that the C_{pk} can be negative.

CONFIDENCE INTERVAL FOR C_{PK}

The degree of confidence we have in the C_{pk} is related to the sample size of the data used to determine the C_{pk} and the amount of risk we are willing to take (α risk). The following equation * can be used to determine the confidence interval for the C_{pk} statistic:

$$C_{pk} \pm Z_{\alpha/2} \sqrt{\frac{1}{9n} + \frac{C_{pk}^2}{2n-2}}$$

Note: For single-sided limits, use Z_α rather than $Z_{\alpha/2}$.

where: n = sample size
α = risk

$$1 - \text{risk} = \text{confidence}$$

Example 1: Two-sided confidence interval:
Determine the 95 percent confidence interval given a C_{pk} of 1.15 and a sample size of $n = 45$.

Another way to phrase this problem is: "I do not know the true C_{pk}, but I am 95 percent confident that it is between _____ and _____."

Select the Z value from the following table:

Risk, %	Confidence, %	$Z_{\alpha/2}$	Z_α
.1	99.9	3.29	3.09
1.0	99.0	2.58	2.33
5.0	95.0	1.96	1.64
10.0	90.0	1.64	1.28

We are assuming a 5 percent risk that our interval will not contain the true C_{pk}; therefore, the level of confidence is 95 percent, and the $Z_{\alpha/2}$ value is 1.96.

$$C_{pk} \pm Z_{\alpha/2} \sqrt{\frac{1}{9n} + \frac{C_{pk}^2}{2n-2}} = 1.15 \pm 1.96 \sqrt{\frac{1}{(9)(45)} + \frac{(1.15)^2}{(2)(45)-2}} = 1.15 \pm 0.26$$

The lower limit is $1.15 - 0.26 = 0.89$.

The upper limit is $1.15 + 0.26 = 1.41$.

An appropriate statement would be: "I do not know the true C_{pk}, but I am 95 percent confident that it is between __0.89__ and __1.41__."

Notice the wide spread in the confidence interval. This spread or error of estimate can be reduced by doing one or both of the following:

1. Increasing the sample size
2. Decreasing the level of confidence

Example 2: Single-sided confidence limit:
Calculating a single-sided confidence *limit* is performed the same way as a *confidence interval,* except the risk is not divided by two. If the limit is a lower limit, we subtract the error term from the C_{pk} estimate. If the limit is an upper limit, we add the error term to the C_{pk} estimate.

A C_{pk} of 1.33 has been determined from a sample of 100 measurements. What is the lower 90 percent confidence?

$n = 100$

$C_{pk} = 1.33$

Confidence $= 0.90$

Risk, $\alpha = 0.10$

$Z_\alpha = 1.28$

This problem can also be phrased: "I do not know the true C_{pk}, but I am 90 percent confident that it is not less than _____."

$$C_{pk} - Z_a \sqrt{\frac{1}{9n} + \frac{C_{pk}^{\,2}}{2n-2}} = 1.33 - 1.28 \sqrt{\frac{1}{(9)(100)} + \frac{(1.33)^2}{(2)(100)-2}} = 1.33 - 0.13 = 1.20$$

An appropriate statement would be: "I do not know the true C_{pk}, but I am 90 percent confident that it is not less than <u>1.20</u>."

Determination of the appropriate sample size for a single-sided confidence limit for C_{pk} can be determined given a level of confidence and a predesired amount of error.

The single-sided error for C_{pk} is given by

$$Z_a \sqrt{\frac{1}{9n} + \frac{C_{\hat{p}k}^{\,2}}{2n-2}}$$

Rearranging and solving for n yields the following quadratic equation:

$$-18n^2 + n\left[18 + \left(\frac{2+9C_{pk}^{\,2}}{\dfrac{E^2}{Z^2}}\right)\right] + \frac{2}{\dfrac{E^2}{Z^2}} = 0$$

which is in the general form of $aX^2 + bX + c = 0$.

$$a = -18$$

$$b = 18 + \left(\frac{2+9Cpk^2}{\dfrac{E^2}{Z^2}}\right)$$

$$c = \frac{2}{\dfrac{E^2}{Z^2}}$$

The sample size n may be determined using the quadratic equation:

$$n = \frac{-b \pm \sqrt{b^2 - 4ac}}{2a}$$

Notice that the b term contains the C_{pk}. A prior estimate for the C_{pk} may be used as a starting point.

Consider the following case:

It is desired to determine the C_{pk} with an acceptable error of 0.166 and a level of confidence of 90 percent. It is estimated that the C_{pk} will be approximately 1.20 based on historical data. What is the appropriate sample size required?

$$\text{Let} \quad C = 90\%$$

$$C_{pk} = 1.20$$

$$E = 0.166$$

$$Z_{\alpha} = 1.28$$

$$a = -18$$

$$b = 18 + \left(\frac{2 + 9C_{pk}^{2}}{\dfrac{E^{2}}{Z^{2}}} \right) \qquad b = 907.35$$

$$c = \frac{2}{\dfrac{E^{2}}{Z^{2}}} = 118.9$$

$$n = \frac{-b \pm \sqrt{b^{2} - 4ac}}{2a}$$

$$n = \frac{-907.35 \pm \sqrt{823283 - 4(-18)(118.8)}}{(2)(-18)}$$

$$n = 50.5 \text{ or } 51$$

CR AND CP INDICES

There are two indices that do not use the process average to determine a process capability; rather, they measure the proportion of the specification tolerance that is being consumed by normal variation. These index numbers are Cr (capability ratio), which is sometimes reported as a %Cr, and Cp.

Normal variation is defined as six standard deviations.

%Cr is the percentage of the tolerance that is being consumed by normal variation. Cp is the reciprocal of Cr:

$$CP = \frac{1}{Cr} \qquad Cr = \frac{6S}{USL - LSL} \qquad \%Cr = \frac{6S}{USL - LSL} \times 100$$

Of these two, the %Cr is more comprehensible as it is a straightforward percentage of consumption. In order to achieve a C_{pk} greater than 1.33 (world-class quality), the %Cr must be less than 75 percent.

Cr and Cp require that the specification be bilateral (have both an upper and lower limit).

Example 1:
The specification for a part is 12.00 ± 0.38, and the standard deviation is $S = 0.08$. What is the %Cr?

$$\%Cr = \frac{6S}{USL - LSL} \times 100 \quad \%Cr = \frac{(6)(0.08)}{12.38 - 11.62} \times 100 \quad \%Cr = 63.2\%$$

With this %Cr, there is potential of achieving a C_{pk} of 1.33.

CONFIDENCE INTERVAL FOR CP AND CR

The confidence interval for Cp can be calculated using the following relationship:

$$Cp\sqrt{\frac{X^2_{n-1,\alpha/2}}{n-1}} > \text{True Cp} > \sqrt{\frac{X^2_{n-1,1-\alpha/2}}{n-1}}$$

Example 1:
A Cp of 2.32 has been calculated based on 29 observations. What is the 95 percent confidence interval for this estimate?

Cp = 2.32

$n = 29$

Confidence = 0.95

Risk, $\alpha = 0.05$

Look up the two required χ^2 values as follows:

For the upper bound, $\chi^2_{n-1,\alpha/2} = \chi^2_{28.025} = 44.4607$

For the lower bound, $\chi^2_{n-1,1-\alpha/2} = \chi^2_{28.0.975} = 15.3079$

$$Cp\sqrt{\frac{X^2_{n-1,\alpha/2}}{n-1}} > \text{True Cp} > Cp\sqrt{\frac{X^2_{n-1,1-\alpha/2}}{n-1}}$$

$$2.32\sqrt{\frac{44.4607}{28}} > \text{True Cp} > 2.32\sqrt{\frac{15.3079}{28}}$$

$$2.92 > \text{True Cp} > 1.72$$

The 95 percent confidence interval is 1.72 to 2.92.

A proper statement would be: "Based on a sample size of 29, I do not know the true Cp, but I am 95 percent confident that it is between 1.72 and 2.92."

The confidence interval for the %Cr can be readily determined by simply taking the inverse of the confidence interval for the Cp and reversing the terms. That is, the lower confidence limit for the Cp is the upper confidence limit for the Cr.

In the previous example, the 95 percent confidence lower bound was 1.72.

$$\frac{1}{1.72} = 0.58 = \text{upper confidence bound for the \%Cr}$$

and

$$\frac{1}{2.92} = 0.34 = \text{lower confidence bound for the \%Cr}$$

If tables of the chi-square distribution are not available, it is necessary to use approximate formulas. In the absence of such tables, one commonly used approximation is

$$X^2{}_{v,\alpha} \cong \left(\sqrt{\left(v - \frac{1}{2} \right)} + \frac{Z}{\sqrt{2}} \right)^2$$

where: v = degrees of freedom = $n - 1$
        Z = Z-score from standard normal distribution, giving α, $\alpha/2$, or $1 - \alpha/2$

From the chi-square table in this example, the upper bound $\chi^2{}_{n-1,\alpha/2} = \chi^2{}_{280025} = 44.4607$. Using the approximation formula:

$$X^2{}_{v,\alpha} \cong \left(\sqrt{\left(28 - \frac{1}{2} \right)} + \frac{1.96}{\sqrt{2}} \right)^2 \qquad X^2{}_{28,0.025} = 43.96$$

The confidence interval for Cr can also be rewritten as follows:

$$\text{Cr} \sqrt{\frac{n-1}{X^2{}_{n-1,\alpha/2}}} < \text{True Cr} < \text{Cr} \sqrt{\frac{n-1}{X^2{}_{n-1,1-\alpha/2}}}$$

CPM INDEX

One of the more recent quality process capability indexes is C_{pm}. A discussion of this index is reported by Chan, Cheng, and Spiring, (July 1988).

The C_{pm} index takes into consideration the proximity to the target value of a specification unlike the C_{pk} that only relates the process variation and average to the nearest specification limit.

$$C_{pm} = \frac{USL - LSL}{6\sigma'}$$

where: $\sigma' = \sqrt{\dfrac{\Sigma(X_i - T)^2}{n - 1}}$

T = target or nominal

$$C_{pm} = \frac{(USL - LSL)}{6\sqrt{\sigma^2 + (\mu - T)^2}} = \frac{1}{Cr\sqrt{1 + \dfrac{(\mu - T^2}{\sigma^2}}}$$

If the process variance σ^2 increases (decreases), the denominator increases (decreases) and C_{pm} will decrease (increase). If the process drifts from its target value (that is, if μ moves away from T), the denominator will again increase, causing C_{pm} to decline. In cases where both the process variance and the process mean change relative to the target, the C_{pm} index reflects these changes as well.

Example 1:
The specification is 15 + 6/–5. The average $\bar{X}$ = 17.0, and the standard deviation S = 1.2. What is C_{pm}?

$$Cr = \frac{6S}{USL - LSL} = 0.65 \quad C_{pm} = \frac{1}{Cr\sqrt{1 + \dfrac{(\mu - T^2}{\sigma^2}}} = \frac{1}{0.65\sqrt{1 + \dfrac{(17.0 - 15.0)^2}{(1.2)^2}}}$$

The following will illustrate the movement of C_{pk} and C_{pm} for a constant specification requirement of 28.00 + 3/–4, where the standard deviation remains a constant 0.43 and the process average changes from 29 to 24. Cr will remain constant (Cr = 0.369).

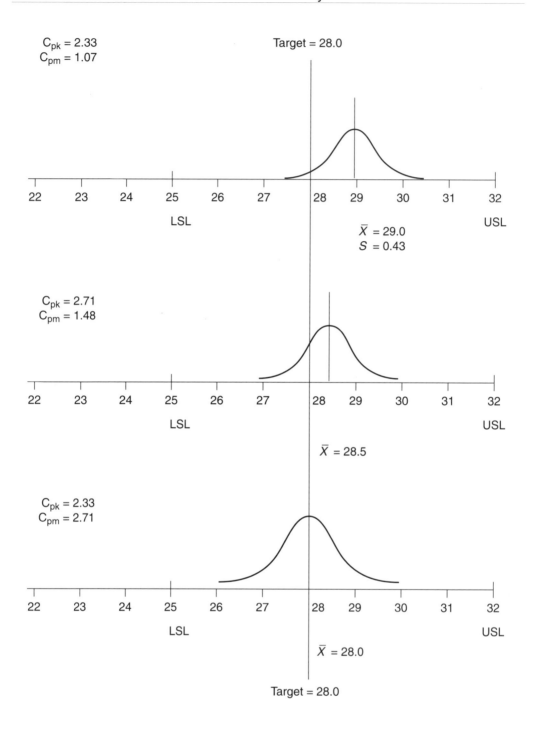

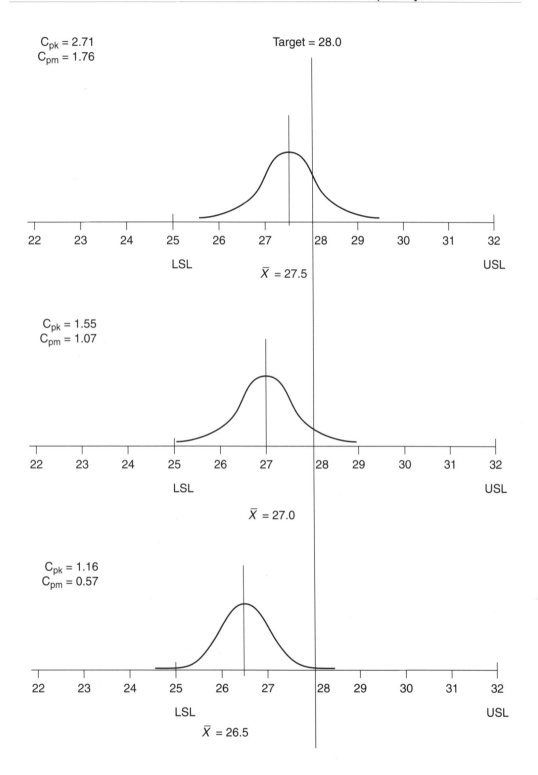

BIBLIOGRAPHY

Bissel, A. F. 1990. How Reliable Is Your Capability Index? *Journal of Applied Statistics* 39:331–340.

Chan, L. K., S. W. Cheng, and F. A. Spiring. *Journal of Quality Technology* 20, no. 3: 162–175.

Chou, Y, D. D. Owen, and S. A. Borrego. 1990. Lower Confidence Limits on Process Capability Indices. *Journal of Quality Technology* July 22: No. 3 98–100.

Kushler, R. H., and P. Hurley. 1992. Confidence Bounds for Capability Indices. *Journal of Quality Technology* Oct 24: No. 4 118–195.

p Chart

Control Charts are selected according to the type of data being taken to monitor a process. Quality data or measurements fall into one of the following two broad categories:

1. *Variable* data are data obtained in which the initial observation is a measurement that can take on any value within the limits that it can occur and within the limits of our ability to make the measurement. Examples of variables are temperature, length, pressure, and mass.

2. *Attribute* data are data obtained from an initial observation of a *yes* or a *no*. Attribute data are normally associated with the compliance to a requirement. Meeting the requirement yields a *yes* response, and failing a requirement yields a *no* response. Variables data can be converted to attribute data by comparison to a requirement. Examples of attribute data would be determining the percentage of observations that do not conform to a requirement. A frequently used attribute would be the percentage of customers not satisfied with the quality of a service or the percentage of voters voting for a specific candidate.

Once we have decided on the type of data being collected, we can narrow the field of available control charts from which to select. Assuming that the data is attribute in nature, we will have the following four basic types available:

1. *p* chart
2. *np* chart
3. *c* chart
4. *u* chart

These charts can be categorized according to:

1. Sample size: fixed or variable
2. Attribute type: defect or defective

The response to these two inquires determines the specific control chart for attributes we will use.

1. Sample size: We may choose to vary the sample size at some time in the future. Our sample size may result from a 100 percent inspection of units being produced over a

fixed period of time such as a shift. The number of units produced over this period may vary from day to day. We may, on the other hand, elect to inspect a fixed or constant number of units for a specified period of time.

2. Attribute type: Attribute data can be related to the total number of defective units (nonconforming) within a sample or the number of defects (nonconformities) in a sample. For example, if we inspect 50 printed circuit boards and find three missing components, two reversed polarity, and three cracks, we would report eight defects. From this information, we do not know the number of defective boards. All of the eight defects could have been found on one board, or eight defective boards could have had one defect each. A defective is a unit of inspection that does not conform to a quality requirement. A defect is the specific incident that caused the unit of inspection to be defective.

Chart name	Sample size	Attribute type
c chart	Constant	Defect
u chart	May be variable	Defect
np chart	Constant	Defective
p chart	May be variable	Defective

If we can define constant as actually being a variable sample size where the amount of variation is defined as zero, then the c chart and np chart can be eliminated from our field. The p chart and u chart can perform very well in place of the c chart and np chart.

We do not have to vary the sample size for the p chart and u chart, but we may do so if we desire. A p chart with a constant sample size performs exactly as an np chart, and a u chart with a constant sample size performs exactly as a c chart.

This module will discuss the application and principles of the p chart.

Control charts for variables are based on the normal distribution as a statistical model. Attribute data is binary in nature in that the initial response of an inspection activity yields only two possible outcomes: The unit of inspection either conforms to a requirement or it does not—*yes* or *no*. The response is discrete. The distribution on which the p chart is based is the binomial distribution.

Where the sample size n is relatively small in proportion to the population N such that n/N is less than 0.10 and the product of the proportion defective and the sample size is greater than five ($np > 5$), the binomial distribution approximates the normal distribution. The average of the binomial distribution is np, and the standard deviation is $\sqrt{npq}$, where n = sample size, p = proportion nonconforming, and $q = 1 - p$.

As with all attribute control charts, the construction and use of a p chart has the following operations or steps:

1. Collection of historical data to characterize the process
2. Calculation of a location statistic (an average)
3. Determination of statistical upper and lower control limits based on ±3 standard deviations
4. Construction of the control chart and plotting of historical data
5. Continuation of monitoring the data, looking for signs of a process change in the future

The following example will illustrate these operations.

Step 1. Collect historical data.

The sample size for a p chart should be sufficiently large as to provide at least one or two defective when a sample is taken. If some prior knowledge is available regarding the expected average proportion defective, the appropriate sample size to provide a positive lower control limit (LCL) can be determined by

$$\text{Minimum sample size } n = \frac{9(1-p)}{p}$$

where: p = the proportion defective for the process

It is suggested that the actual sample size be 10 percent greater than the calculated minimum to provide for a positive LCL.

The number of samples used to characterize the process k is typically a minimum of 25 (more may be used). The 25 minimum will give us a level of confidence that the estimate fairly represents the true nature of the process.

A record of the specific defects observed should be maintained so that appropriate corrective action can be taken and the use of Pareto analysis can be accomplished. A more detailed discussion on Pareto analysis can be found in the module entitled *Pareto Analysis.*

The actual p chart does not depend on the frequency of defects, as it is only concerned with the proportion of defective found. The c chart and u chart focus on defects. See the module entitled *u Chart* for more discussion on the u chart.

The following example of a p chart is based on the responses from a customer-satisfaction survey taken every few days. The sample size is 50. A survey will be sent to each of 50 randomly selected customers. The survey consists of 12 questions with the choice of responses rated from zero to four. The scale rating defines accordingly:

0 = poor

1 = below average

2 = average

3 = above average

4 = excellent

Any response of a zero or a one to any of the 12 questions will be counted as a nonconformity (defect), and the survey will be deemed unsatisfactory (defective unit). This p chart will be based on the proportion of unsatisfied customers.

Data:

n = sample size

np = number of defective

p = proportion defective

Note: For this example, only 16 samples will be used.

Sample #1	Sample #2	Sample #3	Sample #4
8/12/95	**8/15/95**	**8/18/95**	**8/21/95**
$n = 50$	$n = 50$	$n = 50$	$n = 50$
$np = 4$	$np = 2$	$np = 6$	$np = 3$
$p = 0.08$	$p = 0.04$	$p = 0.12$	$p = 0.06$

Sample #5	Sample #6	Sample #7	Sample #8
8/23/95	**8/25/95**	**8/28/95**	**9/2/95**
$n = 50$	$n = 50$	$n = 50$	$n = 50$
$np = 2$	$np = 0$	$np = 5$	$np = 3$
$p = 0.04$	$p = 0.00$	$p = 0.10$	$p = 0.06$

Sample #9	Sample #10	Sample #11	Sample #12
9/5/95	**9/8/95**	**9/11/95**	**9/14/95**
$n = 50$	$n = 50$	$n = 50$	$n = 50$
$np = 4$	$np = 7$	$np = 1$	$np = 4$
$p = 0.08$	$p = 0.14$	$p = 0.02$	$p = 0.08$

Sample #13	Sample #14	Sample #15	Sample #16
9/17/95	**9/20/95**	**9/23/95**	**9/25/95**
$n = 50$	$n = 50$	$n = 50$	$n = 50$
$np = 3$	$np = 6$	$np = 3$	$np = 4$
$p = 0.06$	$p = 0.12$	$p = 0.06$	$p = 0.08$

Record the data on the control chart form.

Step 2. Calculate the location statistic.

The location statistic for the p chart will be the average proportion defective $\bar{p}$ and is determined by dividing the total number of defective found by the total units inspected for the data collected during the historical period.

$$\bar{p} = \frac{4+2+6+3+\cdots+4}{50+50+50+50+\cdots+50} \qquad \bar{p} = \frac{57}{800} \qquad \bar{p} = 0.071$$

Note that $\bar{p}$ is reported one more decimal place than the individual sample proportion defective. This is done to minimize the probability of a single point falling exactly on the average.

Step 3. Determine the control limits based on ±3 standard deviations.

The standard deviation for the binomial distribution where defective are expressed as a proportion is

$$\sqrt{\frac{\bar{p}(1-\bar{p})}{n}}$$

where: $\bar{p}$ = proportional defective
n = sample size

The control limits will be based on the average proportion defective ±3 standard deviations:

$$\bar{p} \pm 3 \sqrt{\frac{\bar{p}(1-\bar{p})}{n}}$$

For this case:

$$0.071 \pm 3 \sqrt{\frac{0.071(1-0.071)}{50}}$$

$$0.071 \pm 0.109$$

The upper control limit (UCL) will be $0.071 + 0.109 = 0.180$.

The LCL will be $0.071 - 0.109 = -0.038$ (when a negative control limit results, default to 0.000).

Based on the historical characterization, we expect to find 99.7 percent of the time between 0.00 and 0.180 proportion defective when selecting a sample of $n = 50$.

The 99.7 percent probability arises from the use of ±3 standard deviations for our statistical limits.

Knowing now that the expected proportion defective is 0.071, we could calculate the proper sample size so as to have a real, positive LCL from the relationship

$$n = \frac{9(1-0.071)}{0.071}$$

$$n = 117.8 = 118$$

If we had taken a sample size of 118, the expected LCL would have been 0.000072.

As a practical matter, the sample size should be somewhat larger than the absolute minimum of 118 (say 125), in which case the LCL would have been 0.002.

Step 4. Construct the control chart, sketch in the average and control limits, and plot the data.

Traditionally, averages are drawn as solid lines (——————), and control limits are drawn as broken lines (- - - - - - - - - -). A vertical plotting scale should be chosen such that part of the chart will be available for those future points that might fall outside of the control limits. A vertical wavy line has been drawn to separate the *history* from the *future* data. Only the process average proportional defective $\bar{p}$ is drawn out into the *future* area of the control chart.

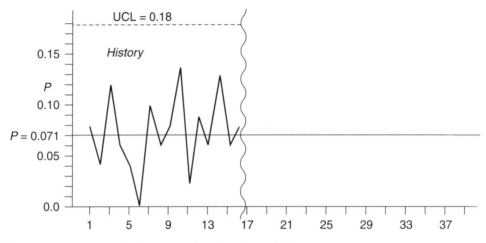

p Chart with Historical Characterization Data Only.

Step 5. Continue to monitor the process, looking for indications that a change has occurred.

Data will be collected as we go into the future. Evidence supporting the conclusion that a change has taken place will be in the form of statistically rare patterns referred to as SPC detection rules. These detection rules have the following two characteristics:

1. All of these SPC patterns or *rules* are statistically rare for a normal, stable process.
2. Most of these SPC detection rules indicate a direction for the process change.

  Rule 1: A lack of control or process change is indicated whenever a single point falls outside of the UCL or LCL.

  Rule 2: A lack of control or process change is indicated whenever seven consecutive points fall on the same side of the process average.

  Rule 3: A lack of control or process change is indicated whenever seven consecutive points are steadily increasing or decreasing.

Anytime a process change is indicated as evident by violation of one of the SPC detection rules, an investigation should be undertaken to identify an assignable cause. There is a small probability that one of these rules will be invoked when, in fact, no change in the process has occurred. This probability, however, is so small that we assume the violation truly represents a process change.

We will continue to monitor the example process to see if a change has occurred relative to our historical period from 8/12 to 9/25. The p chart does allow for a change in the sample size. To demonstrate the effects on our chart, we will change the sample size on the future data from 50 to 150.

A change in sample size will result in a change in the value for three standard deviations, and since we add and subtract three standard deviations to and from the historical process average proportional defective to determine the UCL and LCL, we will expect to see a change in these limits. The larger the sample size, the closer the control limits will be to the average. This decrease in standard deviation as we increase the sample size increases our ability to detect smaller process changes by increasing the sample size.

The new control limits using a sample size of 150 and the historical average proportion of 0.071 will be given by

$$0.071 \pm 3 \sqrt{\frac{0.071(1-0.071)}{150}}$$

$$0.071 \pm 30.063$$

UCL = 0.134

LCL = 0.008

We will sketch in the new control limits and extend the historical average, 0.071. We do not recalculate the average using the future data point, because we want to detect a process change relative to the historical average.

After sketching in the new control limits for $n = 150$ and extending the average, we plot the future data points, looking for a violation of one of the three SPC detection rules.

Future data:

Sample #17	Sample #18	Sample #19	Sample #20
9/28/95	**10/1/95**	**10/4/95**	**10/7/95**
$n = 150$	$n = 150$	$n = 150$	$n = 150$
$np = 18$	$np = 4$	$np = 9$	$np = 15$
$p = 0.12$	$p = 0.03$	$p = 0.06$	$p = 0.10$

Sample #21	Sample #22	Sample #23	Sample #24
10/10/95	**10/13/95**	**10/17/95**	**10/20/95**
$n = 150$	$n = 150$	$n = 150$	$n = 150$
$np = 14$	$np = 14$	$np = 0$	$np = 3$
$p = 0.09$	$p = 0.09$	$p = 0.00$	$p = 0.02$

Notice that rule one has been violated on sample #23, suggesting that the process is performing better than expected (proportional defective has decreased) relative to the historical average of 0.071.

Completed *p* chart:

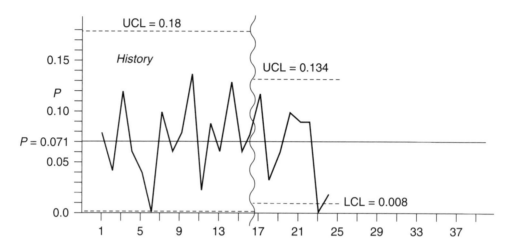

Using the *p* Chart when the proportion defective is extremely low (high-yield processes):

The effects of a very low proportional defective give rise to the problem that very large sample sizes are required and, more critically, that the application of the binomial distribution is not appropriate. If a process were to have a nonconformance rate of, for example 300 ppm and a sample size of 250 were taken, the *np* value would be only $np = (250)$ $(.0003) = 0.075$. This *np* value is significantly below the normal requirement of greater than five; therefore, the binomial would not be appropriate as with the traditional *p* chart. While a *p* chart could be constructed, the resulting control chart could lead to erroneous conclusions. Other problems could also arise. Consider the following case:

$$CL = \overline{P} = 0.0003$$

$$LCL = \overline{P} - 3 \sqrt{\frac{\overline{P}(1-\overline{P})}{n}} \qquad LCL = 0.0003 - 0.00328$$

$$LCL = 0.00$$

$$UCL = \overline{P} - 3 \sqrt{\frac{\overline{P}(1-\overline{P})}{n}} \qquad UCL = 0.0003 + 0.00328$$

$$UCL = 0.00358$$

In this case, if we had one defective in 250, the control chart would be *out of control* with a calculated *p* value of 0.0040, which exceeds the UCL of 0.00358. This process would always be out of control.

The application of a p chart for this case is awkward and is not appropriate for two reasons:

1. No LCL is available.
2. Only samples yielding zero defective will maintain an in-control condition, and we will soon violate the rule of seven points in a row below the average.

Another drawback in using the p chart with very low proportional defective is the slow response to a large shift in the process average.

Consider the previous example, where the process average defective was 300 ppm (0.0003), the sample size is $n = 250$, and the UCL is 0.00358. In this case, finding one defective in a sample of $n = 250$ leads to an out-of-control condition. If the process proportion defective were to double to 600 ppm, the probability of getting one defective in a sample of 250 is

$$P_{x=1} = \left[\frac{n}{x!(n-x)} \right] P^x Q^{n-x} \qquad P_x = (250)(0.0006)(0.9994)^{249} = 0.1292$$

This means that on average 1/0.1292, or about 8, samples must be inspected before a defective is found and the process is judged to be out of control.

Rather than focus on the number of defective found, an alternative approach would be to focus on the cumulative count of nondefective items until a defective item is found. The probability of the nth item being defective is given by

$$P_n = (1-p)^{n-1} p \qquad n = 1,2,3,4,\ldots \text{ (a geometric distribution)}$$

$$\bar{n} = \frac{1}{p} \quad (\bar{n} = \text{ the average run length given a proportion defective of } p.)$$

By summation of the geometric series and expanding:

$$P_n = 1 - \left[1 - np + \frac{n(n-1)}{2!} p^2 + \cdots \right]$$

When n is large, we may substitute $\bar{n}$ for $1/p$ and

$$P_n = 1 - e^{\left(\frac{-n}{\bar{n}} \right)}$$

Solving for n:

$$n = \bar{n} \ln[1 - P_n]$$

By substituting 0.5 for P_n in this equation, we can solve for the median n as follows:

$$\tilde{n} = -\bar{n} \ln[1 - 0.50]$$

$$\tilde{n} = 0.693\bar{n}$$

Control chart for cumulative counts of conforming items, the Σp control chart:

$$\text{Average} = 0.693\bar{n}$$

$$\text{LCL} = -\bar{n}\,\ln\left[1-\left(\frac{\alpha}{2}\right)\right]$$

and since α is small

$$\text{LCL} = \left(\frac{\alpha}{2}\right)\bar{n}$$

$$\text{UCL} = -\bar{n}\,\ln\left[1-\left(1-\frac{\alpha}{2}\right)\right]$$

$$\text{UCL} = -\bar{n}\,\ln\left(\frac{\alpha}{2}\right)$$

If α is set to 0.0027 as with traditional Shewhart control charts, the control limits become

$$\text{LCL} = 0.00135\,\bar{n}$$

$$\text{UCL} = 6.608[1]$$

Sample number	Number of units tested	Cumulative count when count stops	Number of defective units since previous stop	Cumulative count to defective
1	500	500	0	
2	500	1000	0	
3	240	1240	1	1240 (A)
4	800	800	0	
5	1000	1800	0	
6	500	2300	0	
7	900	3200	0	
8	488	3688	1	3688 (B)
9	750	750	0	
10	1000	1750	0	
11	850	2550	0	
12	1000	3550	0	
13	630	4180	1	4180 (C)
14	500	500	0	
15	1000	1500	0	
16	770	2270	0	
17	380	2650	1	2650 (D)

[1]Goh, T. N. A Control Chart for Very High Yield Processes. *Quality Assurance* 13, no. 1 (March 1987).

Example case study:

Light-emitting diodes (LEDs) are 100 percent tested on line in lots that vary from 500 to 1000.

The following test data have been collected:

The average sample size when a defective unit was detected is

$$\bar{n} = \frac{1240 + 3688 + 4180 + 2650}{4} = 2940$$

$$\bar{p} = \frac{1}{2940} = 0.00034 = 340 \text{ ppm}$$

$$\alpha = 0.0027$$

$$CL = 0.693\bar{n} = (0.693)(2940) = 2037$$

$$LCL = 0.003135\bar{n} = (0.00135)(2940) = 4$$

$$UCL = 6.608\bar{n} = (6.608)(2940) = 19,428$$

The cumulative sample sizes are plotted on the vertical axis using a log scale, and the sample number (normal linear scale) is plotted on the horizontal scale. When a point is plotted that represents the detection of a defective, the plotted point is not connected to the next point. The cumulative sample size plotting is restarted. All cumulative points are plotted.

Exceeding the UCL is indicative of the process performing better than expected compared to the historical performance, conversely, points below the LCL indicate a decrease in process performance relative to the past performance.

If multiple cycle log paper is not available, the log of all of the numerical data may be plotted on normal rectangular coordinate graph paper. In this case, the UCL would equal log 19,400 or 4.28, the average (CL) would equal 3.31, and the LCL would equal 0.60.

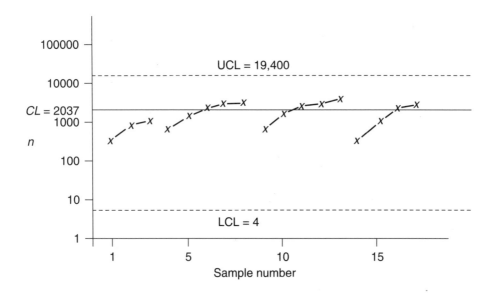

Exercise:

Construct a Σp chart using the following data:
  Set α = 0.05

Sample number	Number of units tested	Cumulative count when count stops	Number of defective units since previous stop	Cumulative count to defective
1	500	500	0	
2	300	800	1	800
3	175	175	1	175
4	600	600	0	
5	500	1100	0	
6	300	1400	0	
7	700	2100	0	
8	1200	3300	1	3300
9	600	600	0	
10	550	1150	1	1150
11	750	750	0	
12	800	1550	0	
13	358	1908	1	1908

EXPONENTIAL MODEL

If we consider the process to be mature and in a steady state of failure rate (but very low) the failures that do occur may be modeled using the traditional reliability model based on the exponential distribution. This model is used extensively in reliability predictions.

$$R = e^{-\left(\frac{n}{\theta}\right)}$$

where: θ = mean time before failure or, for this application, average number of units accumulated when a defective is encountered.
  n = specific sample size for a given level of reliability
  R = the probability of reaching a count of n parts without a failure

By solving this relationship for n at specified levels of reliability, a control chart based on the cumulative sample size to the point of a single failure occurs. If we select an alpha risk of $\alpha = 0.05$, the control chart limits and average change as explained in the following paragraphs.

The LCL is equal to the sample size that represents a reliability of 0.95, or 95 percent probability that this number of units will be found defect free. The UCL will represent the sample size that there will be a 5 percent probability of reaching with no defect being found.

From the exponential relationship for determining reliability, we determine the sample size n for the appropriate reliability.

$$R = e^{-\left(\frac{n}{\theta}\right)}$$

$$\ln R = -\left(\frac{n}{\theta}\right)$$

$$n = \theta \ln R$$

$\theta =$ Mean number of units before failure or average sample size to failure

The centerline for this chart will be the median, which is determined by the point at which the reliability is 50 percent.

Example:
Hypodermic needles are manufactured in a continuous process. The needles are 100 percent inspected by an automatic visual inspection device. The following 18 samples represent the number of units produced before a defective unit is found.

Step 1. Collect historical data:

Date	Sample size n	Date	Sample size n
11/1	120,000	11/24	160,000
11/4	89,000	11/28	76,000
11/7	145,000	11/30	140,000
11/9	78,000	12/1	110,000
11/11	138,000	12/3	98,000
11/13	100,000	12/6	155,000
11/16	180,000	12/9	80,000
11/19	90,000	12/12	105,000
11/22	130,000	12/15	93,000

Estimated MTBF, θ = average sample size n = 115,944

Step 2. Select an alpha risk α and determine the appropriate reliability. Calculate limits.
For this example, the risk will be set at 5 percent ($\alpha = 0.05$).
The LCL will be based on a reliability of 95 percent.

$$LCL = -\theta \ln 0.95 \qquad LCL = -115,944 \ln 0.95 \qquad LCL = (-115,944)(-0.0513)$$

$$LCL = 5948$$

Centerline = Median. Use a reliability of 50 percent for the median.

$$\text{Median} = -\theta \ln 0.50 \quad \text{Median} = -115,944 \ln 0.50 \quad \text{Median} = (-115,944)(-0.693)$$

$$\text{Median} = 80,349$$

Note: The centerline is optional and is not considered in the traditional SPC detection rules, which use the average and the assumption of a normal or approximate normal distribution.

The UCL will be based on a reliability of 5 percent.

$$UCL = -\theta \ln 0.05 \qquad UCL = -115{,}944 \qquad UCL = (-115{,}944)(-3.00)$$
$$UCL = 347{,}832.$$

Step 3. Draw the control chart with limits, and plot the data.

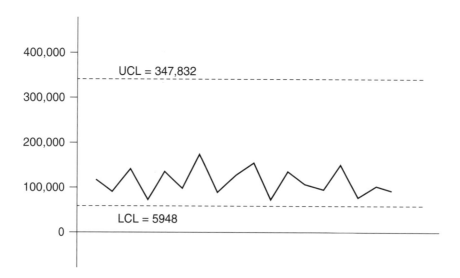

Note on interpretation: The only rule for detecting a process change is exceeding the UCL or LCL. Exceeding the upper limit indicated that the process is performing *better* than the historical period. Having a point below the LCL indicates the process is performing *poorer* than the historical period.

A UCL of $n = 347{,}832$ is the same as having a failure rate λ of 1/347,832, or 0.00000287, which is equivalent to a process defective rate of 2.87 ppm. Exceeding this limit would imply that the process had *improved*.

An LCL of $n = 5948$ is the same as having a failure rate of 1/5948, or 0.000168, which is equivalent to a process defective rate of 168 ppm. Having points below this limit implies that the process has *deteriorated*.

Step 4. Continue to monitor the process looking for signs of a change.

Two indicators will be used to detect a process change:

1. Seven consecutive points increasing or decreasing
2. A single point outside the UCL or LCL

Date	Sample size n
12/18	189,000
12/20	150,000
12/23	220,000
12/26	160,000
12/28	290,000
12/30	360,000

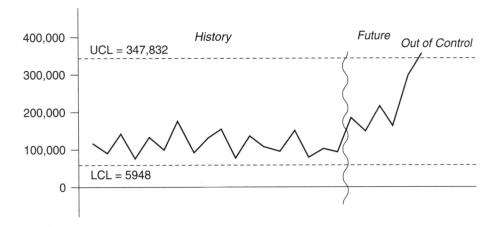

The out-of-control point gives us supporting evidence that the process has change and for the better.

DATA TRANSFORMATION

Lloyd S. Nelson[2] has developed a simple method of control charting processes where the non-conforming rate is extremely low (that is, ppm). This method is based on the fact that the Weibull distribution with a slope parameter (β) of 3.6 has a skewness of zero and a kurtosis of 2.72. This compares favorably to a normal distribution where the skewness is zero and the kurtosis is 3.00.

Therefore, the required transformation is simply to raise the cumulative number of successful samples required to obtain a defective to the 1/3.6 power. This transformed value is then treated as a normally distributed variable and is then monitored using the traditional individual/moving range control chart.

[2]Nelson, L. S. A Control Chart for Parts-Per-Million Nonconforming Items. *Journal of Quality Technology* 26, no. 3 (July 1994).

Example:

A machine is producing 3.5-inch formatted diskettes that are 100 percent certified by writing and the reading data. Each time a diskette fails it is rejected and the count for the interval to the last failure is recorded. The objective is to establish a control chart based on the failure data.

Data:

Sample number	Cumulative count to failure, X	Transformed Data, $Y = X^{0.2777}$	Moving range
1	19,400	15.5	—
2	22,780	16.2	0.7
3	14,000	14.2	2.0
4	12,700	13.8	0.4
5	18,600	15.3	1.5
6	25,000	16.7	1.4
7	15,100	14.5	2.2

$$\text{Process average } \bar{Y} = \frac{15.5 + 16.2 + \ldots + 14.5}{7} = 15.17$$

$$\text{Average moving range } \overline{MR} = \frac{0.7 + 2.0 + \ldots + 2.2}{6} = 1.37$$

$$\text{UCL} = \bar{Y} + 2.66\,\overline{MR} = 15.17 + 1.37 = 16.54$$

$$\text{LCL} = \bar{Y} - 2.66\,\overline{MR} = 15.17 - 1.37 = 13.80$$

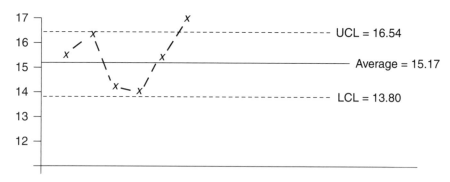

Process Out of control on Sample 6.

BIBLIOGRAPHY

Besterfield, D. H. 1994. *Quality Control.* 4th edition. Englewood Cliffs, NJ: Prentice Hall.

Grant, E. L., and R. S. Leavenworth. 1996, *Statistical Quality Control.* 7th edition. New York: McGraw-Hill.

Montgomery, D. C. 1996. *Introduction to Statistical Quality Control.* 3rd edition. New York: John Wiley and Sons.

Wheeler D. J., and D. S. Chambers. 1992. *Understanding Statistical Process Control.* 2nd edition. Knoxville, TN: SPC Press.

REGRESSION AND CORRELATION

LINEAR LEAST SQUARE FIT

When we attempt to fit a *best* straight line to a collection of paired data, we essentially try to minimize the distance between the line of fit and the data. The formal mathematical procedure that accomplishes the same objective is the *method of least squares.*

This method involves a simple linear relationship for two variables: an independent variable x and a dependent variable y. That is, the value of x is given (independent), and the value of y (dependent) depends on the value of x. This relationship can be written in the general form of

$$\hat{Y} = \hat{\beta}_0 + \hat{\beta}_1 X$$

The values of $\hat{\beta}_0$ and $\hat{\beta}_1$ are such that the sum of the square of the differences between the calculated value of Y and the observed value of Y is minimized.

The least square estimators $\hat{\beta}_0$ and $\hat{\beta}_1$ are calculated as follows

$$\hat{\beta}_0 = \bar{Y} - \hat{\beta}_1 \bar{X} \quad \text{and} \quad \hat{\beta}_1 = \frac{SS_{xy}}{SS_x}$$

where: $SS_x = \Sigma(X_i - \bar{X})^2 = \Sigma X_i^2 - \dfrac{(\Sigma X_i)^2}{n}$

$SS_{xy} = \Sigma(X_i - \bar{X})(Y_i - \bar{Y}) = \Sigma X_i Y_i - \dfrac{(\Sigma X_i)(\Sigma Y_i)}{n}$

Once $\hat{\beta}_0$ and $\hat{\beta}_1$ have been computed, we substitute their values into the equation of a line to obtain the following least squares prediction equation:

$$\hat{Y} = \hat{\beta}_0 + \hat{\beta}_1 X$$

It should be noted that rounding errors can significantly affect the answer you obtain in calculating the SS_x and SS_{xy}. It is recommended that calculations for all sum of squares (SS) be carried to six or more significant figures in the calculation.

Example:
Data have been collected for the life of a deburring tool in hours as a function of the speed in RPMs. Using the least squares method, what is the linear equation relating tool life to speed?

Data table:

Y_i Tool life, hours	X_i Speed, RPM	X_i^2	X_iY_i	Y_i^2
180	1350	1,822,500	243,000	32,400
220	1230	1,512,900	270,600	48,400
130	1380	1,904,400	179,400	16,900
110	1440	2,073,600	158,400	12,100
280	1200	1,440,000	336,000	78,400
200	1300	1,690,000	260,000	40,000
50	1500	2,250,000	75,000	2,500
40	1540	2,371,600	61,600	1,600
Sums 1,210	10,940	15,065,000	1,584,000	232,300

$$SS_x = \Sigma X_i^2 - \frac{(\Sigma X_i)^2}{n} = 15,065,000 - \frac{(10,940)^2}{8} = 104,550$$

$$SS_{xy} = \Sigma X_i Y_i - \frac{(\Sigma X_i)(\Sigma Y_i)}{n} = 1,584,000 - \frac{(10,940)(1210)}{8} = -70,675$$

$$\bar{Y} = \frac{\Sigma Y_i}{n} = \frac{1210}{8} = 151.25 \qquad \bar{X} = \frac{\Sigma X_i}{n} = \frac{10,940}{8} = 1367.5$$

Hence,

$$\hat{\beta}_1 = \frac{SS_{xy}}{SS_x} \qquad \hat{\beta}_1 = \frac{-70,675}{104,550} = -0.676$$

$$\hat{\beta}_0 = \bar{Y} - \hat{\beta}_x \bar{X} \qquad \hat{\beta}_0 = 151.25 - (-0.676)(1367.5) = 1075.67$$

The least squares linear regression equation is

$$\hat{Y} = \hat{\beta}_0 - \hat{\beta}_1 X \qquad \hat{Y} = 1075.67 - 0.676(X)$$

What is the estimated expected tool life at 1400 RPMs?

$$\hat{Y} = 1075.67 - 0.676(1400) = 129.27$$

CONFIDENCE FOR THE ESTIMATED EXPECTED VALUE

Having calculated the expected value for Y (tool life), where X is the tool operating speed, we may determine the confidence limit for the result using the following relationship:

$$\hat{Y} \pm t_{\frac{\alpha}{2}, n-2} S \sqrt{\frac{1}{n} + \frac{(X_p - \bar{X})^2}{SS_x}}$$

where: $\hat{Y}$ = the predicted value of the dependent variable
X_p = the value of the independent variable
n = sample size
SS_x = sum of squares for X

$$SS_x = \Sigma X_i^2 - \frac{(\Sigma X_i)^2}{n}$$

$\overline{X}$ = average of independent variables
S = sample standard deviation
$t_{\alpha/2, n-2}$ = t value for $1 - \alpha$ confidence and $n - 2$ degrees of freedom

Example:
What is the 95 percent confidence interval for the estimated tool life with a speed of 1400 RPMs using the least squares model of $\hat{Y}$ = 1075.67 − 0.676(X)?

$$\hat{Y} = 1075.67 - 0.676(1400) = 129.27$$

$$C = 0.95; \text{ therefore } \alpha/2 = 0.025$$

$$n = 8 \qquad n - 2 = 6 \qquad t_{\frac{\alpha}{2}, n-2} = t_{0.025,6} = 2.447$$

$$S = \text{standard deviation of } Y \text{ for a given value of } X$$

$$S = \sqrt{SS_y - \hat{\beta}_1 SS_{xy}}$$

$$SS_y = \Sigma Y_i^2 - \frac{(\Sigma Y_i)^2}{n} = 232,300 - \frac{(1210)^2}{8} = 49,287.5$$

$$S_{xy} = -70,675 \quad \text{and} \quad \hat{\beta}_1 = \frac{-70,675}{104,550} = -0.676$$

$$S = \sqrt{49,287.5 - (-0.676)(-70,675)} = 38.87$$

$$\hat{Y} \pm t_{\frac{\alpha}{2}, n-2} S \sqrt{\frac{1}{n} + \frac{(X_p - \overline{X})^2}{SS_x}}$$

$$129.27 \pm (2.447)(38.87)\sqrt{\frac{1}{8} + \frac{(1400 - 1367.5)^2}{104,550}} = 129.27 \pm 34.96$$

We are 95 percent confident that based on the least squares model of $\hat{Y}$ = 1075.67 − 0.676(X) with a speed of 1400 RPMs, the tool life will be between 94.31 hours and 164.23 hours.

ASSESSING THE MODEL: DRAWING INFERENCES ABOUT $\hat{\beta}_1$

One method for assessing the linear model is to determine whether a linear relationship actually exists between the two variables. If the true value for β_1 had been equal to zero, no linear relationship would have existed, and the linear model would have been inappropriate.

In general β_1 is unknown, because it is a population parameter. We can, however, use the sample slope $\hat{\beta}_1$ to make inferences about the population slope β_1.

Testing β_1

If we want to test the slope, in order to determine whether some linear relationship exists between X and Y, we test the following hypothesis:

$H_A: \beta_1 \neq 0$

If we want to test for a positive or negative relationship the alternative hypothesis H_A would be $H_A: \beta_1 > 0$ and $H_A: \beta_1 < 0$, respectively.

In all three cases, the null hypothesis H_0 is that $H_0: \beta_1 = 0$.

The test statistic for β_1 is

$$t = \frac{\hat{\beta}_1 - \beta_1}{S_{\hat{\beta}_1}}$$

where: $S_{\hat{\beta}_1}$ = the standard deviation of $\hat{\beta}_1$ (also called the standard error of $\hat{\beta}_1$) and

$$S_{\hat{\beta}_1} = \frac{S_\epsilon}{\sqrt{SS_x}}$$

S_ϵ = standard error of the estimate

$$S_\epsilon = \sqrt{\frac{SS_y - \frac{(SS)^2{}_{xy}}{SS_x}}{n-1}}$$

From the previous example

$$SS_y = 49,287.5$$

$$SS_x = 104,550$$

$$SS_{xy} = -70,675$$

$$S_\epsilon = \sqrt{\frac{49,287.5 - \frac{(-70,675)^2}{104,550}}{8-2}}$$

$$S_\epsilon = 15.87$$

$$S_{\hat{\beta}_1} = \frac{S_\epsilon}{\sqrt{SS_x}} = \frac{15.87}{\sqrt{104,550}} = 0.049$$

Assuming that the error variable ϵ is normally distributed, the test statistic t follows the student t = distribution with $n - 2$ degrees of freedom.

Can we conclude at the 1 percent level of significance that tool wear and speed are linearly related?

We proceed as follows:

H_0: $\beta_1 = 0$

H_A: $\beta_1 \neq 0$

Decision rule:

Reject the null hypothesis H_0 if

$$|t| > t_{\alpha/2,n-1} = t_{0.001,7} = 3.499$$

$$t = \frac{\hat{\beta}_1 - \beta_1}{S_{\hat{\beta}_1}} \qquad t = \frac{-0.676 - 0}{0.049} = -13.795 \qquad |-13.795| > 3,499$$

We reject the null hypothesis and conclude that β_1 is not equal to zero. A linear correlation exists.

Measuring the Strength of the Linear Relationship

In testing β_1, the only question addressed was whether or not there is enough evidence to allow us to conclude that a linear relationship exists. In many cases, it is useful to measure the strength of the linear relationship. We can accomplish this by calculating the coefficient of correlation, ρ.

Its range is

$$-1 \leq \rho \leq +1$$

Because ρ is a population parameter, we must estimate from the sample data. The sample coefficient of correlation is abbreviated r and is defined as follows:

$$r = \frac{SS_{xy}}{\sqrt{(SS_x)(SS_y)}}$$

$$r = \frac{-70,675}{\sqrt{(104,550)(49,287.5)}}$$

$$r = -0.985$$

Testing the Coefficient of Correlation

If $\rho = 0$, the values of x and y are uncorrelated, and the linear model is not appropriate. We can test to determine if x and y are correlated by testing the following hypothesis:

H_0: $\rho = 0$

H_A: $\rho \neq 0$

The test statistic for ρ is

$$t = \frac{r - \rho}{S_r}$$

where: $S_r = \sqrt{\dfrac{1 - r^2}{n - 2}}$

This test statistic is only valid when testing $\rho = 0$. Consequently, the test statistic is simplified to

$$t = \frac{r}{S_r}$$

The complete test follows:

H_0: $\rho = 0$

H_A: $\rho \neq 0$

Test statistic:

$$t = \frac{r}{S_r}$$

Decision rule:

Reject the null hypothesis if $|t| > t_{\alpha/2,n-1} = t_{0.001,7} = 3.499$ (assuming that $\alpha = 0.01$).

Value of the test statistic:

Since $r = -0.985$

$$S_r = \sqrt{\frac{1 - r^2}{n - 2}}$$

$$S_r = \sqrt{\frac{1 - 0.970}{8 - 2}} = 0.0707$$

$$t = \frac{r}{S_r}$$

$$t = \frac{-0.985}{0.0707} = -13.932$$

Conclusion:

$|-13.932| > 3.499$; therefore, we reject H_0 and conclude that $\rho \neq 0$.

There is sufficient evidence to conclude that x and y are correlated and, as a result, a linear relationship exists between them.

The results are identical for both the tests of β_1 and ρ. Because in both cases we are testing to determine whether a linear association exists, the results must agree. In practice, only one of the tests is required to establish whether or not a linear relationship exists.

BIBLIOGRAPHY

Petruccelli, J. D., B. Nandram, and M. Chen. 1999. *Applied Statistics for Engineers and Scientists.* Upper Saddle River, NJ: Prentice Hall.

Sternstein, M. 1996. *Statistics.* Hauppauge, NY: Barron's Educational Series.

Walpole, R. E., and R. H. Myers. 1993. *Probability and Statistics for Engineers and Scientists.* 5th edition. Englewood Cliffs, NJ: Prentice Hall.

RELIABILITY

Reliability can be defined as the *probability* that a product, system, service, component, and so on will *perform* its intended function adequately for a specified period of *time,* operating in a defined *operating environment* without failure. Components of this definition are explained as follows:

- Probability: quantitative measure, likelihood of mission success
- Intended function: for example, to light, cut, rotate, or heat
- Satisfactory: perform according to a specification, degree of compliance
- Specific period of time: minutes, days, months, or number of cycles
- Specified conditions: for example, temperature, speed, or pressure

Stated another way, reliability is

- Probability of success
- Durability
- Dependability
- Quality over time
- Availability to perform a function

Typical reliability statements include: "This car is under warranty for 40,000 miles or 3 years, whichever comes first," or "This mower has a lifetime guarantee."

Concept:

Quality = Does it perform its intended function?

Reliability = How long will it continue to do so?

The reliability or probability of mission success can be determined for individual components of a complex system, or they can be determined for the entire system.

Reliability can be described mathematically by one of several distribution functions, including but not limited to:

1. Normal distribution
2. Exponential distribution
3. Weibull distribution

The concept of reliability requires an understanding of the following terms and relationships:

Failure: Not performing as required. For example, a successful push-up is defined as lying in an extended prone position and using your arms to fully raise your body vertically off the floor while keeping the length of your body rigid and your toes on the floor.

Failure rate, λ: The number of failures that occur in a number of cycles or time period. For example, there is one fatality per million due to parachutes failing to open, or there are six car wrecks every seven months on interstate 95.

Failure rate is determined by dividing the total number of failures by the total time or cycles accumulated for the failures.

Five motors are run on a test stand until they all fail. Given the following data, calculate the failure rate:

Motor number	Hours at failure
1	600
2	1000
3	850
4	1100
5	960

$$\text{Failure rate, } \lambda = \frac{\text{Number of failures}}{\text{Total test time}} = \frac{5}{4510} = 0.0011$$

Mean time before failure (MTBF), θ: The average time or cycles a unit runs before failure. MTBF can be written as a function of failure rate, $1/\lambda$.

For the previous example

$$\text{MTBF} = \frac{4510}{5} = 902 \text{ hours}$$

The failure rate for most things changes over time. The relationship of the failure rate with respect to time is depicted by the *Bathtub curve.*

Consider the following data giving the failure rates for an automatic dishwasher. The failure rates were determined on dishwashers at different ages, ranging from one hour old to 20 years old.

At one hour old, the failure rate is relatively high at one failure per 714 units, or a failure rate of 0.0014. After the units have reached an age of one day, the failure rate decreases to one failure in 10,000, or a failure rate of 0.0001. By an age of one month, the failure rate is one in 555,555, or 0.0000018.

During this initial period of the "life" of the dishwasher, the period from one hour to one month, the failure rate steadily decreased. This period in the life, where the failure rate is decreasing, is referred to as the *burn-in* period, or the period of *infant mortality*. Failures in this area are typically due to poor workmanship or defective components and raw materials.

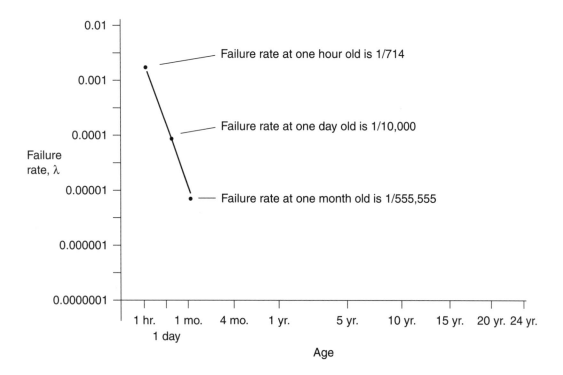

During the period from four months to about 12 years, the failure rate has stabilized to a constant failure rate of slightly less than one failure per million units—1/1,000,000 or 0.000001.

This portion of the life of the dishwasher is called the *useful life* and is characterized by a constant failure rate.

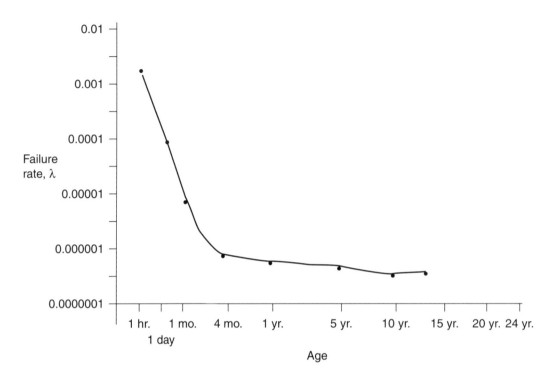

During the period from 12 years to 24 years, the failure rate increases. The group of dishwashers that are 24 years old are failing at a rate of one in 100, or a failure rate of 0.01. The area of increasing failure is the *burn-out* period. It is during this time that individual components are wearing out as a result of of metal fatigue, stress, oxidation, and corrosion.

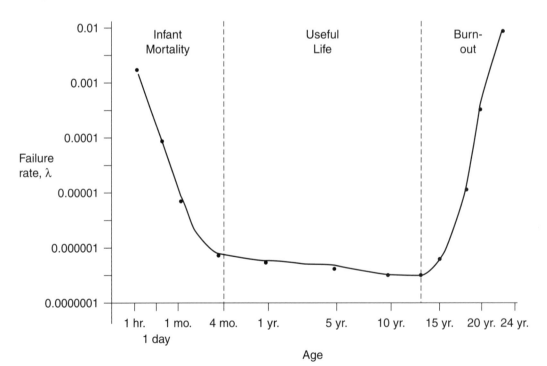

Summary:

The period life of a product where the failure rate is decreasing is defined as the *infant mortality* or *burn-in* phase. Where the failure rate is constant defines the period of *useful life*. Where the failure rate is increasing, the product is in a *burn-out* phase.

When the product is in a state of constant failure, rate predictions regarding reliability may be made using the *exponential distribution or exponential model.*

EXPONENTIAL DISTRIBUTION

Given a *constant* failure rate λ or MTBF θ, the probability of mission success or reliability based on the exponential model is given by

$$R_t = e^{-\left(\frac{t}{\theta}\right)} \qquad \text{or} \qquad R_t = e^{-t\lambda}$$

where: t = time or cycles associated with the reliability
θ = MTBF
λ = failure rate

Example calculation:
The MTBF of an electric motor is 660 hours. What is the probability that the motor will last or run for 400 hours?

$$R_t = e^{-\left(\frac{t}{\theta}\right)} \qquad \begin{array}{l} t = 400 \\ \theta = 660 \end{array}$$

$$R_t = e^{-0.61}$$

$$R_t = 0.54 \text{ or } 54\%$$

There is only a 54 percent chance that the motor will survive 400 hours of running time.

Supplemental problem:

The failure rate during the useful life of a product is 0.00018. What is the probability that the unit will survive 1500 hours of service?

Note: There is an assumption that the failure rate is calculated based on the same time unit as the reliability calculation objective. That is, the failure rate of 0.00018 means that one failure is occurring every 5556 *hours.*

Answer: 76 percent

Determination of Failure Rate, λ

Failure rates are determined by testing units until failure. There are three modes of testing:

1. Complete data: All units are tested until all units fail.
2. Time-censored data (type I testing): Multiple units are started at the same time, and the test is discontinued upon reaching a predetermined time.
3. Failure-censored data (type II testing): Multiple units are started, and the test is discontinued upon reaching a predetermined number of failures.

In all cases, the failure rate λ is determined by the general relationship:

$$\lambda = \frac{\text{number of failures}}{\text{total test time}}$$

1. Complete data

Eight electric motors are tested. All fail at the following times. What is the estimate for the failure rate, λ?

Failure times, hours: 900, 600, 1100, 1350, 1000, 750, 1800, 2890

$$\lambda = \frac{n}{\sum_{i=1}^{n} t_i} \qquad \lambda = \frac{8}{10{,}390} \qquad \lambda = 0.00077$$

2. Time-censored failure data (type I)

In a life test, seven batteries were run for 140 hours, at which time the test was terminated. During the test time, three batteries failed at 28, 49, and 110 hours respectively. The remaining four batteries were functioning properly when the test was discontinued. What is the estimated failure rate?

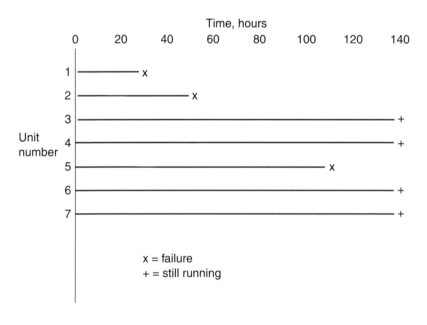

x = failure
+ = still running

$$\lambda = \frac{r}{\displaystyle\sum_{i=1}^{n} t_i + (n-r)T}$$

where: r = total number of failures = 3
   n = number of units on test = 7
   t = time at which the test is terminated = 140

$$\lambda = \frac{3}{(28+49+110)+(7-3)140}$$

$\lambda = 0.0040$ failures per hour

$\theta = 1/0.0040 = 250$ hours

Supplemental problem:

The gloss is measured on a painted test panel each day. A gloss reading of less than 25 microlumens is considered a failure. Nine test panels are tested, and each panel is measured automatically every five daylight hours. The test was run a total of 1500 hours, at which time the test was terminated. Four test panels failed at 120, 345, 655, and 1100 hours respectively. What is the failure rate?

3. Failure-censored data (type II)

Precision fluid metering devices are designed to deliver 5.00 ml. of liquid. Nine random units were selected for reliability testing. The specification for acceptance is that the unit deliver 5.00 ± 0.10 ml. The nine selected units are cycled over and over, checking the delivery to confirm that the specification is met. The test is terminated when r failures have occurred. The total number of cycles at failure are recorded. Given the following data, what is the estimated failure rate?

This calculation is performed the same as with the time censored type I data. Note that in this example, the number of cycles accumulated at failure rather than the total amount of time is used for the calculation.

$$\lambda = \frac{r}{\displaystyle\sum_{i=1}^{n} t_i + (n-r)T}$$

where: r = number of failures = 6
   n = number of units on test = 9
   T = time when the test was terminated (when r failures occur) = 12,600

Unit	Cycles on failure	Status
5	250	F
4	800	F
7	1100	F
3	1800	F
2	4800	F
1	12,600	F
8	12,500	+
9	12,600	+
6	12,600	+

F = failed
+ = still running at termination

$$\lambda = \frac{r}{\displaystyle\sum_{i=1}^{n} t_i + (n-r)T} \qquad \lambda = \frac{6}{21,350 + (3)(12,600)} \qquad \lambda = 0.000101$$

$$\theta = \frac{1}{0.000101} = 9901$$

Confidence Interval for λ from Time-Censored (Type I) Data

The $100(1 - \alpha)$ percent confidence interval for λ is given by

$$\lambda\left(\frac{\chi^2\left(2r, 1-\frac{\alpha}{2}\right)}{2r}\right) < \text{True } \lambda < \lambda\left(\frac{\chi^2\left(2r+2, \frac{\alpha}{2}\right)}{2r}\right)$$

where: λ = estimated failure rate
r = number of failures
α = risk ($\alpha = 1.00 -$ confidence)
χ^2 = chi-square

Example:
Ten units have been tested for 250 days. At the end of the 250-day period, seven units had failed. What is the estimated failure rate, and what is the 90 percent confidence interval for the estimate?

Data:

Unit no.	Day at failure
1	145
2	200
3	250+
4	188
5	250+
6	250+
7	220
8	120
9	227
10	150

$$\lambda = \frac{7}{2000} = 0.0035$$

$$\lambda\left(\frac{\chi^2\left(2r,1-\frac{\alpha}{2}\right)}{2r}\right) < \text{True } \lambda < \lambda\left(\frac{\chi^2\left(2r+2.\frac{\alpha}{2}\right)}{2r}\right)$$

where: $r = 7$
$\alpha = 0.10$
$\lambda = 0.0035$

$$\lambda\left(\frac{\chi^2\left(2r,1-\frac{\alpha}{2}\right)}{2r}\right) < \text{True } \lambda < \lambda\left(\frac{\chi^2\left(2r+2.\frac{\alpha}{2}\right)}{2r}\right)$$

$$\lambda\left(\frac{\chi^2{}_{14,0.95}}{14}\right) < \text{True } \lambda < \lambda\left(\frac{\chi^2{}_{16,0.05}}{14}\right)$$

Looking up the appropriate chi-square value using a chi-square table.

$$\chi^2{}_{14,0.95} = 6.57$$

$$\chi^2{}_{16,0.05} = 26.3$$

$$0.0035\left(\frac{6.57}{14}\right) < \text{True } \lambda < 0.0035\left(\frac{26.3}{14}\right)$$

$$0.0164 < \text{True } \lambda < 0.0066$$

An appropriate statement would be: "We do not know the true failure rate, but we are 90 percent confident that it is between 0.0164 and 0.0066."

Confidence Interval for λ from Failure-Censored (Type II) Data

Failure rates that are determined from reliability tests that are terminated when a specified number of failures have been reached are referred to as failure-censored or failure-truncated tests. The confidence interval for this type of data is determined using the following relationship:

$$\lambda\left(\frac{\chi^2\left(2r,1-\frac{\alpha}{2}\right)}{2r}\right) < \text{True } \lambda < \lambda\left(\frac{\chi^2\left(2r.\frac{\alpha}{2}\right)}{2r}\right)$$

Note that the only difference between this relationship and the time-censored confidence interval is in the upper limit equation, where the degrees of freedom for the chi-square is $2r$ for the failure-censored data and $2r + 2$ for the time-censored data.

Once the confidence interval for λ has been obtained, the confidence interval for the MTBF can be determined by simply taking the reciprocal of the limits for the failure and reversing the order of the intervals. The MTBF resulting from taking the reciprocal of the lower limit for the failure value yields the upper confidence limit for the MTBF.

Example:
A sample of eight panel bulbs was tested, and after five failures, the estimate for λ was determined to be 4.6×10^{-4} failures per hour. Determine the 95 percent confidence interval for this failure-censored data and the corresponding interval for the MTBF.

$$\lambda = 4.6 \times 10^{-4} \text{ failures per hour}$$

$2r = 2 \times 5 = 10$

$1 - \alpha = 0.95$

$\alpha = 0.05$

$\alpha/2 = 0.025$

$1 - \alpha/2 = 0.975$

From the chi-square table, look up the required chi-square values for $1 - \alpha/2$ and $\alpha/2$ degrees of freedom.

$$\chi^2_{(2r, 1-\alpha/2)} = \chi^2_{(10, 0.975)} = 3.25$$
$$\chi^2_{(2r, \alpha/2)} = \chi^2_{(10, 0.025)} = 20.48$$

The 95 percent confidence interval for λ is

$$\frac{4.6 \times 10^{-4} \times 3.25}{10} < \text{True failure rate} < \frac{4.6 \times 10^{-4} \times 20.48}{10}$$

$$0.000149 < \text{True failure rate} < 0.000942$$

and

$$1062 < \text{True MTBF} < 6711$$

An alternative equation for calculating the confidence interval for the failure rate using the total time on all test units when the test is terminated and the total number of failed units is determined is given by:

$$\frac{2T}{\chi^2_{1-\frac{\alpha}{2}, 2r}} < \text{True } \lambda < \frac{2T}{\chi^2_{\frac{\alpha}{2}, 2r}}$$

where: T = total test time on all units when the test is terminated
r = total number of failed units

Example:
A set of seven units is tested until all seven have failed. The total time when the seventh unit failed was 6000 hours. Calculate the single-sided lower confidence limit at 95 percent confidence for the failure rate.

$$\alpha = 0.10$$

$$\frac{\chi^2_{1-\alpha, 2r}}{2T} < \text{True } \lambda$$

Note that since we are asking for only the lower confidence limit, the risk α is not divided by 2. All of the risk is assigned to the single-sided confidence limit.

$$\chi^2_{1-\alpha,2r} = \chi^2_{0.95,14} = 6.57$$

$$\frac{\chi^2_{1-\alpha,2r}}{2T} = \frac{6.57}{12,000} = 0.00055$$

This lower 95 percent confidence limit for the failure rate is the recripocal of the upper 95 percent confidence limit for the MTBF, θ.

The upper 95 percent confidence limit is 18,182.

Success run theorem

There are some cases where no failures have occurred during a reliability demonstration test. For those cases, a concept known as the success run theorem may be used to report the reliability at a given level of reliability. The relationship of the reliability, level of confidence, and number of failure free samples is given in the formula.

$$R = (1 - C)^{1/n}$$

where R is the reliability, C is the level of confidence, and n is the sample size or number of failure-free test items.

Most often we use the success run theorem to determine the number of failure-free test items required to demonstrate a minimum reliability at a given level of confidence. By rearranging the original formula, we may solve for n.

$$n = \frac{\ln(1-C)}{\ln R}$$

It is important to realize that time is not the criteria. Items are simply tested to a predetermined time. If the items operate to the prescribed time, success is demonstrated.

Sample application:

Success for an electric relay is defined as 150,000 cycles without failure. How many relays must pass this test in order to demonstrate a reliability of 95 percent with a confidence of 90 percent?

$$n = \frac{\ln(1-C)}{\ln R}$$

$$n = \frac{\ln(1-0.90)}{\ln 0.95} \qquad n = \frac{-2.303}{-0.051} = 45$$

Table 1 may be used to determine the appropriate sample size for several levels of reliability and confidence. The intersection of the row where $R = 0.95$ with the column where $C = 0.90$ gives a sample size of $n = 45$. Forty-five relays must successfully operate for

Table 1. Success Run Sample Size.

Minimum reliability, R	Confidence level, C						
	0.80	**0.85**	**0.90**	**0.95**	**0.99**	**0.995**	**0.9999**
0.80	8	9	11	14	21	24	31
0.85	10	12	15	19	29	33	43
0.90	16	19	22	29	44	51	66
0.95	32	37	45	59	90	104	135
0.99	161	189	230	299	459	528	688
0.995	322	379	460	598	919	1058	1379
0.999	1609	1897	2303	2995	4603	5296	6905

150,000 cycles to demonstrate a reliability of 95 percent with a level of confidence of 90 percent. We are 90 percent confident that the reliability is not less than 95 percent.

Reliability Modeling

The reliability of more complex systems can be modeled using the reliability of individual components and block diagrams. Components can be diagramed using series and parallel configurations.

Series Systems

The system reliability for components that function in series is determined by the product of the reliability of each component.

$$R_S = R_1 \times R_2 \times R_3 \times \ldots \times R_K$$

The system reliability for a series of components is always less reliable than the least reliable component.

Using a reliability block diagram (RBD), calculate the reliability of the following system:

$$R_S = 0.90 \times 0.95 \times 0.99$$
$$R_S = 0.85$$

If the failure rate or MTBF is given, the reliability must be determined in order to perform the system reliability model. Failure rates and MTBFs cannot be combined directly.

Example:
Find the system at $t = 200$ hours for a series system configuration given one component has a failure rate of 0.002 and the other component has an MTBF of 700 hours.

Given $\theta = 700$ hours, the reliability at $t = 200$ is

$$R_{t=200} = e^{-\left(\frac{t}{\theta}\right)}$$

where: $t = 200$
          $\theta = 700$

$$R_{t=200} = e^{-(0.29)}$$
$$R_{t=200} = 0.75$$

Given $\lambda = 0.002$, the reliability at $t = 200$ is

$$R_{t=200} = e^{-\lambda t}$$

where: $t = 200$
          $\lambda = 0.002$

$$R_{t=200} = e^{-(0.002 \times 200)}$$
$$R_{t=200} = 0.67$$

The RBD can now be constructed for the individual components:

$$R_S = 0.75 \times 0.67$$
$$R_S = 0.50$$

Parallel Systems

The reliability of systems that have components that function in parallel are determined by subtracting the product of the component unreliabilities $(1 - R)$ from 1. The system reliability of a parallel system will always be more reliable than the most reliable component.

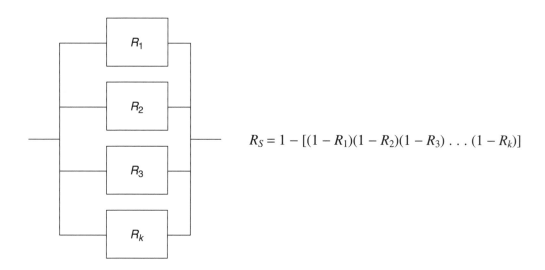

$$R_S = 1 - [(1 - R_1)(1 - R_2)(1 - R_3) \ldots (1 - R_k)]$$

If all of the components are of equal reliability, then the system reliability is given by

$$R_S = 1 - (1 - R)^n$$

where: n = number of components in parallel

Example:
What is the system reliability for three components having individual reliabilities of 0.85, 0.90, and 0.80?

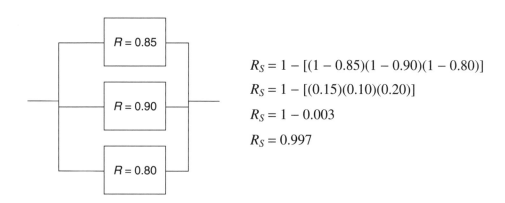

$$R_S = 1 - [(1 - 0.85)(1 - 0.90)(1 - 0.80)]$$
$$R_S = 1 - [(0.15)(0.10)(0.20)]$$
$$R_S = 1 - 0.003$$
$$R_S = 0.997$$

Complex Systems

Block diagrams of more complex arrangements utilizing a mixture of both serial and parallel components can be developed. Parallel components are reduced to an equivalent single component and the completion of the reliability calculation based on a serial model.

Example:
Determine the reliability of the following RBD:

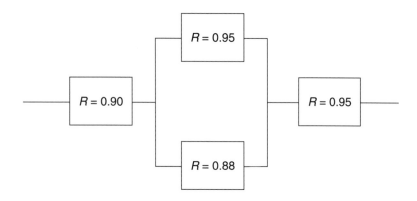

Reduce the two parallel components to the equivalent of one component as follows:

$$R = 1 - [(1 - 0.95)(1 - 0.88)]$$
$$R = 1 - [(0.05)(0.12)]$$
$$R = 1 - 0.006$$
$$R = 0.994$$

Determine the system reliability as a series configuration:

$$R_S = 0.90 \times 0.994 \times 0.95$$
$$R_S = 0.85$$

Shared-Load Parallel Systems

In the previous calculation, it is assumed that only one of the parallel components is required for success and that the failure of one of the parallel components does not increase the load on the remaining parallel component. Problems where the failure rate of the surviving component is different than it would have been in the active redundancy state are more complex.

Shared-load problems with two parallel components where the failure rate changes with the number of surviving components are determined by

$$R_t = e^{-2\lambda_1} + \left(\frac{2\lambda_1}{2\lambda_1 - \lambda_2} \right)(e^{-\lambda_1 t} - e^{-2\lambda_1 t})$$

where: $\lambda_1 t$ = failure rate when both components are in operable condition
$\lambda_2 t$ = failure rate of the surviving component when one component has failed

Example:
A mobile refrigeration unit is served by two cooling systems in parallel. Operational data have shown that one unit will provide adequate cooling, but the additional stress on the surviving unit increases its failure rate. The failure rate of the single surviving unit λ_2 is 0.00025 per hour. When cooling is provided by both units operating, the individual failure rate for each λ_1 is 0.000075 per hour. What is the 1000 hour reliability for this system?

$$R_t = e^{-2\lambda_1 t} + \left(\frac{2\lambda_1}{2\lambda_1 - \lambda_2} \right)(e^{-\lambda_2 t} - e^{-2\lambda_1 t})$$

where: $t = 1000$
$\lambda_1 = 0.000075$
$\lambda_2 = 0.00025$

$$R_t = e^{-2(0.000075)(1000)} + \left(\frac{2(0.000075)}{(2)(0.000075) - (0.00025)} \right)(e^{-(0.00025)(1000)} - e^{-(2)(0.000075)(1000)})$$

$R_t = e^{-0.15} + (-1.5)(e^{-.25} - e^{-0.75})$ $R_t = 0.8607 + (-1.5)(0.7788 - 0.8607)$

$R_t = 0.993$

Standby Parallel Systems

Parallel configurations can have a unit in a standby mode, ready to switch in when the active component fails. The system reliability depends on the reliability of the switch and the reliability of the standby component. The more complex case is where the two components have different failure rates and the switching is imperfect. The system reliability is determined by

$$R_t = e^{-\lambda_1} + R_{sw} \left(\frac{\lambda_1}{\lambda_2 - \lambda_1} \right)(e^{-\lambda_1 t} - e^{-\lambda_2 t})$$

Example:
A pump with a failure rate λ_1 of 0.0002 per hour is backed up with a standby pump with a failure rate, λ_2 of 0.0015 per hour. The backup pump is switched into place upon failure of the primary pump. The reliability of the switch is 0.95. What is the reliability of the system at $t = 400$ hours?

$$R_t = e^{-\lambda_1 t} + R_{sw} \left(\frac{\lambda_1}{\lambda_2 - \lambda_1} \right)(e^{-\lambda_1 t} - e^{-\lambda_2 t})$$

$$R_t = e^{-(0.0002)(400)} + 0.95 \left(\frac{0.0002}{0.0015 - 0.0002} \right)(e^{-(0.0002)(400)} - e^{-(0.0015)(400)})$$

$R_t = 0.9231 + 0.95(0.1538)(0.9231 - 0.5488)$

$R_t = 0.9778$

BIBLIOGRAPHY

Dovich, R. A. 1990. *Reliability Statistics.* Milwaukee, WI: ASQC Quality Press.

Krishnamoorthi, K. S. 1995. *Reliability Methods for Engineers.* Milwaukee, WI: ASQC Quality Press.

O'Connor, P. D. T. 1995. *Practical Reliability Engineering.* 3rd edition, New York: John Wiley and Sons.

SEQUENTIAL SIMPLEX OPTIMIZATION

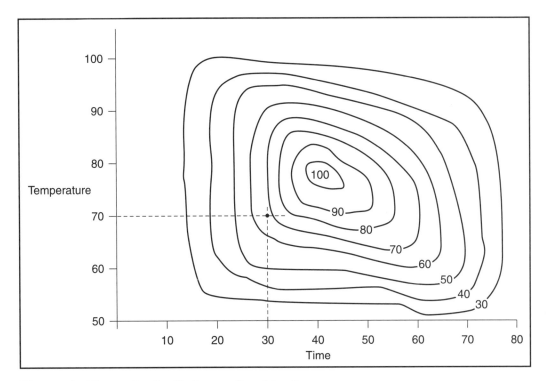

Figure 1. Theoretically Optimum Combination.

THEORETICAL OPTIMUM

The intersection of the two broken lines in Figure 1 shows the location of the result of a single experiment performed with the conditions or combinations of temperature (70°C) and time (30 minutes) for a chemical reaction. The response for this one set of conditions is 68 percent as indicated by the contour of the experimental space, which is, in reality, unknown. The response surface is unknown, and the shape can only be determined by a series of experiments. One experiment, as is the case with Figure 1, gives no information regarding the response surface.

Shape is a differential quality, dy/dx. In order to define a shape or surface, there must be a dx. This requires at least two different values for variable x. If we are to understand how a factor affects a response, we must vary it.

351

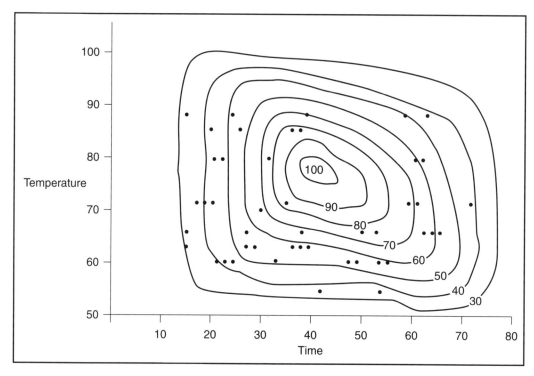

Figure 2. *Shotgun* Approach to Optimization.

STRATEGIES FOR OPTIMIZATION

Shotgun Approach

Randomly changing one or two variables and measuring the response is one approach to optimization. One advantage to this approach is that if enough experiments are performed, they will cover the entire response surface. This is important if information is needed for the total experimental space. A totally random approach does lead to the possibility of repeating an experiment already covered. This shotgun approach is sometimes referred to as a *stochastic* or *probabilistic* approach to optimization. Figure 2 illustrates the shotgun approach. The absolute optimum may or may not be attained.

Approach of a Single Factor at a Time

Figure 3 shows the possible result when varying one factor at a time, a frequently used strategy that is also called the *multifactor* strategy. Consider the seemingly logical statement: "If you want to find out the effect of a factor on a response, you must hold all other factors constant and vary the factor you are investigating." While this approach sounds good, it fails in that the answer you get is conditional on the values of the other factors and their settings. You are totally ignoring the potential of interactions. Consider the experiments in Figure 3.

Initially, an evaluation of the time variable is conducted by keeping the temperature fixed at the theoretical "best" setting of 70° C and varying the time around the theoretical point of 30 minutes. This gives the three experimental conditions of *A, B,* and *C.* The response of

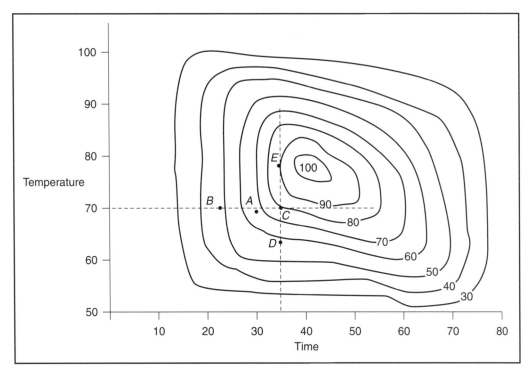

Figure 3. Approach of One Factor at a Time.

experiment *C* indicates that an improvement in yield from 68 percent (condition *A*) to a yield of 78 percent occurs for condition *C*. The experimenter now holds the time constant at 34 minutes and varies temperature around the theoretical "best" setting of 70° C by running experiments *D* and *E*. The results indicate that the higher temperature of 79° C is the best condition or optimal setting with a yield of 88 percent. Based on this knowledge, the experimenter might be lead to increase the temperature a little more and hold time constant at 34 minutes. Doing this would lead to a decrease in yield. What has happened in Figure 3 is that the experimenter has become stranded on the oblique ridge. The only way to move up the ridge toward the optimum is to change both factors simultaneously. The strategy of a single factor at a time will not accomplish the optimum setting. A true multifactor optimization strategy is required.

Basic Simplex Algorithm

A simplex is a geometric figure that has a number of vertexes (corners). If there are *k* number of factors, then the simplex is defined by *k* + 1 points. A two-factor simplex is defined by three points of a triangle. If there are three factors, then the simplex is defined by four points in space (the corners of a tetrahedron). Two and three factors can be visualized easily, but four factors are difficult to illustrate.

Simplex Terminology:

1. A *simplex* is a geometric figure defined by number of points.
2. A *vertex* is a corner of a simplex. It is one of the points that define a simplex.

3. A *face* is that part of the simplex that remains after one of the vertexes is removed.
4. A *hyperface* is the same as a face, but it is used when the simplex is of four dimensions or greater.
5. A *centroid* is the center of mass of all of the vertexes in that simplex.
6. The *centroid of the remaining hyperface* is the center of mass of the vertexes that remain when one vertex is removed from the simplex.

Simplex Calculations

Problem:

Two variables will be considered for the optimization of a process. The response variable will be percent yield. The two independent variables or factors will be time (X_1) and temperature (X_2). An initial simplex will be formed by three vertexes around an initial set of conditions located at 60° C and 20 minutes. At this operating condition, the yield is 36 percent. Figure 4 shows the simplex worksheet for the first simplex. The three vertexes are labeled *W* for the vertex giving the worst response, *B* for the best response, and *N* for the next best response. When two factors are considered, there will always be three vertexes describing the simplex.

This initial simplex is defined by the following three vertexes. The yield for each of these conditions is determined and ranked as best (*B*), worst (*W*) and next best (*N*).

Simplex 1 coordinates:

Time	Temperature	Yield	Rank
20	64	42	B
23	58	37	N
17	58	31	W

Simplex number: 1

		Factor			
		X_1	X_2	Response	Rank
Coordinates of retained vertexes		20	64	42	B
		23	58	37	N
Σ		43	122		
$\bar{P} = \Sigma/k$		21.5	61		
W		17	58	31	
$(\bar{P} - W)$		4.5	3		
$R = \bar{P} + (\bar{P} - W)$		26	64	54	

Figure 4.

Having established an initial set of conditions, running the experiments, and ranking the three responses according to **b**est, **n**ext best, and **w**orst, the reflected vertex is calculated using the worksheet. This reflected vertex for simplex #1 is: Time $(X_1) = 26$ minutes, and temperature $(X_2) = 64°$ C. The result of this experiment gives a yield of 54 percent.

Of these current four responses, we discard the current **w**orst of 31 percent and retain the remaining vertexes. These responses are reranked B, N, and W, and the data are recorded on the simplex #2 worksheet. The new reflected vertex is determined and run. The resulting response is 50 percent.

Simplex number: 2

	Factor			
	X_1	X_2	Response	Rank
Coordinates of retained vertexes	26	64	54	B
	20	64	42	N
Σ	46	128		
$\bar{P} = \Sigma/k$	23	64		
W	23	58	37	
$(\bar{P} - W)$	0	6		
$R = \bar{P} + (\bar{P} - W)$	23	70	50	

Figure 5.

Discard the current worst vertex, rerank the remaining three, and record on the worksheet for simplex #3.

Simplex number: 3

	Factor			
	X_1	X_2	Response	Rank
Coordinates of retained vertexes	26	64	54	B
	23	70	50	N
Σ	49	134		
$\bar{P} = \Sigma/k$	24.5	67		
W	20	64	42	
$(\bar{P} - W)$	4.5	3		
$R = \bar{P} + (\bar{P} - W)$	29	70	66	

Figure 6.

The response for this reflected vertex is 66 percent. Discarding the current worst vertex corresponding to the yield of 42 percent and reranking generates simplex #4:

Simplex number: 4

		Factor			
		X_1	X_2	Response	Rank
Coordinates of retained vertexes		29	70	66	B
		26	64	54	N
Σ		55	134		
$\bar{P} = \Sigma/k$		27.5	67		
W		23	70	50	
$(\bar{P} - W)$		4.5	−3		
$R = \bar{P} + (\bar{P} - W)$		32	64	57	

Figure 7.

The response for the reflected vertex of 32 minutes and 64° C is 57 percent. Discard the current worst, rerank, and calculate the reflected vertex for simplex #5:

Simplex number: 5

		Factor			
		X_1	X_2	Response	Rank
Coordinates of retained vertexes		29	70	66	B
		32	64	57	N
Σ		61	134		
$\bar{P} = \Sigma/k$		30.5	67		
W		26	64	54	
$(\bar{P} - W)$		4.5	3		
$R = \bar{P} + (\bar{P} - W)$		35	70	80	

Figure 8.

Simplex number: 6

	Factor			
	X_1	X_2	Response	Rank
Coordinates of retained vertexes	35	70	80	B
	29	70	66	N
Σ	64	140		
$\bar{P} = Σ/k$	32	70		
W	32	64	57	
$(\bar{P} - W)$	0	6		
$R = \bar{P} + (\bar{P} - W)$	32	76	80	

Figure 9.

Discarding the current worst and reranking will establish simplex #7. The reflected vertex generated in simplex #6 gave a yield of 80 percent, the same as the original best for simplex #6. The choice for the vertex for the best and next best does not matter as long as both of the 80 percent yields are used.

Simplex number: 7

	Factor			
	X_1	X_2	Response	Rank
Coordinates of retained vertexes	35	70	80	B
	32	76	80	N
Σ	67	146		
$\bar{P} = Σ/k$	33.5	73		
W	29	70	66	
$(\bar{P} - W)$	4.5	3		
$R = \bar{P} + (\bar{P} - W)$	38	76	98	

Figure 10.

Simplex number: 8

	Factor			
	X_1	X_2	Response	Rank
Coordinates of retained vertexes	38	76	98	B
	32	76	80	N
Σ	70	152		
$\bar{P} = \Sigma/k$	35	76		
W	35	70	80	
$(\bar{P} - W)$	0	6		
$R = \bar{P} + (\bar{P} - W)$	35	82	89	

Figure 11.

Simplex number: 9

	Factor			
	X_1	X_2	Response	Rank
Coordinates of retained vertexes	38	76	98	B
	35	82	89	N
Σ	73	158		
$\bar{P} = \Sigma/k$	36.5	79		
W	32	76	80	
$(\bar{P} - W)$	4.5	3		
$R = \bar{P} + (\bar{P} - W)$	41	82	95	

Figure 12.

Simplex number: 10

	Factor			
	X_1	X_2	Response	Rank
Coordinates of retained vertexes	38	76	98	B
	41	82	95	N
Σ	79	158		
$\bar{P} = \Sigma/k$	39.5	79		
W	35	82	89	
$(\bar{P} - W)$	4.5	−3		
$R = \bar{P} + (\bar{P} - W)$	44	76	(100)	

Figure 13.

The resulting vertex for simplex #9 gives a yield of 100 percent, and the optimization is complete. See Figure 13 for the complete map of the 10 simplexes (or simplices).

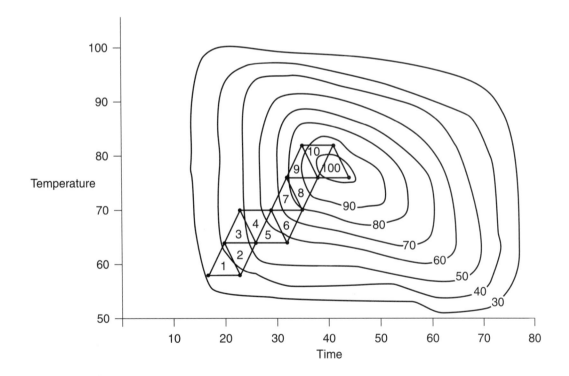

See the following table for a summary of the simplex movements toward optimization:

	Vertex	X_1 (time)	X_2 (Temp.)	Response, percent yield
1.	Initial	20	64	42 Best
2.	Initial	23	58	37 Next best
3.	Initial	17	58	31 Worst
4.	Reflected	26	64	54
5.	Reflected	23	70	50
6.	Reflected	29	70	66
7.	Reflected	32	64	57
8.	Reflected	35	70	80
9.	Reflected	32	76	80
10.	Reflected	38	76	98
11.	Reflected	35	82	89
12.	Reflected	41	82	95
13.	Reflected	44	76	100

BIBLIOGRAPHY

Walters, F. H., L. R. Parker, Jr., S. L. Morgan, and S. N. Deming. 1991. *Sequential Simplex Optimization.* Boca Raton, FL: CRC Press.

Short-Run Attribute Charts

Traditional control charts for attributes include the p chart and np chart for defective where the sample size is variable and constant, respectively, and the $\bar{u}$ chart and C chart for defects where the sample size is variable and constant, respectively.

In all of these cases, it is assumed that the process average is unique to a particular process or product and is used to provide a historical basis for detecting a process change. If the process is one, for example, that photocopies paper $8.5'' \times 11''$ on one side where the typical run size is 100 sheets, the average number of defects $\bar{u}$ might be 0.02. This information could be used to construct a U chart.

$$\text{UCL} = \bar{U} + 3 \sqrt{\frac{\bar{U}}{n}} \qquad \text{and} \qquad \text{LCL} = \bar{U} - 3 \sqrt{\frac{\bar{U}}{n}}$$

where: $\bar{U}$ = the average defects per unit

If the size of the paper changes and a more complex operation such as two-sided copying and collating is required, then the process average number of defects per unit might be different. If this occurs, then another control chart will need to be developed for the redefined process using its unique process average. If $\bar{U}$ changes, then the entire control chart changes.

Rather than plot the U value for each sample taken, the corresponding Z-score may be plotted. By plotting the Z value for each sample, we are standardizing the data. Z-scores are determined by calculating the magnitude of the difference in a reported value and its expected average in units of standard deviation. In the case of variables data, a Z-score is determined by:

$$Z = \frac{X_i - \bar{X}}{\sigma}$$

where: X_i = individual value
X = average
σ = standard deviation

Replacing the terms for a u chart, the corresponding Z-score becomes

$$Z = \frac{U - \bar{U}}{\sqrt{\dfrac{\bar{U}}{n}}}$$

where: U = individual sample defects / unit

$\bar{U}$ = average defects per unit

n = sample size

The $\bar{U}$ value is unique to a specific product. The $\bar{U}$ value for $8.5'' \times 11''$ paper was 0.02, and for $11'' \times 17''$ paper, 2-sided the $\bar{U}$ value might be 0.12.

If the $\bar{U}$ value is known for each product, then the Z-scores may be determined and plotted on a universal chart. Since the objective of all control charts is to determine if the process average changes relative to the historical process average or target process average, we will refer to the process average as the target average or $\bar{U}_t$.

Development of the short-run u chart

$$LCL < U < UCL$$

$$\bar{U} - 3\sqrt{\frac{\bar{U}}{n}} < U < \bar{U} + 3\sqrt{\frac{\bar{U}}{n}}$$

$$-3\sqrt{\frac{\bar{U}}{n}} < U - \bar{U} < +3\sqrt{\frac{\bar{U}}{n}}$$

$$-3 < \frac{U - \bar{U}}{\sqrt{\frac{\bar{U}}{n}}} < +3$$

The Lower Control Limit, LCL = –3

The Upper Control Limit, UCL = +3

The process average = 0

The individual plot points are Z-scores, or

$$\frac{U - \bar{U}}{\sqrt{\frac{\bar{U}}{n}}}$$

Each $\bar{U}$ has a unique value of $\bar{U}_t$.

$\bar{U}_t$ may come from

1. Previous control charts for that product or item
2. An assumed level of defects/unit

Case study:

The Ionic Corporation manufactures depth-finding units and has several models that are produced in small lots. The product line is broad, ranging from relatively simple units for consumer use to sophisticated commercial units that are much more complex. Data from the final inspection audits for four production models give the following historical process average defects/unit $\overline{U}_t$.

Part	Target $\overline{U}_t$
A	0.20
B	0.10
C	0.13
D	0.17

As quality engineer, you would like to establish a u chart using the short-run technique. You will use a short-run method, because there is no real long run of any specific model as would be required for a tradition u chart, and the process for manufacturing is more or less the same. Only the complexity of the unit produced changes.

Step 1. Collect historical data.

All of the units produced are inspected and tested to a rigorous format. The sample size represents 100 percent of the units produced. The following information is available for the past 13 production runs:

Date	Product	Sample size n	Number of defects c	Defects per unit U
11/2	A	18	5	0.28
11/3	C	23	2	0.09
11/4	D	7	1	0.14
11/4	A	11	2	0.18
11/5	C	38	4	0.11
11/5	D	16	3	0.19
11/6	A	7	0	0.00
11/6	D	23	2	0.09
11/7	D	11	3	0.27
11/8	B	42	2	0.05
11/8	C	17	3	0.28
11/9	A	13	1	0.08
11/9	B	26	2	0.08

Step 2. Calculate a Z-score for each sample collected.

$$Z = \frac{U - \overline{U}_t}{\sqrt{\dfrac{\overline{U}_t}{n}}}$$

where: $n =$ sample size
$\overline{U}_t =$ target average defects / unit for specific part or model

First sample:

$$Part = A$$

$$\overline{U}_t = 0.20$$

$$n = 18$$

$$c = 5$$

$$U = \frac{c}{n} = \frac{5}{18} = 0.28$$

$$Z = \frac{U - \overline{U}_t}{\sqrt{\dfrac{\overline{U}_t}{n}}} \qquad Z = \frac{0.28 - 0.20}{\sqrt{\dfrac{0.28}{18}}} = +1.28$$

Second sample:

$$Part = C$$

$$\overline{U}_t = 0.13$$

$$n = 23$$

$$c = 2$$

$$U = \frac{c}{n} = \frac{2}{23} = 0.09$$

$$Z = \frac{U - \overline{U}_t}{\sqrt{\dfrac{U_t}{n}}} \qquad Z = \frac{0.09 - 0.13}{\sqrt{\dfrac{0.13}{23}}} = -0.53$$

Continue calculation for all 13 samples.

Date	Product	Sample size n	Number of defects c	Defects per unit U	Z-score
11/2	A	18	5	0.28	+1.32
11/3	C	23	2	0.09	−0.53
11/4	D	7	1	0.14	+0.16
11/4	A	11	2	0.18	−0.15
11/5	C	38	4	0.11	−0.34
11/5	D	16	3	0.19	+0.19
11/6	A	7	0	0.00	−1.18
11/6	D	23	2	0.09	−0.93
11/7	D	11	3	0.27	+0.80
11/8	B	42	2	0.05	−1.02
11/8	C	17	3	0.28	+1.72
11/9	A	13	1	0.08	+0.97
11/9	B	26	2	0.08	−0.32

Step 3. Record and plot historical data on the control chart.

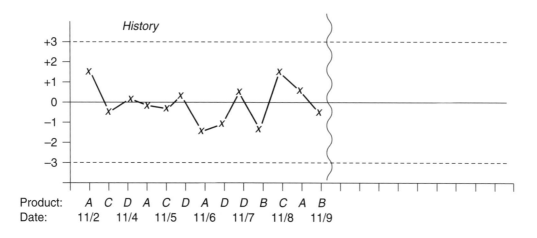

Product:	A	C	D	A	C	D	A	D	D	B	C	A	B
Date:	11/2		11/4		11/5		11/6		11/7		11/8		11/9

Step 4. Continue to collect and plot data, looking for evidence of a process change.

Date	Product	Sample size n	Number of defects c	Defects per unit U	Z-score
11/10	C	40	5	0.13	0.00
11/10	A	19	8	0.42	+2.14
11/11	B	23	6	0.26	+2.43
11/11	C	15	3	0.20	+0.75
11/12	D	30	8	0.27	+1.33
11/12	B	25	2	0.08	−0.32
11/13	A	9	3	0.33	+0.87
11/14	B	18	4	0.22	+1.61
11/14	C	22	9	0.41	+3.64

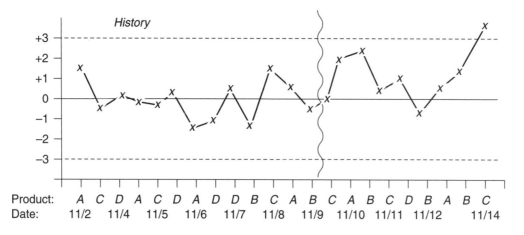

Product:	A	C	D	A	C	D	A	D	D	B	C	A	B	C	A	B	C	D	B	A	B	C
Date:	11/2		11/4		11/5		11/6		11/7		11/8		11/9		11/10		11/11		11/12			11/14

A process change is indicated with sample #22 occurring on 11/14 with product C.

SHORT-RUN P CHART

The same principle used to develop the u chart is used for the p chart, only the standard deviation calculation changes. The calculation for the plotting characteristic Z using the appropriate estimate for the standard deviation is given by

$$Z = \frac{P - \bar{P}_t}{\sqrt{\dfrac{\bar{P}_t(1 - \bar{P}_t)}{n}}}$$

A value for the target process average proportion $\bar{P}_t$ must be given for each product.

Example:
A manufacturer of noninterruptible power supplies has an operation that makes wiring harnesses to be used on its product lines. There are several different styles of harnesses, and the department manager wants to use a short-run p chart to monitor the process. In some cases, the lot size for orders may be as high as 100, but there are many runs of different styles. The harness production line consists of four workers that normally work in other departments. This production is assembled once or twice a week to produce harnesses that will be used for the next production period. The manager wants only one chart to monitor the department performance. Given the following information, develop the short-run p chart:

Model number	Process average proportional defective, $\bar{P}_t$
A	0.03
B	0.07
C	0.11
D	0.05
E	0.09

Production data:

Date	Product	Lot size	Number defective	Proportional defective P
4/5	A	45	1	0.022
4/5	D	123	8	0.065
4/5	B	88	4	0.045
4/11	C	28	2	0.071
4/11	B	35	3	0.086
4/11	E	58	3	0.052
4/11	D	75	1	0.013
4/15	E	95	11	0.116
4/15	C	42	7	0.167
4/15	B	65	5	0.077
4/15	D	27	1	0.037
4/19	C	55	1	0.018
4/19	E	40	4	0.100

Calculate the Z-scores:

Sample #1

$$Z = \frac{P - \overline{P}_t}{\sqrt{\dfrac{\overline{P}_t(1 - \overline{P}_t)}{n}}}$$

$$Z = \frac{0.022 - 0.020_t}{\sqrt{\dfrac{0.03(1 - 0.03)}{45}}} = -0.31$$

Sample #2

$$Z = \frac{P - \overline{P}_t}{\sqrt{\dfrac{\overline{P}_t(1 - \overline{P}_t)}{n}}}$$

$$Z = \frac{0.022 - 0.020}{\sqrt{\dfrac{0.03(1 - 0.03)}{45}}} = -0.76$$

Continue to calculate the Z-scores for the remaining data and record.

Date	Product	Lot size	Number defective	Proportional defective P	Z_P
4/5	A	45	1	0.022	−0.31
4/5	D	123	8	0.065	0.76
4/5	B	88	4	0.045	−1.08
4/11	C	28	2	0.071	−0.66
4/11	B	35	3	0.086	0.37
4/11	E	58	3	0.052	−1.01
4/11	D	75	1	0.013	−1.47
4/15	E	95	11	0.116	0.89
4/15	C	42	7	0.167	1.18
4/15	B	65	5	0.077	0.22
4/15	D	27	1	0.037	−0.31
4/19	C	55	1	0.018	−2.18
4/19	E	40	4	0.100	0.22

Plot the Z_P scores:

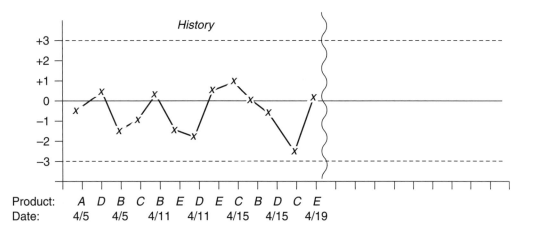

Continue to collect data, plot points, and look for signs of a process change:

Date	Product	Lot size	Number defective	Proportional defective P	Z_P
4/20	B	85	10	0.118	+1.73
4/20	E	30	4	0.133	+0.82
4/21	A	75	7	0.093	+3.20
4/22	C	32	3	0.094	−0.29
4/23	B	48	4	0.083	+0.35

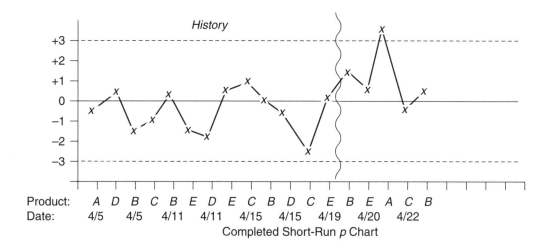

Completed Short-Run *p* Chart

BIBLIOGRAPHY

Besterfield, D. H. 1994. *Quality Control.* 4th edition. Englewood Cliffs, NJ: Prentice Hall.

Montgomery, D. C. 1996. *Introduction to Statistical Quality Control.* 3rd edition. New York: John Wiley and Sons.

Wheeler, D. J. 1991. *Short Run SPC.* Knoxville, TN: SPC Press.

SHORT-RUN AVERAGE/RANGE CHARTS

The traditional implementation of statistical process control (SPC) requires that a minimal number of sample observations be obtained to characterize the process or set up a historical reference against which future changes in the process are monitored. The recommended number of subgroups required for such historical characterization varies. For the average/range control chart, a minimum of 25 subgroups are suggested, each of a subgroup size of four or five. In many manufacturing processes, the opportunity to measure 125 units of production would be impractical if not impossible due to low production volumes in a *just-in-time* operation.

By consolidating the measurements of a universal characteristic that are common to several different products into one population, we may monitor this characteristic as if it originated from a single process.

The concept for short-run average/range charts is based on a normalization of the statistic we will use to monitor changes in the location (central tendency) and variation. We will be measuring the amount of change for a specific characteristic relative to the expected value for that characteristic. This will be done for several different products, all having the same characteristic. Each of the products, however, will have a unique expected value for both the location statistic and the variation statistic. These historical expected values are established with historical data.

The following derivations are made to generate the new short-run plotting characteristic that will replace the traditional average and range used in the $\overline{X}/R$ chart.

FOR THE SHORT-RUN AVERAGE CHART

Initial premise:

$$\text{LCL}_{\overline{X}} < \overline{X} < \text{UCL}_{\overline{X}}$$

Replace LCL and UCL with defining equation:

$$\overline{\overline{X}} - A_2 \overline{R} < \overline{X} < \overline{\overline{X}} + A_2 \overline{R}$$

Subtract $\overline{\overline{X}}$ from all terms:

$$-A_2 \overline{R} < \overline{X} - \overline{\overline{X}} < +A_2 \overline{R}$$

Divide all terms by $\bar{R}$:

$$\frac{-A_2 \bar{R}}{\bar{\bar{R}}} < \frac{\bar{X} - \bar{\bar{X}}}{\bar{\bar{R}}} < \frac{+A_2 \bar{R}}{\bar{\bar{R}}}$$

$$-A_2 < \frac{\bar{X} - \bar{\bar{X}}}{\bar{\bar{R}}} < +A_2$$

Rather than plotting the average $\bar{X}$ as in the case of a normal average/range chart, we will plot

$$\frac{\bar{X} - \bar{\bar{X}}}{\bar{\bar{R}}}$$

where: $\bar{X}$ and $\bar{R}$ will have unique values for each product

The control limits will be based on $\pm A_2$ for this normalized chart.

Control limits for $\dfrac{\bar{X} - \bar{\bar{X}}}{\bar{\bar{R}}}$ are: UCL $= +A_2$ and LCL $= -A_2$.

The value for A_2 depends on the subgroup sample size.

FOR THE SHORT-RUN RANGE CHART

Initial premise:

$$\text{LCL}_R < R < \text{UCL}_R$$

Replace LCL and UCL with formula:

$$D_3 \bar{R} < R < D_4 \bar{R}$$

Divide all terms by $\bar{R}$:

$$\frac{D_3 \bar{R}}{\bar{\bar{R}}} < \frac{R}{\bar{\bar{R}}} < \frac{D_4 \bar{R}}{\bar{\bar{R}}}$$

$$D_3 < \frac{R}{\bar{\bar{R}}} < D_4$$

Rather than plotting the range R, we will plot $\dfrac{R}{\bar{R}}$. There will be a unique $\bar{R}$ value for each product.

Control limits are: UCL $= D_4$ and LCL $= D_3$.

Each of the plotting parameters requires a unique value for the $\bar{X}$ and $\bar{R}$ for each part or unit of inspection. The following sections explain the several sources for these values.

FOR $\bar{\bar{X}}$

1. Use $\bar{\bar{X}}$ from the previous control chart data.
2. Calculate the average from any available data.
3. Let the nominal = $\bar{\bar{X}}$.

Note: Method #1 is the preferred choice. The decision to use method #3 over method #2 is determined by comparing the absolute value of the difference between the nominal and calculating the average from m observations to a critical difference value F_cS calculated from the sample standard deviation S.

If $|\text{Nominal} - \text{Average}| \geq F_cS,^*$ The average of M observations should be used rather than the nominal where F_c is a factor determined by the sample size m from which S, the sample standard deviation was determined.

F_c Factors:

Sample size m	F_c	Sample size m	F_c
4	1.1765	12	0.5185
5	0.9535	13	0.4942
6	0.8226	14	0.4733
7	0.7344	15	0.4547
8	0.6699	16	0.4383
9	0.6200	17	0.4235
10	0.5796	18	0.4101
11	0.5463	19	0.3978

Example 1:

Five measurements are available, and the specification nominal is 10.000. Should we use the nominal or average of the five measurement for the target $\bar{\bar{X}}$?

Data:

10.101, 10.008, 9.993, 10.102, 10.007

The sample average $\bar{X} = 10.042$, and the sample standard deviation $S = 0.0054$

$m = 5$ and the table value for $F_c = 0.9535$.

Critical difference $C_d = F_cS = 0.9535 \times 0.0054 = 0.0515$

The absolute difference between the nominal and sample average is
$|10.000 - 10.042| = 0.0042$

Since 0.0042 is not greater than the critical difference of 0.0515, we can use the nominal in place of the average of the five observations.

Rule: **If the difference between the average and nominal is greater than or equal to the critical difference, use average rather than nominal for the target $\bar{\bar{X}}$.**

FOR $\bar{R}$

1. Use $\bar{R}$.
2. Calculate $\bar{R}$ from an assumed C_{pk}:

$$\bar{R} = \frac{(USL - LSL)d_2}{6C_{pk}}$$

where: d_2 = factor dependent on the selected subgroup sample size
USL = upper specification limit
LSL = lower specification limit

For a unilateral specification, we may use

$$\bar{R} = \frac{(USL - \bar{X}_t)d_2}{3C_{pk}} \quad \text{or} \quad \bar{R} = \frac{(\bar{X}_t - LSL)d_2}{3C_{pk}}$$

3. Calculate $\bar{R}$ from the sample standard deviation S.
$\bar{R}$ is derived from S as follows:

$$\sigma = \frac{\bar{R}}{d_2} \quad \text{and} \quad \sigma = \frac{\bar{S}}{c_4}$$

Therefore,

$$\frac{\bar{R}}{d_2} = \frac{\bar{S}}{c_4}$$

$$\bar{R} = \left(\frac{d_2}{c_4}\right)\bar{S}$$

The value for d_2 depends on the selected subgroup sample size.

The value for c_4 depends on the sample size n from which the sample standard deviation S is calculated. The values for d_2 and c_4 can be found in the table of control chart constants.

Case study:

Product is routinely sampled and brought into a testing laboratory for evaluation. The products are all different models, but the characteristic being measured is common to all products. Your objective is to establish a short-run control chart. The production schedule consists of several product changes and is relatively short for a given product design. You will be working with three different product models. The subgroup has a sample size of $n = 3$.

Model A:

This is a new design, and a temporary specification of 30 ± 24 has been issued. The expected process capability index (C_{pk}) is 1.00.

Model B:

The specification is 150 ± 20, and you have test data for five units tested last month.

Model C:

Another new model is added with an expected C_{pk} of 1.20 and a specification of 400 ± 20.

Step 1. Establish working values for target $\overline{X}$ and $\overline{R}$ for each model.

Model A:

$\overline{\overline{X}}_A$ Determination

Based on an expected $C_{pk} = 1.00$ and a specification requirement of 30 ± 24, the expected target average $\overline{\overline{X}}$ will be set at the nominal.

$$\overline{\overline{X}}_A = 30.0$$

$\overline{R}_A$ Determination

Use an expected $C_{pk} = 1.00$.
    We are using a subgroup sample size of $n = 3$; therefore $d_2 = 1.693$.

$$\overline{R} = \frac{(USL - LSL)d_2}{6C_{pk}} \quad \overline{R} = \frac{(54.0 - 6.0)1.693}{(6)(1.00)} \quad \overline{R} = \frac{81.264}{6.00} \quad \overline{R} = 13.5$$

Model B:

$\overline{\overline{X}}_B$ Determination

Five measurements have been obtained and yield an average of $\overline{X} = 160.56$ and a sample standard deviation of $S = 4.79$. You may consider using the specification nominal of 150.0 for the target $\overline{X}$, but you first must test to see if this is statistically acceptable. If it is not, you will use the average of the five available data point instead of the nominal.

Critical difference, $C_d = F_c S = (0.9535)(4.79) = 4.57$

If the absolute difference between the average of the five measurements exceeds the critical difference C_d, then use the average of five measurements rather than the nominal for $\overline{\overline{X}}_B$.

$$\text{Difference} = \left| \text{Nominal} - \text{Average} \right| = \left| 150.0 - 160.56 \right| = 10.6$$

$10.6 > 4.57$; Therefore, use $\overline{\overline{X}}_B = 160.56$.

$\overline{R}_B$ Determination

The standard deviation S from the five measurements will be used to estimate the expected range for model B when subgroups of $n = 3$ are taken.

$d_2 = 1.693$ for subgroup size $n = 3$

$c_4 = 0.9400$ for a sample standard deviation S calculated from five measurements

$S = 4.79$

$$\overline{R} = \left(\frac{d_2}{c_4} \right) \overline{S} \quad \overline{R}_B = \left(\frac{1.693}{0.9400} \right) 4.79 \quad \overline{R}_B = 8.63$$

Model C:

$\overline{\overline{X}}_c$ Determination

The specification is 400 ± 20, and no prior data are available. Therefore, we will set the target $\overline{\overline{X}}$ to equal the nominal of 400:

$$\overline{\overline{X}}_C = 400.0$$

$\overline{R}_c$ Determination

Using the specification of 400 ± 20, the expected $C_{pk} = 1.20$, and the subgroup sample size of $n = 3$:

$$\overline{R} = \frac{(\text{USL} - \text{LSL})d_2}{6C_{pk}} \quad \overline{R} = \frac{(420 - 380)1.693}{(6)(1.20)} \quad \overline{R} = \frac{67.72}{7.20} \quad \overline{R} = 9.41$$

Summary of $\overline{\overline{X}}$ and $\overline{R}$ by Model

Model	$\overline{\overline{X}}$	$\overline{R}$
A	30.00	13.50
B	160.56	8.63
C	400.00	9.41

These values are recorded in the upper-left portion of the short-run $\overline{X}/R$ chart form.

Step 2. Collect data.

Model	Sample 1 A	Sample 2 A	Sample 3 A	Sample 4 B	Sample 5 B
Measurement					
(1)	37.1	41.9	44.4	161.6	163.4
(2)	37.3	43.1	34.1	157.9	160.3
(3)	30.1	26.1	33.1	158.4	161.7
Average	34.8	37.0	37.2	159.3	161.8
Range	7.2	17.0	11.3	3.7	3.1

Model	Sample 6 B	Sample 7 B	Sample 8 B	Sample 9 C	Sample 10 C
Measurement					
(1)	163.0	161.2	153.9	411.0	395.0
(2)	161.7	154.2	161.9	400.0	399.0
(3)	160.0	162.4	156.3	405.0	416.0
Average	161.6	159.3	157.4	405.3	403.3
Range	3.0	8.2	8.0	11.0	21.0

Model	Sample 11 A	Sample 12 B	Sample 13 B
Measurement			
(1)	44.9	165.8	161.8
(2)	37.6	161.7	157.6
(3)	35.4	159.2	161.9
Average	39.3	162.2	160.4
Range	9.5	6.6	4.3

Step 3. Calculate average, range, average plot point, and range plot point, and plot the points on the chart.

Subgroup 1: Model A

The plotting points are calculated, recorded on the chart form, and plotted.

Measurements:	37.1
	37.3
	30.1
Average $\overline{X}$:	34.8
Range R:	7.2

The plot point for the average is

$$\frac{\overline{X} - \overline{\overline{X}}_A}{\overline{R}_A} = \frac{34.8 - 30.00}{13.50} = 0.36$$

The plot point for the range is

$$\frac{R}{\overline{R}_A} = \frac{7.2}{13.50} = 0.53$$

Subgroup 2: Model A

The second subgroup metrics are calculated and plotted, and the points are connected to the first subgroup point.

Measurements:	41.9
	43.1
	26.1
Average, $\overline{X}$:	37.0
Range, R:	17.0

The plot point for the average is

$$\frac{\overline{X} - \overline{\overline{X}}_A}{\overline{R}_A} = \frac{37.0 - 30.00}{13.50} = 0.52$$

The plot point for the range is

$$\frac{R}{\overline{R}_A} = \frac{17.0}{13.50} = 1.26$$

Subgroup 3: Model A

Calculate, plot, and connect to subgroup point 3.

Measurements:	44.4
	34.1
	33.1
Average $\overline{X}$:	37.2
Range $\overline{R}$:	11.3

The plot point for the average is

$$\frac{\overline{X} - \overline{\overline{X}}_A}{\overline{R}_A} = \frac{37.2 - 30.00}{13.5} = 0.53$$

The plot point for the range is

$$\frac{R}{\overline{R}_A} = \frac{11.3}{13.50} = 0.84$$

Subgroup 4: Model B

Note than new target values for $\overline{\overline{X}}$ and $\overline{R}$ must be used.

Measurements:	161.6
	157.9
	158.4
Average $\overline{X}$:	159.3
Range $\overline{R}$:	3.7

The plot point for average is

$$\frac{\overline{X} - \overline{\overline{X}}_B}{\overline{R}_B} = \frac{159.3 - 160.56}{8.63} = 0.15$$

$$\frac{R}{\overline{R}_B} = \frac{3.7}{8.63} = 0.43$$

The remaining subgroup samples 5 through 13 are calculated and plotted on the chart. The process exhibits good statistical control with no points outside the upper or lower control limits for both the average and range portions.

See the following completed control chart with all of the historical data points plotted.

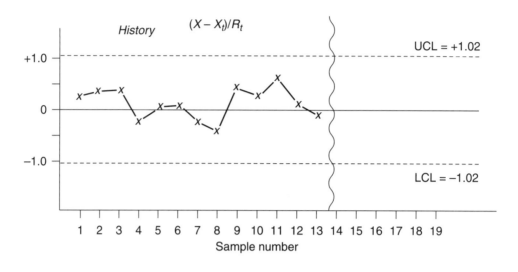

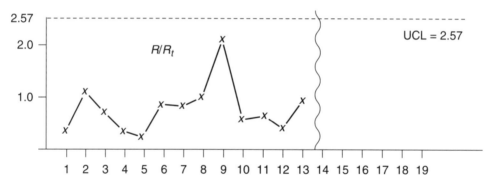

Step 5: Continue to collect data into the future, looking for signs of a process change.

During the next few days, additional data are collected. Based on these observations, do you feel that a process change has occurred? If so, was there a change in the average or in the variation?

Model	Sample 14 *B*	Sample 15 *A*	Sample 16 *A*	Sample 17 *B*	Sample 18 *C*
Measurement					
(1)	150.3	26.4	26.0	149.0	390.4
(2)	159.4	29.2	21.0	158.2	388.7
(3)	53.7	23.0	28.2	152.0	386.9
Average	154.5	26.2	25.1	153.1	388.7
Range	9.1	6.2	7.2	9.2	3.5

Model:	Sample 19 *C*
Measurement:	
(1)	394.9
(2)	392.3
(3)	390.2
Average	384.9
Range	9.4

A summary of all subgroup calculations follows:

Subgroup number	Model	Average $\bar{X}$	Range $\bar{R}$	$\dfrac{\bar{X} - \bar{\bar{X}}_T}{\bar{R}_T}$	$\dfrac{R}{\bar{R}_T}$
1	A	34.8	7.2	0.36	0.53
2	A	37.0	17.0	0.52	1.26
3	A	37.2	11.3	0.53	0.84
4	B	159.3	3.7	−0.15	0.43
5	B	161.8	3.1	0.14	0.36
6	B	161.6	3.0	0.14	0.35
7	B	159.3	8.2	−0.15	0.95
8	B	157.4	8.0	−0.37	0.93
9	C	405.3	11.0	0.56	1.17
10	C	403.3	21.0	0.35	2.23
11	A	39.3	9.5	0.69	0.70
12	B	162.2	6.6	0.19	0.76
13	B	160.4	4.3	−0.02	0.50
14	B	154.5	9.1	−0.70	1.05
15	A	26.2	6.2	−0.28	0.46
16	A	25.1	7.2	−0.36	0.53
17	B	153.1	9.2	−0.86	1.07
18	C	388.7	3.5	−1.20	0.37
19	C	384.9	9.4	−1.60	0.99

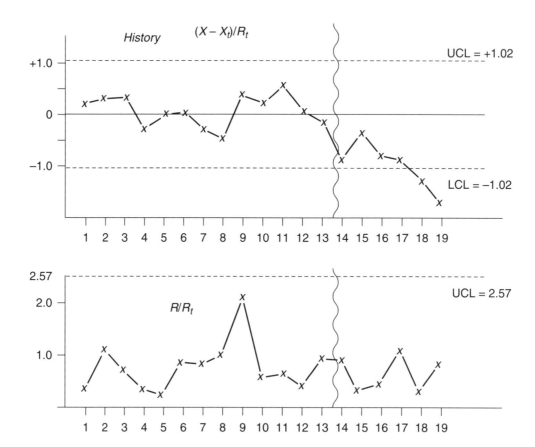

The control chart indicated an out-of-control condition beginning with sample 18.

BIBLIOGRAPHY

Besterfield, D. H. 1994. *Quality Control.* 4th edition. Englewood Cliffs, NJ: Prentice Hall.
Montgomery, D. C. 1996. *Introduction to Statistical Quality Control.* 3rd edition. New York: John Wiley and Sons.
Wheeler, D. J. 1991. *Short Run SPC.* Knoxville, TN: SPC Press.

SHORT-RUN INDIVIDUAL/ MOVING RANGE CHART

Individual/moving range control charts are useful when the opportunity for data is limited and the data is from a family of products that have the same characteristic but the target, tolerance or the average and standard deviation for different products vary. Examples of this might include the angle of a mounting bracket for several different brackets. One model might have a mounting angle of 55 degrees, and another would have a requirement of 60 degrees.

Another example might be a batch manufacturing process for the manufacture of hot melt adhesive. The measured characteristic will be the melt index. The specification for four product's requirements are as follows:

Product	Melt index
A	25–75
B	88–120
C	18–36
D	133–150

A traditional individual/moving range control chart would require between 60 to 100 observations to characterize the process. The time to acquire this number of individual data points for several different products would make it very difficult to establish a traditional individual/moving range control chart.

By deriving a few relationships from the traditional methods and by making a few assumptions, we can place related data from several products on a single control using short-run techniques.

The following derivations will be used to develop a format for short-run statistical process control (SPC).

DERIVATION OF THE PLOTTING CHARACTERISTIC FOR THE LOCATION STATISTIC

In a traditional individual/moving range control chart, the location statistic is the individual observation and its relationship to the average of all of the individuals. Control limits for the individuals are based on the overall average ± 3 standard deviations. The three standard deviations are calculated from $3S = 2.66\,\overline{MR}$, where $\overline{MR}$ is the average moving range.

For any individual/moving range control:

1. $\text{LCL}_x < X < \text{UCL}_x$
2. $\overline{X} - 2.66 \, \overline{MR} < X < \overline{X} + 2.66\overline{MR}$
3. Subtracting $\overline{X}$ from all terms:

$$-2.66\overline{MR} < X - \overline{X} < +2.66\overline{MR}$$

4. Dividing all terms by $\overline{MR}$:

$$-2.66 < \frac{X - \overline{X}}{\overline{MR}} < +2.66$$

Using this transformation, the new plot point for the individuals will be $X - \dfrac{X - \overline{X}}{MR}$, and the new upper and lower control limits for this plot point will be ±2.66.

For each product, there will be a unique average and average moving range. This unique average and moving range will be identified as the target average X_T and the target moving range $\overline{MR}_T$.

DERIVATION OF THE PLOTTING CHARACTERISTIC FOR THE VARIATION STATISTIC

The traditional characteristic for the location statistic is the moving range and how it behaves relative to the upper control limit (UCL). Recall that there is no lower control limit (LCL) for the moving range when using the individual/moving range control chart.

For all cases of the moving range ($n = 2$):

$$\text{Moving range} = |X_1 - X_2|$$

For the plot points X_1 and X_2 as transformed to the short-run format, we have a moving range calculated as

$$\text{Moving range} = \left| \frac{X_1 - \overline{X}_T}{\overline{MR}_T} - \frac{X_2 - \overline{X}_T}{\overline{MR}_T} \right|$$

The traditional UCL for the moving range is given by $\text{UCL}_R = 3.268\overline{MR}$. Therefore, any given moving should be less than the UCL:

$$3.268\overline{MR} > MR$$

Dividing by $\overline{MR}$, we have

$$3.268 > \frac{MR}{\overline{MR}}$$

But since the individual plot points have already been divided by an expected or target average moving range $\overline{MR}_T$, we do not have to divide again. The plot point for the short-run moving range simply becomes the moving range, MR, and the UCL becomes 3.268 with an average moving range of 1.00.

SELECTING A VALUE FOR $\overline{X}_T$ AND $\overline{MR}_T$

Each product must have an expected average $\overline{X}_T$

We may let the nominal $= \overline{X}_T$, or we may determine it from a collection of data such as a collection of measurements or control chart data.

For an expected value of the average moving range $\overline{MR}_T$, we may:

1. Use historical control chart data
2. Estimate $\overline{MR}_T$ from an expected or measured Cpk
   a. For unilateral specifications:

   $$\overline{MR}_T = \frac{1.128\,|\,\text{spec. limit} - \overline{X}_T\,|}{3C_{pk}}$$

   b. For bilateral specifications:

   $$\overline{MR}_T = \frac{1.128\,|\,\text{upper spec.} - \text{lower spec.}\,|}{6C_{pk}}$$

3. Calculate $\overline{MR}_T$ from a collection of data using the relationship of $\overline{MR}_T = f_{MR}\, S,$

   where: f_{MR} = a constant dependent on the number of data points used to calculate the sample standard deviation
   S = the sample standard deviation

Sample size n	f_{MR}
5	1.200
6	1.185
7	1.180
8	1.169
9	1.164
10	1.160
11	1.156
12	1.154
13	1.152
14	1.150
15	1.148
20	1.143
25	1.140

Case Study

The Acme Chemical Co. produces several bioplasmic double regenerative phalangic processing aids (the famous Glycomul line). Each of these products has specific hydroxyl value. These products are produced in a batch operation requiring approximately four hours to produce a 2000-gallon batch of product. The specification from the various products is as follows:

Product	Specification
Glycomul S	235–260
Glycomul L	330–350
Glycomul T	58–69

Step 1. Determine values for $\overline{X}_T$ and $\overline{MR}_T$ for each product.

For Glycomul S:

We have 20 previous batch records. The average of the 20 batches is 249.8, and the standard deviation of the data points is $S = 2.19$. Using the relationship $\overline{MR}_T = f_{MR} S$, the expected average moving range is

$$\overline{MR}_T = (1.143)(2.19)$$

$$\overline{MR}_T = 2.50$$

We will let the average of $249.8 = \overline{X}_T$.

For Glycomul L:

We have 12 batch records for glycomul L. The average and standard deviation for these data points give $\overline{X}_T = 335.2$ and $S = 3.79$. Calculate $\overline{MR}_T$ from $f_{MR} = 1.154$ and $S = 3.79$.

$$\overline{MR}_T = (1.154)(3.79)$$

$$\overline{MR}_T = 4.37$$

For Glycomul T:

Since this is a new product, we have no previous data. The development pilot plant trials indicate that there will be some difficulty in manufacturing this product. We will assume a C_{pk} of 1.00 and let the process average equal the specification nominal.

$$\overline{X}_T = \text{Nominal} = 58 - \frac{(69 - 58)}{2} = 63.5$$

For $\overline{MR}_T$:

$$\overline{MR}_T = \frac{1.128 \,|\, \text{upper spec.} - \text{lower spec.}\,|}{6C_{pk}}$$

$$\overline{MR}_T = \frac{1.128\,|\,69 - 58\,|}{6(1.00)} \qquad \overline{MR}_T = 2.07$$

Summary of product parameters:

Product	Code	$\overline{X}_T$	$\overline{MR}_T$	Specification
Glycomul S	A	249.8	2.50	235–260
Glycomul L	B	335.2	4.37	330–350
Glycomul T	C	63.5	2.07	58–69

This information should be recorded on the control chart.

Step 2. Collect historical data to characterize the process.

Data is collected for the next 18 batches of production. This information is recorded on the control chart.

Date	Product	Hydroxyl value	Date	Product	Hydroxyl value
11/3/96	A	250	11/6/96	C	65
11/3/96	A	248	11/7/96	C	62
11/3/96	A	253	11/7/96	A	252
11/3/96	B	334	11/7/96	B	335
11/4/96	B	336	11/8/96	A	247
11/5/96	C	63	11/8/96	A	250
11/5/96	A	249	11/8/96	C	65
11/5/96	A	253	11/9/96	C	60
11/6/96	C	64	11/9/96	C	66

Step 3. Calculate the location and variation plot for each data point. Plot points on the chart.

Sample #1

Each observation must have a transformed value to plot for the individual. This sample is the hydroxyl value for a batch of Glycomul S identified as product A. The target average is 249.8, and the target moving range is 2.50. The calculated short-run plotting value for this sample is

$$\text{Individual plot point} = \frac{(X - \overline{X}_T)}{\overline{MR}_T} = \frac{(248.0 - 249.8)}{2.50} = \frac{1.80}{2.50} = 0.72$$

The variation plot point for this first sample is undefined as there is no moving range possible with only one data point. For this reason, the first moving range cell is blackened in.

Sample #2

Another batch of Glycomul S is run with a reported hydroxyl value of 248.0. The target value is again 249.8 with a target moving range of 2.50. The calculated short-run plot point is

$$\text{Individual plot point} = \frac{(X - \overline{X}_T)}{\overline{MR}_T} = \frac{(248.0 - 249.8)}{2.50} = \frac{1.80}{2.50} = 0.72$$

The moving range is the algebraic difference between the largest and the smallest consecutive values.

$$\text{Moving range} = 0.08 - (-0.72) = 0.80$$

Note: All moving ranges are positive.
Plot the points and connect to the previous point.

Sample #3

A third batch of Glycomul S is run with a reported hydroxyl value of 253.0. The location plot point is

$$\frac{(X - \overline{X}_T)}{\overline{MR}_T} = \frac{(253.0 - 249.8)}{2.50} = \frac{3.20}{2.50} = 1.28$$

$$\text{Moving range} = 1.28 - (-0.72) = 2.00$$

Sample #4

With the fourth sample, we have a product change. The next batch run is Glycomul L, which has a different set of short-run parameters. For this product, $X_T = 335.2$ and $\overline{MR}_T = 4.37$. The measured hydroxyl value for this batch is 334.0.

$$\frac{(X - \overline{X}_T)}{\overline{MR}_T} = \frac{(334.0 - 335.2)}{4.37} = \frac{-1.20}{4.37} = -0.27$$

$$\text{Moving range} = 1.28 - (-0.27) = 1.55$$

Sample #5

Another batch of Glycomul is run with a reported hydroxyl value of 336.0.

$$\frac{(X - \overline{X}_T)}{\overline{MR}_T} = \frac{(336.0 - 335.2)}{4.37} = \frac{0.80}{4.37} = 0.18$$

$$\text{Moving range} = 0.18 - (-0.27) = 0.45$$

Sample #6

A new product Glycomul T, which was never produced except in a pilot operation, is run for the first time. We are using the nominal for the target average $\overline{X}_T = 63.5$. The expected average moving range, $\overline{MR}_T$ is based on an assumed C_{pk} of 1.00. The estimate for $\overline{MR}_T$ is 2.07.

$$\frac{(X - \overline{X}_T)}{\overline{MR}_T} = \frac{(63.0 - 63.5)}{2.07} = \frac{-0.50}{2.07} = -0.24$$

$$\text{Moving range} = 0.18 - (-0.24) = 0.42$$

Continue calculating the plot points, and plot the results. The completed control chart showing all of the plotted historical data points follows:

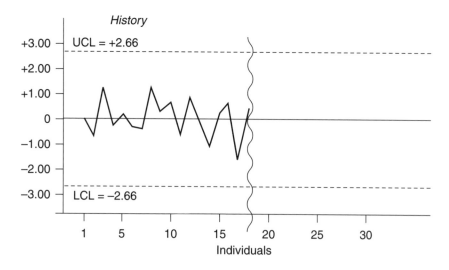

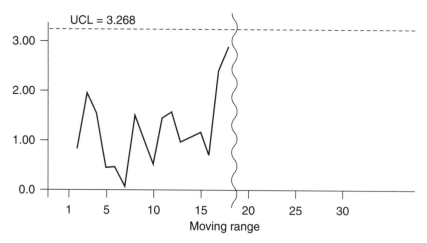

Step 4. Continue to collect future data, looking for signs of a process change.

A change in the supplier of a key raw material was made beginning with sample #19. Using the following sample data, do you feel that this change in supplier changed the process?

Sample	Product	Hydroxyl value	Sample	Product	Hydroxyl value
19	A	248	23	B	333
20	A	250	24	C	59
21	B	331	24	C	59
22	B	329	25	A	243

The completed short-run individual/moving range control chart follows. It appears that the process has changed with sample number 25.

Step 5.

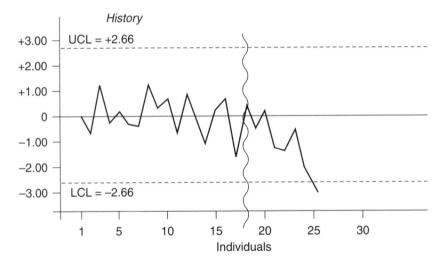

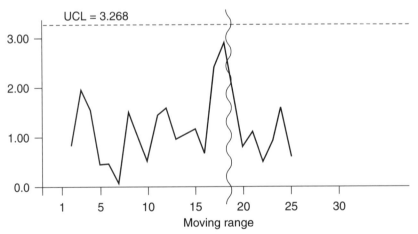

BIBLIOGRAPHY

Besterfield, D. H. 1994. *Quality Control.* 4th edition. Englewood Cliffs, NJ: Prentice Hall.

Montgomery, D. C. 1996. *Introduction to Statistical Quality Control.* 3rd edition. New York: John Wiley and Sons.

Wheeler, D. J. 1991. *Short Run SPC.* Knoxville, TN: SPC Press.

SPC CHART INTERPRETATION

A statistical process control (SPC) chart is essentially a set of statistical control limits applied to a set of sequential data from samples chosen from a process. The data comprising each of the plotted points are a location statistic such as an individual, an average, a median, a proportion, and so on. If the control chart is one to monitor variables data, then an additional associated chart for the process variation statistic can be utilized. Examples of variation statistics are the range, standard deviation, or moving range.

By their design, control charts utilize unique and statistically rare patterns that we can associate with process changes. These relatively rare or unnatural patterns are usually assumed to be caused by disturbances or influences that interfere with the ordinary behavior of the process. These causes that disturb or alter the output of a process are called *assignable causes.* They can be caused by:

1. Equipment
2. Personnel
3. Materials

Most of these patterns can be characterized by several features such as:

1. Degree of statistical rarity for processes where no change has occurred (rate of false alarm)
2. Direction in which the process has changed (increased or decreased)

Not all of the SPC detection rules indicate a direction of process change. For that reason, it is suggested that these rules or patterns be given a low priority for consideration and application.

Several of these patterns or *SPC detection rules* will be discussed. These SPC rules will be numbered as they are presented, and the example chart diagrams will be that of an average. The detection rules however, apply to all control charts (individual/moving range, p chart, and so on). The example control chart diagrams will have the following defined areas:

Zone A: The area defined by the average $\pm 1S$

Zone B: The area defined by the limits of $\bar{X} - 1S$ to $\bar{X} - 2S$ and $\bar{X} + 1S$ to $\bar{X} + 2S$

Zone C: The area defined by the limits of $\bar{X} - 2S$ to $\bar{X} - 3S$ and $\bar{X} + 2S$ to $\bar{X} + 3S$

Rule 1. A single point outside of the control limits (outside of zone C, $\bar{X} \pm 3S$)

Violation of rule 1 will be invoked once out of 370 times when, in fact, there has been no shift or change in the location statistic of a process. Control limits are based on an average $\pm 3S$, in which case we expect to have values inside of the limits 99.73 percent of the time and outside of the limits 0.27 percent of the time (0.135 percent above the upper control limit [UCL] and 0.135 percent below the lower control limit [LCL]).

Some texts refer to rule 1 as indicating an *out-of-control* condition and the others as *rule violations*. In fact, however, the occurrence of any of the rules indicates a *rule violation*, which we interpret as a sign of process change with the full understanding that there is some small probability that the rule violation is a false alarm and no process change has really occurred.

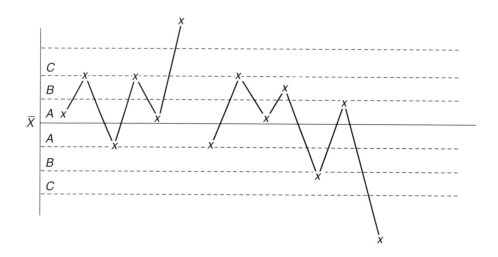

Rule 2. Seven points in a row in zone A or beyond and none outside of the control limits

The probability of getting a single point above the average is 0.50, and the probability of getting a single point above the UCL is 0.00135. Therefore, the probability of getting a point between the average and the UCL is

$$
\begin{array}{r}
0.500000 \\
-0.00135 \\
\hline
0.49865
\end{array}
$$

The probability of getting seven points in a row on the same side of the average but not exceeding the control limits is:

$$2(0.499325)^7 = (0.49865)^6 = 0.015, \text{ or } 1 \text{ in } 67$$

The number of consecutive points in a row varies from reference to reference. All are correct; only the false alarm rate (and, hence, the average run length, ARL) differ. See the following sources for various selections for the number of points on one side of the average:

Number of consecutive points

8
9
7
9

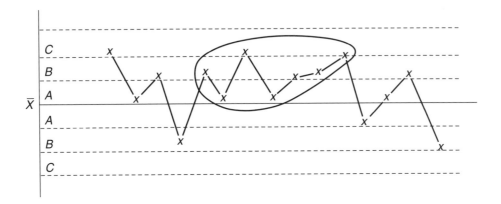

When using rule 2 to detect changes in the range portion of a average/range control chart, caution should be exercised, as the range is not normally a distributed variable.

A comparison of the normal distribution, the range distribution for $n = 2$, and the distribution for $n = 5$ is presented in the following table and figure:

Limit region	Normal	Range, $n = 2$	Range, $n = 5$
>Avg. + 3S	0.00135	0.0094	0.0046
Avg. + 2S to Avg. + 3S	0.02135	0.0360	0.0294
Avg. + 1S to Avg. + 2S	0.1360	0.1162	0.1231
Avg. to Avg. + 1S	0.3413	0.2622	0.3120
Avg. To Avg. − 1S	0.3413	0.1724	0.3396
Avg. − 1S to Avg. − 2S	0.1360	0.1910	0.1735
Avg. − 2S to Avg. − 3S	0.02135	0.2128	0.0178
<Avg. − 3S	0.00135	0.0000	0.0000

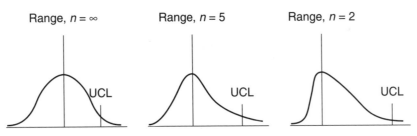

Distribution of Range as a Function of Sample Size *n*.

As *n* increases, the distribution approaches a normal distribution. Small sample sizes ($n < 7$) tend to be skewed to the higher values.

Rule 2 is applicable for attribute control charts. In certain cases for attribute control charts, however, the LCL will be calculated and will default to a zero. This will inhibit the ability to have single values less than LCL (rule 1) but will not affect the probability of rule 2.

Rule 3. Seven points consecutive increasing or decreasing (six intervals)

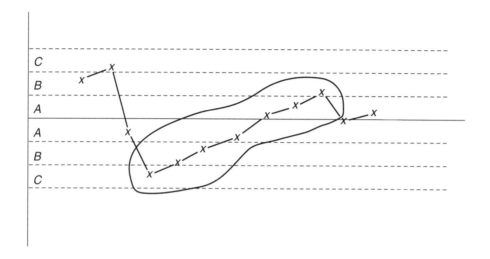

The probability of getting seven consecutive point increasing (or decreasing) due to random chance is $2/7! = 0.0004$, or 1 in 2500.

The International Standard ISO-8258 (Shewhart control charts) uses only six (instead of seven), in which case the probability changes to $0.0028 (= 2/6!)$, or 1 in 360.

Rule 4. Two out of three consecutive points falling in zone C

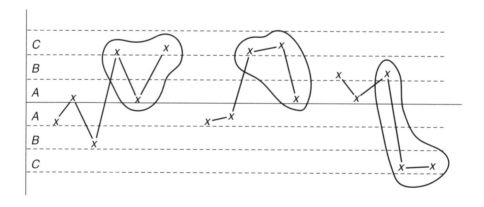

The probability of getting two consecutive points in zone C or beyond is determined by squaring the probability of getting a single point, or $0.0227 \times 0.0227 = 0.000516$ (1 in 1941).

The probability of getting a third point in the sequence in another area but not in zone C is 0.9773 (= 1 − 0.0227). The probability of getting three consecutive points, two of which are in zone C or beyond and one that is not is given by

$$0.0227 \times 0.0227 \times 0.9773 \times 3 = 0.00151$$

The probabilities are multiplied by three because there are three independent events.

We multiply this probability by two to accommodate rule violations above and below the average to obtain the overall probability of this rule violation, which is 0.00302, or 1 in 331.

Rule 5. Four out of five consecutive points falling in zone B or zone C and none exceeding the upper or lower control limit

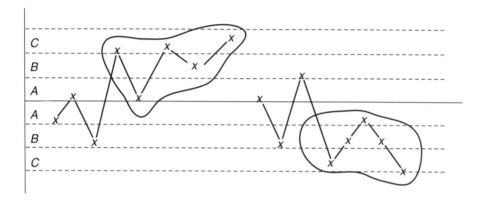

The probability of rule 5 occurring randomly when there has been no change is given by

$$(0.1574)^4(0.840)(5)(2) = 0.0051, \text{ or } 1 \text{ in } 196$$

The probability of getting a single point in zone B or zone C is 0.1574 [0.1360 for zone B + 0.02135 for zone C). The probability of getting a point not in zone B or zone C but inside of the UCL and the LCL is 0.8399 (1 − 0.1574 − 2(0.00135)].

Rule 6. Five consecutive points outside of zone A but none exceeding the upper or lower control limit

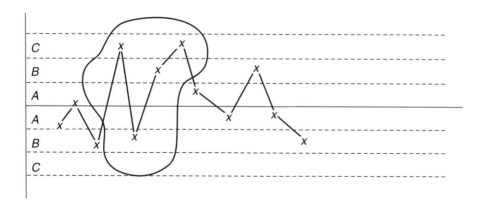

The probability of getting five consecutive points outside of zone A but inside of the control limits is

$$\left(\frac{15}{16}\right)(0.31738)^5 = 0.00302, \text{ or } 1 \text{ in } 331$$

This particular rule does not indicate a direction of process change but perhaps reflects a condition of mixed distributions where samples are drawn from different machines, processes, or raw materials.

BIBLIOGRAPHY

Montgomery, D. C. 1996. *Introduction to Statistical Quality Control.* 3rd edition. New York: John Wiley and Sons.

Wheeler, D. J. 1995. *Advanced Topics in SPC.* Knoxville, TN: SPC Press.

Nelson, L. S. *Journal of Quality Technology* 16, no. 4 (October 1984).

Grant, E. L., and Leavenworth, R. S. 1996. *Statistical Quality Control.* 7th edition. New York, NY: McGraw-Hill.

ISO-8258: 1991. *Shewhart Control Charts.* 1st edition.

TAGUCHI LOSS FUNCTION

Taguchi defines *quality* as the avoidance of loss to society in terms of economic loss or dollars. This loss is a result of the noncompliance to manufacturing or service specification targets. He also suggests that this loss can be mathematically modeled as a function. The Taguchi loss function is quadratic in nature in that Taguchi hypothesizes that the loss is proportional to the square of the deviation from the target value.

Taguchi loss function:

$$Cx = K(X - Tg)^2$$

where: Cx = cost, dollars
K = proportionally constant (loss parameter or loss constant)
X = observed or measured value
Tg = Target or nominal value (perfection), sometimes abbreviated τ

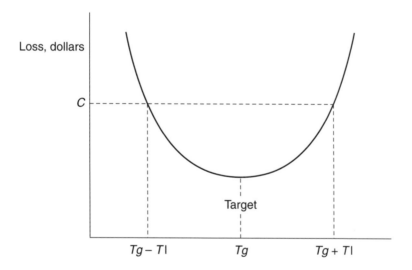

The key to utilizing the Taguchi loss function is in the determination of the proportionally constant K.

The specification for a shaft is 1.00 ± 0.05. The cost for nonconformance (Cx) is \$25.00. What is the value for K?

$$Cx = K(X - Tg)^2 \qquad K = \frac{Cx}{(X - Tg)^2} \qquad K = \frac{25.00}{(1.05 - 1.00)^2} = 10,000$$

Knowing K, we may now calculate the loss for any shaft diameter. For example, given $K = 10,000$ what is the expected loss when the shaft diameter is 1.08?

$$Cx = K(X - Tg)^2 \qquad Cx = 10,000(1.08 - 1.00)^2 \qquad Cx = 10,000(0.08)^2 \qquad Cx = \$64.00$$

Even if the shaft is within the specification limits, there will be an associated loss. The Taguchi philosophy is such that *any* deviation from perfection (target) will result in a loss (to society). The loss for a shaft measuring 0.98 is

$$Cx = 10,000(0.98 - 1.00)2 \qquad Cx = \$4.00$$

A customer's specification for a particular part is 10.00 ± 0.15 ($Tg \pm Tlc$), and the associated cost of nonconformance is \$35.00, Cc. In an effort to avoid producing nonconforming parts, the manufacturer sets an internal specification of 10.00 ± 0.05 ($Tg \pm Tlm$). The internal cost to the manufacturer for failing to meet this requirement is \$12.00.

The customer will realize a cost of Cc whenever the product lies at or exceeds the specification of $Tg \pm Tlc$. The customers loss will be

$$Cc = K(Tg + Tlc - Tg)^2$$

$$Cc = K(Tlc)^2$$

Likewise, it can be shown that the loss to the manufacturer when exceeding the manufacturing specification is

$$Cm = K(Tlm)^2$$

The value for the loss parameter K can be found by solving for K in either relationship.

$$\text{Loss parameter } K = \frac{Cc}{(Tlc)^2} = \frac{Cm}{(Tlm)^2}$$

In this example where $Cm = \$12.00$ and $Tlm = \pm 0.05$, the value for the loss parameter K is given by

$$K = \frac{12.00}{(0.05)^2} = \$4800$$

We can now determine the required manufacturing specification tolerance:

$$K = \frac{Cc}{(Tlc)^2} = \frac{Cm}{(Tlm)^2}$$

$$Tlm = Tlc \sqrt{\frac{Cm}{Cc}} \qquad Tlm = 0.15 \sqrt{\frac{12}{35}} \qquad Tlm = \pm 0.088$$

The manufacturer's tolerance can be determined when the *ratio* of the consumer's loss to the manufacturer's loss is known.

Example:

Customer's tolerance, $Tlc = \pm 0.015$

Customer's loss, $Cc = \$25.00$

Manufacturer's loss, $Cm = \$15.00$

$$\text{Manufacturer's tolerance, } Tlm = Tlc \sqrt{\frac{Cm}{Cc}}$$

$$\text{Manufacturer's tolerance, } Tlm = 0.015 \sqrt{\frac{15}{25}}$$

$$\text{Manufacturer's tolerance, } Tlm = \pm 0.012$$

In some unique cases, the loss incurred when the product is below the lower specification may be different than that when the upper specification is exceeded. An example would be the diameter of a hole or shaft. If a shaft is oversized, it is relatively easy or less costly to reduce by removing more material; however, if the shaft is undersized, it is more complex and costly to add material. From a practical point, the undersized part is simply scrapped.

The shape of the Taguchi loss function is nonsymmetrical due to these differences.

Consider the following case where the specification is 10.00 ± 0.20. The cost of being at the lower specification is $120.00 (*LCx*), and the cost of being at the upper specification is $22.00 (*UCx*).

These values will be used to calculate a unique K value for those measurements falling below the target and another unique K for those measurements falling above the target. These two K values will be designated as K_L and K_U.

K_L determination:

Loss at lower Specification, $LCx = \$120.00$

$$120.00 = K_L (X - Tg)^2 \qquad K_L = \frac{120.00}{(X - Tg)^2} \qquad K_L \frac{120.00}{(0.2)^2} \qquad K_L = \$3000$$

K_U determination:

Loss at upper specification, $UCx = \$22.00$

$$22.00 = K_U (Tg - X)^2 \qquad K_U = \frac{22.00}{(Tg - X)^2} \qquad K_U = \frac{22.00}{(10.0 - 9.8)^2} \qquad K_U = \$550$$

The loss function for the parts measuring less than the target is given by

$$LCx = 3000(Tg - X)^2$$

The loss function for the parts measuring greater than the target is given by

$$UCx = 550 (X - Tg)^2$$

Using these loss functions, the following losses are determined for various measurement observations:

	Measurement	Loss, dollars	Measurement	Loss, dollars
	9.5	750	10.3	49.5
	9.6	480	10.4	88
	9.7	270	10.5	137.5
LS	9.8	120	10.6	198
	9.9	30	10.7	269.5
Tg	10.0	0	10.8	352
	10.1	5.5	10.9	446
US	10.2	22	11.0	550

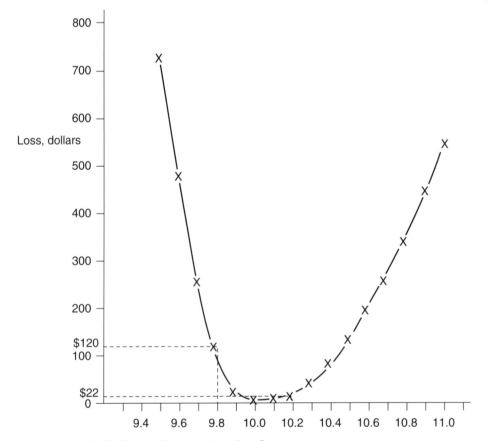

Taguchi Loss with Differing Proportionality Constants.

LOSSES WITH NONSYMMETRIC SPECIFICATIONS

Consider the follow case. A process variable has a specification requirement of 8.50 (+0.2/–0.5) and a loss of $18.00 C_L associated with falling below the lower specification and a loss of $24.00 C_U associated with exceeding the upper specification. Given the following 25 observations, what is the average expected total loss *ATL* for this process?

Data:

8.6	8.35	8.4	8.4	8.5
8.4	8.4	8.65	8.6	8.2
8.15	8.35	8.6	8.4	8.4
8.2	8.3	8.4	8.5	8.5
8.3	8.4	8.35	8.5	8.5

$$\text{Upper mean square, UMS} = \frac{1}{n}\Sigma(X_U - tgt)^2$$

where: X_U = data below the target or nominal
n = total samples
tgt = target or nominal

$$\text{UMS} = \frac{1}{25}[3(8.6 - 8.5)^2 + 2(8.65 - 8.5)^2] = 0.0021$$

$$\text{Lower mean square, LMS} = \frac{1}{n}\Sigma(X_L - tgt)^2$$

where: X_L = data below target or nominal

$$\text{LMS} = \frac{1}{25}[(8.15 - 8.5)^2 + 2(8.2 - 8.5)^2 + 2(8.3 - 8.5)^2$$
$$+ 3(8.35 - 8.5)^2 + 8(8.4 - 8.5)^2)] = 0.0212$$

$$K_L = \frac{C_L}{(tgt - \text{lower spec.})^2} = \frac{18.00}{(0.5)^2} = 72$$

$$K_U = \frac{C_U}{(\text{Upper spec.} - tgt)^2} = \frac{24.00}{(0.2)^2} = 600$$

Average loss lower, $AL_L = K_L \text{ (LMS)} = (72)(0.0212) = \$1.53$

Average loss upper, $AL_U = K_U \text{ (UMS)} = (600)(0.0021) = \$1.26$

Average total loss, $\text{ATL} = AL_L + AL_U = \$1.53 + \$1.26 = \$2.79$

BIBLIOGRAPHY

Grant, E. L., and R. S. Leavenworth. 1996. *Statistical Quality Control.* 7th edition. New York: McGraw-Hill.

Wheeler, D. J. 1995. *Advanced Topics in Statistical Process Control.* Knoxville, TN: SPC Press.

TESTING FOR A NORMAL DISTRIBUTION

In many statistical applications, there is an assumption that the data come from a normally distributed universe. Applications such as process capability can be affected by nonnormal distributions. This module discusses four methods for testing for a normal distribution. The methods presented here are graphical and statistical. The graphical methods are more subjective in the interpretation but are generally easier to use. There are advantages and disadvantages for both the analytical and graphical methods.

NORMAL PROBABILITY PLOT

Normal probability plots are a method to test for a normal distribution. They are based on the fact that data plotted on normal probability graph paper will yield a straight line when the data are normally distributed. The normal probability paper is scaled in the vertical axis based on the normal probability function and is scaled in the horizontal axis as a linear scale.

A normal probability plot has two advantages:

1. It can be accomplished with a small sample size ($n \geq 15$).
2. It is relatively easy to perform and requires simple calculations.

However, the one disadvantage is that the criteria for normalcy is the degree to which a straight line is obtained. This judgement is sometimes very subjective.

The normal plotting technique will be illustrated using the following example.

The head of radiology in a midwestern hospital wants to determine if the length of time required to perform a series of tests is normally distributed. Fifteen samples have been obtained with the time recorded in minutes.

Step 1. Collect data and arrange in ascending order.

Sample number I	Sample value X
1	14.0
2	15.8
3	16.0
4	17.3
5	17.7
6	18.3
7	19.0

Sample number I	Sample value X
8	19.0
9	20.2
10	20.4
11	21.0
12	21.9
13	22.3
14	22.8
15	24.6

Step 2. Calculate the median rank MR for each data value.

There are several methods to estimate the median rank for data. One method (Hazen's) is based on the relationship:

$$\% MR = \frac{(i - 0.5)}{n} \times 100$$

Another relationship is that of Benard's median rank. Benard's approximation is accurate to 1 percent for samples of $n = 5$ and 0.1 percent for samples of $n = 50$.

$$\% MR = \frac{(i - 0.3)}{(n + 0.4)} \times 100$$

where: n = total sample size
i = order of data

Both the Hazen and Benard formulas are derived from the general form of

$$MR = \frac{i - c}{n - 2c + 1} \text{ for the Hazen } c = 0.5 \text{ and for the Benard } c = 0.3$$

Another significantly more complex median rank determination is that defined by

$$MR = \frac{n - i + (0.5)^{1/n}(2i - n - 1)}{n - 1}$$

For the example given, Benard's median rank will be used.

1st median rank: $\% MR = \dfrac{(i-0.3)}{(n+0.4)} \times 100$ $\% MR = \dfrac{(1-0.3)}{(15+0.4)} \times 100$ $\% MR = \dfrac{(0.7)}{(15.4)} \times 100$

$\% MR = 4.55\%$

2nd median rank: $\% MR = \dfrac{(i-0.3)}{(n+0.4)} \times 100$ $\% MR = \dfrac{(2-0.3)}{(15+0.4)} \times 100$ $\% MR = \dfrac{(1.7)}{(15.4)} \times 100$

$\% MR = 11.0\%$

In a similar manner continue to calculate the percent median rank for all 15 observations.

Sample number i	Sample value X	% median rank
1	14.0	4.55
2	15.8	11.00
3	16.0	17.53
4	17.3	24.03
5	17.7	30.52
6	18.3	37.01
7	19.0	43.51
8	19.0	50.00
9	20.2	56.49
10	20.4	62.99
11	21.0	69.48
12	21.9	75.97
13	22.3	82.47
14	22.8	88.96
15	24.6	95.45

Step 3. Plot the data vs. the percent median ranks.

Determine a horizontal plotting scale. Locate the value 14.0 on the horizontal axis and the corresponding percent median rank on the vertical axis. Caution should be used in locating the vertical percent median rank as the scale is not linear and changes as the median rank number increases in magnitude.

1st plotted value:

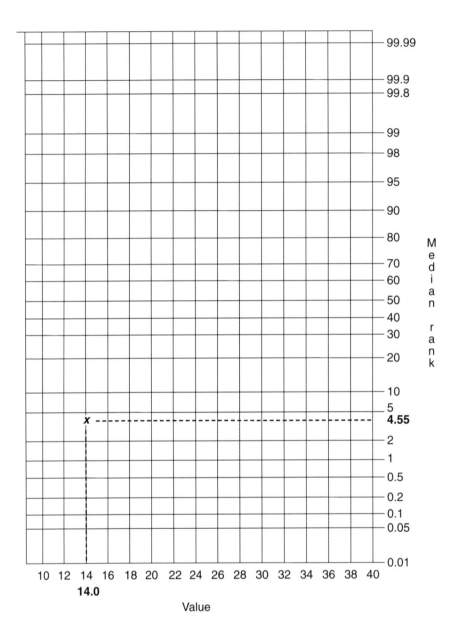

Value

The second data point is plotted; $X_2 = 15.8$ and median rank position = 11.00%.

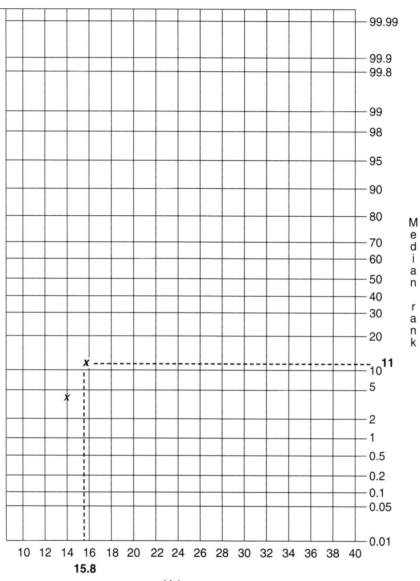

Continue to plot the remaining points in a similar manner. When all of the points have been plotted, draw the best straight line possible through the points. Be less influenced by the first and last points of the data, as this is the region where the tails of the distribution influence the distribution of the normal probability the most. The degree to which a single straight line can be drawn though all of the points is proportional to the degree that a normal distribution is present.

The following figure shows the completed normal probability plot. A normal distribution is supported due to the straightness of the fitted line to the plot.

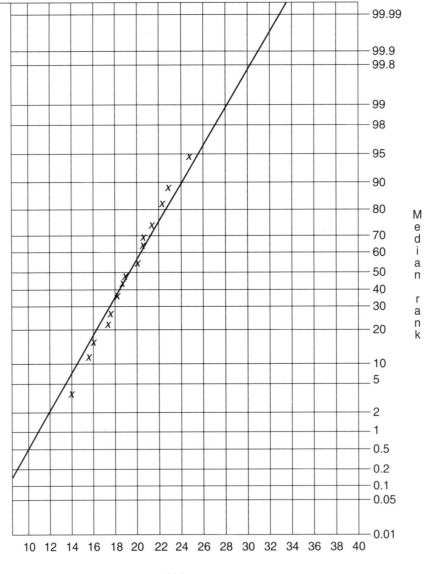

Completed Normal Probability Plot.

The degree to which a straight line can be drawn through all of the points reflects the degree to which a normal distribution is appropriate to describe the data. This example exhibits normalcy.

If normal probability paper is not available, the median ranks (not %MR but MR) may be transformed and plotted using ordinary rectangular graph paper. The transformation equation is given by

$$Y_i = 4.91[(MR)^{0.14} - (1 - MR)^{0.14}]$$

Normal Probability Plots Using Transformed data

When normal probability plotting paper is available, one may still test for a normal distribution using conventional rectangular coordinate paper by transposing the median rank values or P_i values to Y_i values according to the following:

$$Y_i = 4.91[(P_i)^{0.14} - (1 - P_i)^{0.14}]$$

No transformation of the individual data values, X_i, is required.
Consider the following example: Fourteen data values, X_i, are given. The Benard's median rank values are determined and transformed.

Order, i	Data value, X_i	Median rank, P_i	Transformed median rank, Y_i
1	25.08	0.0486	−1.661
2	28.99	0.1181	−1.184
3	29.60	0.1875	−0.885
4	30.54	0.2569	−0.651
5	30.75	0.3264	−0.448
6	32.85	0.3958	−0.263
7	33.51	0.4653	−0.087
8	34.10	0.5347	+0.087
9	34.62	0.6042	+0.263
10	35.90	0.6736	+0.448
11	36.07	0.7431	+0.651
12	36.10	0.8125	+0.885
13	39.36	0.8819	+1.184
14	42.03	0.9514	+1.661

Details of computation for first data point.

Note: The Benard's median rank is calculated as a proportion, not as a percent.

$$\text{Median rank, } P_i = \frac{i - 0.3}{n + 0.4} = \frac{1 - 0.3}{14 + 0.4} = 0.0486$$

Transformation of P_i:

$$Y_i = 4.91[P_i)^{0.14} - (1 - P_i)^{0.14}]$$

$$Y_i = 4.91[(0.0486)^{0.14} - (1 - 0.0486)^{0.14}] = -1.661$$

The Y_i and X_i values are plotted on rectangular coordinate graph paper. The resulting Y_i values represent the vertical plot positions, and the individual data values, X_i, represent the horizontal positions. The degree to which a straight line is obtained is proportional to the degree of normalcy for the data. The completed plot is indicative of normally distributed data because the line is relatively straight. The X_i intercept is an estimate of the average of the data, and the slope is inversely proportional to the standard deviation.

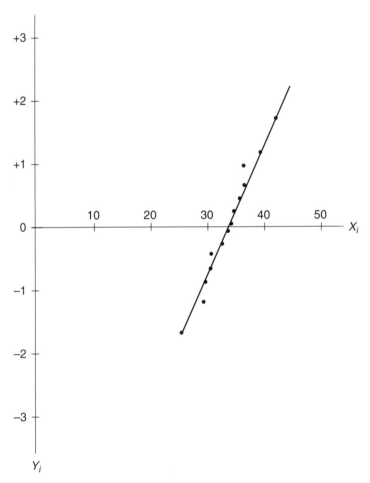

Normal Probability Plot Using Rectangular Coordinates.

CHI-SQUARE GOODNESS-OF-FIT TEST

The chi-square (χ^2) test is derived from expected values with an assumed distribution compared to the distribution of the data being tested.

The chi-square test statistic is calculated by summing up the square of the differences between the observed data and the expected data and dividing by the expected value for a given interval.

$$\chi^2 = \Sigma \frac{(O_i - E_i)^2}{E_i}$$

where: O_i = the observed frequency for a given interval
E_i = the expected frequency for a given interval

The chi-square goodness-of-fit test will be illustrated with the following example.

Step 1. Collect approximately 300 data points, and determine the average and standard deviation.

For convenience, the data have been arranged into 12 column and 25 rows.

Data:

	1	2	3	4	5	6	7	8	9	10	11	12
1	53	56	48	47	56	47	47	51	53	49	50	55
2	46	46	50	44	49	51	45	50	49	51	50	54
3	50	49	52	45	51	56	57	47	48	48	51	44
4	49	46	47	48	51	54	48	45	55	49	46	49
5	48	51	50	52	48	49	51	50	45	53	50	50
6	50	54	50	53	49	51	49	50	49	48	48	50
7	52	50	57	51	54	51	49	55	49	50	53	53
8	54	49	51	52	53	48	54	42	54	49	45	51
9	51	50	51	47	51	48	48	55	58	50	50	49
10	48	51	48	48	47	55	48	45	52	50	50	46
11	54	50	50	50	51	47	52	47	48	49	48	53
12	54	46	50	51	53	50	52	57	47	48	55	49
13	51	56	51	47	49	50	48	48	47	53	50	50
14	47	52	53	52	51	49	47	51	53	49	51	42
15	50	56	50	46	50	48	48	50	51	51	50	54
16	50	52	48	46	48	46	50	43	54	52	51	48
17	44	44	45	49	52	46	49	52	49	48	49	52
18	50	47	51	54	50	47	47	55	53	57	51	53
19	50	49	52	48	54	52	56	47	47	51	48	56
20	47	53	47	53	49	49	53	48	48	54	53	49
21	52	51	45	48	50	53	50	58	55	49	47	53
22	53	48	52	51	50	52	50	50	49	53	51	47
23	51	49	51	53	47	51	53	54	45	45	50	56
24	48	47	57	42	47	48	51	54	52	52	44	50
25	43	53	49	50	49	47	46	53	53	51	53	51

Step 2. Calculate the average and sample standard deviation of the data.

The average, $\bar{X} = 50.5$ and the standard deviation, $s = 3.0$.

Step 3. Sort the data into six discrete intervals.

The intervals are determined as a function of the average and standard deviation. The data are tallied into the intervals as a frequency of occurrence.

Interval definition	Interval value	Frequency
$\leq \bar{X} - 2S$	<44	5
$\bar{X} - 2S$ to $\bar{X} - 1S$	44 to 47	27
$\bar{X} - 1S$ to $\bar{X}$	47 to 50	95
$\bar{X}$ to $\bar{X} + 1S$	50 to 53	106
$\bar{X} + 1S$ to $\bar{X} + 2S$	53 to 56	52
$\geq \bar{X} + 2S$	>56	15

Step 4. Determine the expected frequency of occurrence in the defined intervals given the average and standard deviation for a normal distribution.

If we are dealing with a normal distribution, we can calculate the expected number of data values falling inside the defined intervals. The expected percentage of the data falling inside the intervals and the actual observed data values inside the intervals is calculated as follows:

Interval definition	% of data for normal distribution	Expected data points	Observed data points
$\leq \bar{X} - 2S$	2.5	7.5	5
$\bar{X} - 2S$ to $\bar{X} - 1S$	13.5	40.5	27
$\bar{X} - 1S$ to $\bar{X}$	34.0	102	95
$\bar{X}$ to $\bar{X} + 1S$	34.0	102	106
$\bar{X} + 1S$ to $\bar{X} + 2S$	13.5	40.5	52
$\geq \bar{X} + 2S$	2.5	7.5	15

Step 5. Calculate the chi-square statistic for the data.

The difference between the observed and expected frequency is squared and divided by the expected frequency for each interval. The sum of these terms give the chi-square statistic:

$$\chi^2 = \Sigma \frac{(O_i - E_i)^2}{E_i}$$

For the first interval, we have

$$\frac{(5.0 - 7.5)^2}{7.5} = 0.80$$

For the second interval, we have

$$\frac{(27-40.5)^2}{40.5} = 4.50$$

All intervals are determined and presented in the following table. The total of the individual terms gives the chi-square statistic.

Interval definition	Expected E data points	Observed O data points	$\dfrac{(O-E)^2}{E}$
$\leq \bar{X} - 2S$	7.5	5	0.8
$\bar{X} - 2S$ to $\bar{X} - 1S$	40.5	27	4.5
$\bar{X} - 1S$ to $\bar{X}$	102	95	0.5
$\bar{X}$ to $\bar{X} + 1S$	102	106	0.2
$\bar{X} + 1S$ to $\bar{X} + 2S$	40.5	52	2.7
$\geq \bar{X} + 2S$	7.5	15	7.5

$$X^2 = \Sigma \frac{(O_i - E_i)^2}{E_i} = 16.2$$

Step 6. Compare the calculated chi-square to a critical chi-square value.

The critical chi-square value is determined by looking up the chi-square value for $n - 1 - f$ degrees of freedom and an α risk of 10 percent (0.10). The degrees of freedom are based on a sample size n (number of intervals) of six minus three degrees of freedom. We subtract one more degree of freedom than the total parameters we are estimating, f. There are two parameters being estimated: the average and the standard deviation. The total degrees of freedom is three.

$$\chi^2_{(3,.10)} = 6.251$$

The criterion for rejection of the assumption of a normal distribution is if the calculated χ^2 is greater than the critical χ^2.

If χ^2 calc. $> \chi^2$ crit, reject the assumption of normalcy.

$16.2 > 6.251$; therefore, we reject the hypothesis that the distribution is normal.

SKEWNESS AND KURTOSIS

Skewness Sk and Kurtosis Ku are measures of the normalcy of a distribution. A perfect normal distribution will have a $Sk = 0$ and a $Ku = 0$. Depending on the method of calculation, Ku for a normal distribution may be $Ku = 3$.

Ku and *Sk* are based on calculations using moments of the mean. The *k*th moment of the mean is given by

$$\mu_k = \frac{\Sigma(X_i - \overline{X})^k}{n}$$

$$\mu_k = \text{the } k\text{th moment of the mean}$$

where: $\overline{X}$ = average
X_i = an individual observation

The first moment about the mean is always zero, and the second moment about the mean adjusted for bias is the variance.

$$\mu_i = 0$$

$$\mu_2\left(\frac{n}{n-1}\right) = s$$

A measure of skewness often used is

$$Sk = \sqrt{\frac{\mu_3^2}{\mu_2^3}}$$

and a measure of kurtosis is

$$Ku = \frac{\mu_4}{\mu_2^2} - 3$$

The standard deviations for the skewness and kurtosis are

$$S_{SK} = \sqrt{\frac{6}{n}} \qquad S_{KU} = \sqrt{\frac{24}{n}}$$

The confidence intervals for the skewness and kurtosis are given by

$$\hat{Sk} \pm Z_{\alpha/2}\sqrt{\frac{6}{n}} \quad \text{and} \quad \hat{Ku} \pm Z_{\alpha/2}\sqrt{\frac{24}{n}}$$

$Z_{\alpha/2}$ = 1.96 for 95 percent confidence and 1.645 for 90 percent confidence.

Any *Sk* or *Ku* within 0 ± 1.96 is assumed to come from a normal distribution with a level of confidence of 95 percent. Similarly, if they are within 0 ± 1.645, the level of confidence is 90 percent.

The following data will be used to illustrate calculation of *Sk* and *Ku* and testing for normalcy of the data:

X	Frequency f	$(X - \bar{X})^2$	$(X - \bar{X})^3$	$(X - \bar{X})^4$
6	1	13.67	−50.65	187.40
7	6	7.29	−19.68	53.14
8	15	2.87	−4.91	8.35
9	13	0.49	−0.34	0.24
10	11	0.09	0.03	0.01
11	8	1.69	2.20	2.86
12	6	5.29	12.17	27.98
13	4	10.89	35.94	118.59
14	2	18.49	79.51	341.88
15	1	28.09	148.88	789.05

$$\mu = \frac{\Sigma(X - \bar{X})^2}{n} \qquad \mu_2 = \frac{(1)(13.67) + (6)(7.29) + (15)(2.87) + \ldots + (1)(28.09)}{1 + 6 + 15 + \ldots + 1} = 261.72$$

$$\mu_3 = \frac{\Sigma(X - \bar{X})^3}{n} \qquad \mu_3 = \frac{(1)(-50.65) + (6)(-19.68) + (15)(-4.91) + \ldots + (1)(148.88)}{1 + 6 + 15 + \ldots + 1} = 295.81$$

$$\mu_4 = \frac{\Sigma(X - \bar{X})^4}{n} \qquad \mu_4 = \frac{(1)(187.40) + (6)(53.14) + (15)(8.35) + \ldots + (1)(789.05)}{1 + 6 + 15 + \ldots + 1} = 2772.65$$

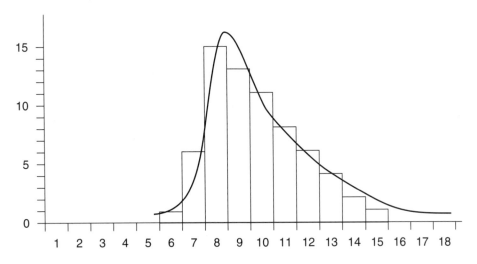

Frequency Histogram of Data.

$$Sk = \sqrt{\frac{\mu_3^2}{\mu_2^3}} = \sqrt{\frac{\left(\dfrac{295.81}{67}\right)^2}{\left(\dfrac{261.72}{67}\right)^3}} = 0.57$$

90 percent confidence interval:

$$0.57 \pm 1.645\left(\sqrt{\frac{6}{67}}\right) \rightarrow 0.08 \text{ to } 1.06$$

$$Ku = \frac{\mu_4}{\mu_2^2} - 3 = \frac{\dfrac{41.383}{67}}{\left(\dfrac{261.72}{67}\right)^2} - 3 = -0.29$$

90 percent confidence interval:

$$-0.29 \pm 1.645\left(\sqrt{\frac{24}{67}}\right) \rightarrow -1.27 \text{ to } 0.69$$

This distribution is skewed to the left as indicated by the shape of the frequency histogram and the positive *Sk*. The distribution is also flatter than a perfectly normal distribution as indicated by the slightly negative *Ku*. The distribution is normal with respect to the amount of kurtosis since the confidence interval for the kurtosis contains zero (which is expected for a kurtosis from a normal distribution). However, the confidence interval for the skewness does not contain zero, and we conclude that the data comes from some distribution other than a normal one.

BIBLIOGRAPHY

Betteley, G., N. Mettrick, E. Sweeney, and D. Wilson. 1994. *Using Statistics in Industry.* New York: Prentice Hall.

Dovich, R. A. 1992. *Quality Engineering Statistics.* Milwaukee, WI: ASQC Quality Press.

Grant, E. L., and R. S. Leavenworth. 1996. *Statistical Quality Control.* 7th edition. New York: McGraw-Hill.

Hayes, G. E., and H. G. Romig. 1982. *Modern Quality Control.* 3rd edition. Encino, CA: Glencoe.

Montgomery, D. C. 1996. *Introduction to Statistical Quality Control.* 3rd edition. New York: John Wiley and Sons.

Petruccelli, J. D., B. Nandram, and M. Chen. 1999. *Applied Statistics for Engineers and Scientists.* Upper Saddle River, NJ: Prentice Hall.

Walpole, R. E., and R. H. Myers. 1993. *Probability and Statistics for Engineers and Scientists.* 5th edition. Englewood Cliffs, NJ: Prentice Hall.

WEIBULL ANALYSIS

The Weibull distribution is one of the most used density functions in reliability. It is named after Waloddi Weibull who invented it in 1937. Weibull's 1951 paper, "*A Statistical Distribution Function of Wide Applicability,*" led to a wide use of the distribution. Weibull demonstrated that this distribution is very flexible for a wide variety of data distributions.

Though the initial reaction to Weibull's distribution varied from skepticism to outright rejection, his distribution has passed the test of time and today is widely accepted as a viable model for application in reliability engineering.

The Weibull probability density function is defined as follows:

$$F(t) = \frac{\beta(t-\delta)^{\beta}-1}{\eta^{\beta}} \; e^{\left[-\left(\frac{t-\delta}{\eta}\right)^{\beta}\right]}$$

where: β (beta) = the shape parameter or Weibull slope
η (eta) = the scale parameter or characteristic life
δ (delta) = the location parameter or minimum life characteristic

Note: If $\beta \neq 1$, we abbreviate the characteristic life as η. When $\beta = 1$, we refer to the characteristic life as the mean time before failure (MTBF) and abbreviate it as θ.

The Weibull distribution can be used as a tenable substitute for several other distributions by varying the shape parameter.

Shape parameter, β	Distribution
1	Exponential
2	Rayleigh
2.5	Lognormal
3.6	Normal

The Weibull reliability function describes the probability of survival as a function of time.

$$R_t = e^{-\left(\frac{t-\delta}{\eta}\right)^{\beta}}$$

When $\beta < 1$, the rate of change in the reliability is decreasing; when $\beta = 1$, the failure rate is constant; and when $\beta > 1$, the failure rate is increasing. These three areas are frequently referred to as infant mortality, useful life, and burn out.

Example of a Weibull reliability calculation:
Calculate the reliability of a component at $t = 80$, where $\beta = 1.2$, $\eta = 230$, and $\delta = 5$.

$$R_t = e^{-\left(\frac{t-\delta}{\eta}\right)^{\beta}}$$

$$R_{80} = e^{-\left(\frac{80-5}{230}\right)^{1.2}}$$

$$R_{80} = 0.77 \text{ or } 77\%$$

The hazard function represents the instantaneous failure rate and can be used to characterize failures in accordance with the bathtub curve.

$$h(t) = \frac{\beta(t-\delta)^{\beta-1}}{\eta^{\beta}}$$

Example:
What is the instantaneous failure rate at $t = 80$, where $\beta = 1.2$, $\eta = 230$, and $\delta = 5$?

$$h(t) = \frac{1.2(80-5)^{0.2}}{230^{1.2}} = 0.0042$$

While the Weibull is very versatile, its drawback is the difficulty in estimating its parameters. There are several methods for estimating the Weibull parameters.

PROBABILITY PLOTTING

Taking the logarithm of the Weibull cumulative distribution twice, the following relationship is obtained:

$$\ln\left[\ln\left(\frac{1}{1-f(t)}\right)\right] = \beta \ln(t) - \beta \ln \theta$$

This is a linear equation with $\ln(t)$ as the independent variable, β as the slope, the dependent variable as $\ln\left[\ln\left(\frac{1}{1-f(t)}\right)\right]$, and $\beta \ln \theta$ as the y-intercept. Plotting $\ln(t)$ against $\ln\left[\ln\left(\frac{1}{1-f(t)}\right)\right]$, we may graphically determine the Weibull parameters β and θ.

For an estimate of $f(t)$, we may use Benard's median rank.

$$f(t) = \frac{i - 0.3}{n + 0.4}$$

where: i = the failure order
             n = the sample size

Example calculation:
Ten identical tools were run until all had failed, giving the following time-to-failure data:
118, 190, 221, 333, 87, 175, 137, 203, 264, and 270

Step 1. Assume $\delta = 0$, and order the data in ascending order.

Time to failure
87
118
137
175
190
203
221
264
270
333

Step 2. Calculate the median rank for each observation.
For the first observation, the median rank (MR) is

$$f(t) = \frac{i - 0.3}{n + 0.4} = \frac{i - 0.3}{10 + 0.4} = 0.067$$

For the seventh observation, the MR is

$$f(t) = \frac{i - 0.3}{n + 0.4} = \frac{7 - 0.3}{10 + 0.4} = 0.644$$

Step 3. Determine the natural logarithm for the time to failure for each observation.

Step 4. Determine the natural logarithm of the natural logarithm of the inverse of $1 - \text{MR}$ for each observation.
For the first ordered observation:

$$\ln\left(\ln\frac{1}{1 - f(t)}\right) = \ln\left(\ln\frac{1}{1 - 0.067}\right) = -2.669$$

For the seventh ordered observation:

$$\ln\left(\ln\frac{1}{1 - f(t)}\right) = \ln\left(\ln\frac{1}{1 - 0.644}\right) = -0.0323$$

The following table summarizes the calculations for steps 1–4:

Order, i	Time to failure	$f(t)$	$\ln(t)$	$\ln[\ln(1/1 - f(t))]$
1	87	0.067	4.47	−2.669
2	118	0.163	4.77	−1.726
3	137	0.259	4.92	−1.205
4	175	0.355	5.16	−0.824
5	190	0.452	5.25	−0.508
6	203	0.548	5.31	0.231
7	221	0.644	5.40	0.032
8	264	0.740	5.58	0.298
9	270	0.837	5.60	0.596
10	333	0.933	5.81	0.994

Step 5. Plot the $\ln(t)$ on the x-axis and $\ln\left[\ln\left(\dfrac{1}{1-f(t)}\right)\right]$ on the y-axis.

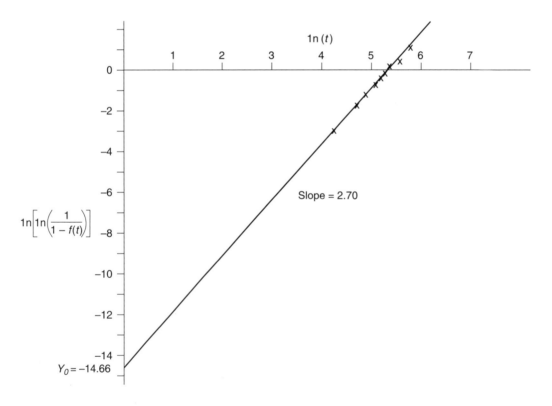

The Weibull slope, $\beta = 2.70$. The Weibull characteristic life η is calculated from the equation

$$\eta = e^{\left(\frac{-y_0}{\beta}\right)}$$

where: Y_0 = the y-intercept
        β = the Weibull slope

The characteristic life is $\eta = e^{\left(\frac{14.66}{2.70}\right)} = 228$.

PLOTTING DATA USING WEIBULL PAPER

A less rigorous graphical technique relies on the use of a special graph paper called Weibull plotting paper. Failure times plotted as a function of the %MR will yield a straight line (with a slope $\beta = 1$) if the probability density function follows that of an exponential model. Graphically we can determine the slope directly from the special paper. The characteristic life η may be determined by locating the failure time that corresponds to a cumulative percent failure of 63 percent. If the plotted data do not yield a single straight line, there may be the influence of a positive minimum life parameter γ. Subtracting the estimated value for the minimum life characteristic, the original data may be adjusted somewhat to yield a straight line, and then an estimate of the characteristic life and Weibull slope may be determined. Consider the following data:

Order number, i	Failure time, hours	%MR
1	28	7.5
2	39	12.7
3	50	20.1
4	60	27.6
5	70	35.1
6	85	42.5
7	100	50.0
8	115	57.5
9	135	64.9
10	160	72.4
11	200	79.9
12	250	87.3
13	370	94.8

A plot of the failure times as a function of the %MR yields a slightly curved line. Discretion must be used in the interpretation of the underlining cause for the nonlinearity. This could be due to the exixtence of mixed distributions, or there could be a failure-free time or a minimum life characteristic.

The minimum life characteristic is time that, when subtracted from all of the failure times, will yield a somewhat straight line. Essentially by subtracting the minimum life characteristic from all of the failure and by replotting the data, we are reducing the Weibull plot to a two-parameter model using the characteristic life and Weibull slope.

We begin by plotting the original data as a function of the %MR. The data are ordered in decreasing order. The MR is calculated using Benard's approximation. The *i*th MR is given by

$$\%MR = \frac{i - 0.3}{n + 0.4} \times 100$$

The original data give the following plot:

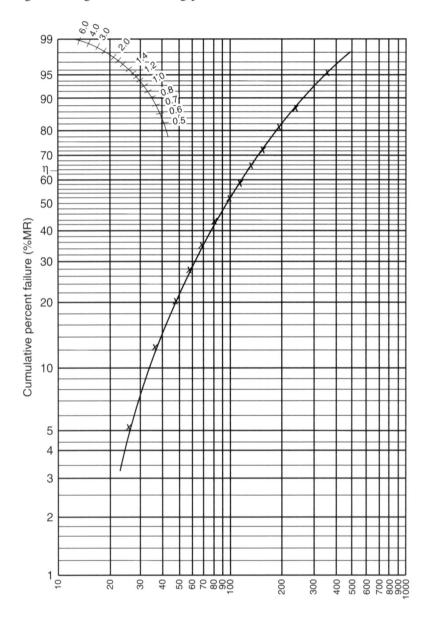

Estimation of Minimum Life Characteristic γ

Equally spaced horizontal lines covering the entire area of the curve are drawn. The midpoint of these two lines is designated as t_2, the maximum failure point as t_3, and the minimum failure point as t_1.

The minimum life characteristic is calculated by

$$\lambda = t_2 - \frac{(t_3 - t_2)(t_2 - t_1)}{(t_3 - t_2) - (t_2 - t_1)}$$

For this example, $t_1 = 28$, $t_2 = 70$, and $t_3 = 370$

$$\gamma = 21$$

Subtracting 21 from all of the data gives the new adjusted data from which the Weibull slope and characteristic life may be determined.

Order number, i	Failure time, hours	%MR	Adjusted failure times
1	28	7.5	7
2	39	12.7	18
3	50	20.1	29
4	60	27.6	39
5	70	35.1	49
6	85	42.5	64
7	100	50.0	79
8	115	57.5	94
9	135	64.9	114
10	160	72.4	139
11	200	79.9	179
12	250	87.3	229
13	370	94.8	349

Replotting the adjusted data gives a nearly straight line.

The characteristic life η is determined by following the 63 percent MR position across until intercepting the adjusted straight line and then going down this position vertically until reaching the horizontal scale. The value for η is read as approximately 112.

The Weibull slope is determined by drawing a perpendicular line from the straight line (from the data) to the reference point on the left designated by the dot (60% MR). This is line L_1. Another perpendicular line is drawn from this reference point to intersect the curved arc from where the Weibull slope is read. This line is designated as L_2. The estimated value for the Weibull slope β is 1.15.

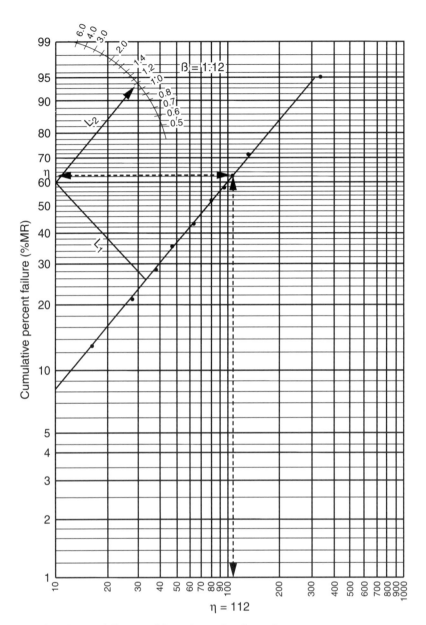

Weibull Plot of Adjusted Data with values for β and η.

BIBLIOGRAPHY

Abernethy, R. B. 1994. *The New Weibull Handbook*. North Palm Beach, FL: Robert B.
    Abernethy.

Dodson, B. 1994. *Weibull Analysis*. Milwaukee, WI: ASQC Quality Press.

Dovich, R. A. 1992. *Reliability Engineering Statistics*. Milwaukee, WI: ASQC Quality Press.

O'Connor, P. D. T. 1991. *Practical Reliability Engineering*. New York: John Wiley and Sons.

Weibull, W. 1951. A Statistical Distribution Function of Wide Applicability. *Journal of
    Applied Mechanics*. 18:293–297.

ZONE CHARTS

The successful implementation of statistical process control (SPC) into the work environment requires that the people most closely related to the process maintain the control chart. In many cases and for a variety of reasons, these individuals may feel uncomfortable with the task of plotting data and having to remember a number of the traditional "rules" that indicate an out-of-control condition or evidence that a change in the process characteristic being monitored has occurred.

Zone charting offers an effective way to overcome many of these types of obstacles. Zone charting employs the use of a summation rule based on the probability of getting values within certain discrete areas as defined by the process average and increments of ± 1, ± 2, and ± 3 standard deviations.

Traditional Shewhart control charts define a location statistic and a variation statistic. The variation statistic is based on ± 3 standard deviations and is derived of some type of range measurement.

The individual/moving range control chart limits for the individual portion are determined by

$$\text{Upper Control Limit (UCL)} = \overline{X} + 2.66\,\overline{MR}$$

and

$$\text{Lower Control Limit (LCL)} = \overline{X} - 2.66\,\overline{MR}$$

where: $\overline{MR}$ = moving range
$\overline{X}$ = average of individuals

The average/range control chart limits for the averages portion is determined by

$$\text{Upper Control Limit} = \overline{\overline{X}} + A_2\overline{R}$$

and

$$\text{Lower Control Limit} = \overline{\overline{X}} - A_2\overline{R}$$

where: $\overline{R}$ = range
$\overline{\overline{X}}$ = average of averages
A_2 = a factor dependent upon the subgroup sample size

In both cases, the control chart limits are based on an average response ±3 standard deviations.

The performance of a control chart is determined by the average amount of time required to detect a change in the process. This performance is measured in terms of the average run length (ARL). The ARL for a Shewhart control chart using a single detection rule of a single point outside the upper or lower control limit where a $1 - \Sigma$ shift has occurred is ARL = 42.

By adding additional rules of detection, we can reduce the ARL for a given amount of shift in the process. For example, by adding the rule of seven points in a row on the same side of the average, we reduce the ARL of 42 to an ARL of 11 (for a $1 - \Sigma$ shift). Other rules are available, but the task of remembering all of the rules becomes burdensome in application.

Zone charts are based on the probability of getting values in each of eight defined zones or areas. The weighting factors are proportional to the inverse of the probability of the event occurring. The zone factors are also adjusted to allow for various ARLs for a given amount of process change.

CONSTRUCTION OF ZONE CHARTS

I   Collect sufficient process data to characterize by determination of the process average and variation. The variation statistic will either be the average range or average moving range depending on the variable chart selected to monitor the process.

II Compute the process average and standard deviation of the process s. Determine the average ±1, ±2, and ±3 values. Each zone area will be identified as A, B, C, and D.

Zone A = average ±1S. The probability of getting a single point inside one of these zones is approximately 34 percent.

Zone B = average + 1S to average + 2S and the average − 1S to average − 2S. The probability of getting a value in one of these zones is approximately 13.5 percent.

Zone C = average + 2S to average + 3S and the average − 2S to average − 3S. The probability of getting values in one of these zones is approximately 2.35 percent.

Zone C = zones above the average + 3S and below the average − 3S. The probability of getting values in one of these zones is approximately 0.15 percent.

ARLs for Selected Zone Scales Deviation from Historical Average in units of Σ

Zone scores					Process change				
A	*B*	*C*	*D*	**CRS**	**0.0**	**0.5**	**1.0**	**1.5**	**2.0**
0	1	3	4	4	49	19	7	4	2.5
0	1	2	4	4	100	26	8	4	2.9
0	1	2	5	5	211	45	11	5	3.3
0	1	2	6	6	320	63	14	6	3.8
0	1	5	12	12	358	86	17	7	3.8
0	1	2	4	5	311	44	11	5	3.6
1	2	4	8	8	41	16	6	4	2.8
0	2	4	8	8	100	26	8	4	2.8
1'	2	6	8	8	27	13	6	3	2.4
1	3	6	12	12	62	21	7	4	2.9

Shewhart 3 Σ limits only:					370	140	42	13	6.5
Shewhart 3 Σ + 7 in a row:					100	33	12	7	3

III Select an appropriate zone-weighting scale. These weights are cumulatively summed until a critical run sum (CRS) has been reached or exceeded. Upon reaching or exceeding the CRS, the process is deemed out of control, a condition that indicates a process change relative to the historical characterization. The CRS value triggers the alarm system for the process.

Several zone-weighting systems and CRS values are available depending on the ARL desired for a given amount of process change.

IV Collect data and plot individual averages or individuals depending on the chart type. Use the following steps:

a. Locate the zone area for the first value to be plotted, and record the proper zone score.

b. Locate the next point. If on the same side of the average as the preceding value, add the zone score of this value to the plotted zone score of the previous value.

c. If the next point is on the opposite side of the average as the previous one, do not accumulate the zone score; start the counting process over.

d. Upon reaching the CRS value, discontinue the accumulation process and start over. Never connect a value that exceeds the CRS to the next point.

The following data will be used to develop an average/range control chart using the zone format.

Collect Historical Data

#1	#2	#3	#4	#5	#6
7.14	7.17	7.13	7.15	7.11	7.19
7.03	7.14	7.23	7.16	7.25	7.20
7.33	7.09	7.23	7.08	7.25	7.28
7.35	7.29	7.37	7.18	7.33	7.27
7.11	7.13	7.08	7.15	7.15	7.31
$\bar{X} = 7.19$	$\bar{X} = 7.16$	$\bar{X} = 7.21$	$\bar{X} = 7.14$	$\bar{X} = 7.22$	$\bar{X} = 7.25$
$R = 0.32$	$R = 0.20$	$R = 0.29$	$R = 0.10$	$R = 0.22$	$R = 0.12$

#7	#8	#9	#10	#11	#12
7.37	7.01	7.42	7.23	7.14	7.26
7.05	7.14	7.22	7.05	7.08	7.36
7.09	7.27	6.98	7.34	7.26	7.12
7.11	7.23	7.12	7.11	7.25	7.24
6.86	7.22	7.24	7.36	7.18	7.16
$\bar{X} = 7.10$	$\bar{X} = 7.17$	$\bar{X} = 7.20$	$\bar{X} = 7.22$	$\bar{X} = 7.18$	$\bar{X} = 7.23$
$R = 0.51$	$R = 0.26$	$R = 0.44$	$R = 0.31$	$R = 0.18$	$R = 0.24$

#13	#14	#15	#16	#17	#18
7.17	7.15	7.37	7.27	7.20	7.28
7.13	7.21	7.23	7.06	7.18	7.20
7.15	7.30	7.16	7.34	7.00	7.25
7.15	7.18	7.13	7.05	7.39	7.41
7.12	7.25	7.06	7.19	7.05	7.17
$\bar{X} = 7.14$	$\bar{X} = 7.22$	$\bar{X} = 7.19$	$\bar{X} = 7.18$	$\bar{X} = 7.16$	$\bar{X} = 7.28$
$R = 0.05$	$R = 0.15$	$R = 0.31$	$R = 0.29$	$R = 0.39$	$R = 0.24$

Samples 1–18 will be used to determine the historical characterization of the process. The location statistic will be the grand average of the averages $\bar{\bar{X}}$.

$$\overline{\overline{X}} = \frac{\overline{X}_1 + \overline{X}_2 + \overline{X}_3 + \cdots + \overline{X}_k}{k} \qquad \overline{\overline{X}} = \frac{7.19 + 7.16 + 7.21 + \cdots + 7.28}{18} \qquad \overline{\overline{X}} = 7.191$$

The variation statistic will be the average range $\overline{R}$.

$$\overline{R} = \frac{R_1 + R_2 + R_3 + \cdots + R_k}{k} \qquad \overline{R} = \frac{0.32 + 0.20 + 0.29 + \cdots + 0.24}{18} \qquad \overline{R} = 0.257$$

The control limits are based on an average ± 3 standard deviations. Three standard deviations for the distribution of averages are calculated using the following relationship:

$$3S_{\overline{X}} = A_2 \overline{R}$$

$$3S_{\overline{X}} = (0.577)(0.257) = 0.148$$

Note: The range portion for this example of the zone format average/range chart will not be presented, as the range portion is performed exactly as with a traditional range chart. See the module entitled *Average/Range Control Chart*.

One standard deviation for the distribution of averages is determined by dividing $3S_{\overline{X}}$ by three.

$$3S_{\overline{X}} = 0.049$$

Zones in increments of one standard deviation about the average are determined, and a zone-weighting scale is assigned for each of the ± 1, ± 2, and ± 3 standard deviation zones. Table 1 lists the zone-definition area and the corresponding zone score.

Table 1.

Zone location	Zone value (1)	Zone score
$> \overline{X} + 3S$	>7.338	8
$\overline{X} + 2S$ to $\overline{X} + 3S$	7.289 to 7.338	4
$\overline{X} + 1S$ to $\overline{X} + 2S$	7.240 to 7.289	2
$\overline{X}$ to $\overline{X} + 1S$	7.191 to 7.240	1
$\overline{X} - 1S$ to $\overline{X}$	7.142 to 7.191	1
$\overline{X} - 1S$ to $\overline{X} - 2S$	7.093 to 7.142	2
$\overline{X} - 2S$ to $\overline{X} - 3S$	7.093 to 7.044	4
$< \overline{X} - 3S$	<7.044	8

These zone values and the zone-weighting factors are recorded on the following zone format $\overline{X}/R$ chart:

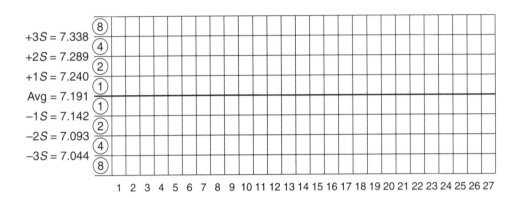

Each of the averages is plotted in the appropriate zone location. The following rules of protocol are used with the zone charting technique:

1. Plotting points are circled with the appropriate zone score placed inside the circle position.
2. Continue to accumulate with the current zone score the previous circled zone score as long as the positions remain on the same side of the average line. Connect the circled values with a line.
3. If the average line in crossed going from one zone area to another, start the accumulation process over again.
4. An indication of an out-of-control condition is evident when a cumulated zone score of greater than or equal to eight is obtained.

The first few values will be plotted to illustrate the concept.

Sample #1

An average of 7.19 is located in zone 1 below the average. A zone score of 1 is placed in this area:

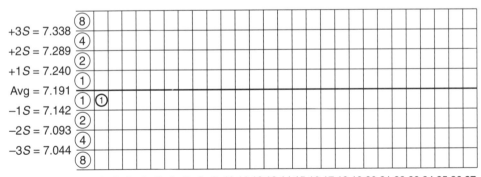

Sample #2

An average of 7.16 is located in zone 1 below. The average has not been crossed; there-fore, we add the previous value of 1 to the current zone area of 1 for a total zone score of 2. The 2 is circled and connected to the preceding value.

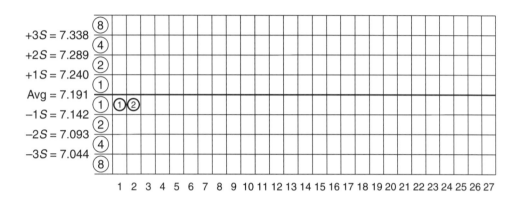

Sample #3

An average of 7.21 is located in zone 1 above. The average has been crossed going from sample #2 to the current position for sample #3; therefore, the accumulation process is started over. The correct value for this zone position is 1.

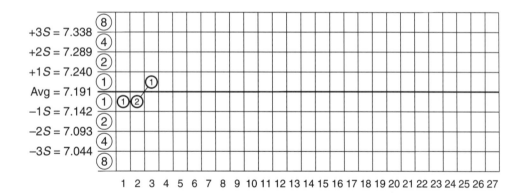

Sample #4

An average of 7.14 is located in zone 2 below. The average has been crossed going from the previous value to the current sample #4; therefore, the accumulation process is started over.

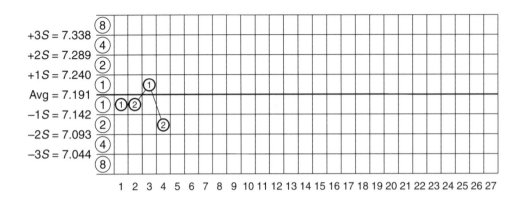

Continue to plot the remaining points for samples 5–18.

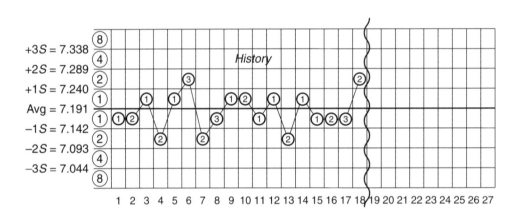

A change has been made in the operation of the process. Based on the next nine sets of data, do you have evidence to support the conclusion that the process data reflect any change?

#19	#20	#21	#22	#23	#24
7.20	7.42	7.14	6.94	7.06	7.13
7.02	7.22	7.02	7.06	7.04	7.24
7.17	6.98	7.27	7.18	7.07	6.97
7.05	7.24	7.22	7.10	7.16	7.23
7.06	6.94	7.22	7.02	7.94	7.41
$\bar{X} = 7.10$	$\bar{X} = 7.16$	$\bar{X} = 7.17$	$\bar{X} = 7.06$	$\bar{X} = 7.05$	$\bar{X} = 7.20$
$R = 0.18$	$R = 0.48$	$R = 0.25$	$R = 0.24$	$R = 0.22$	$R = 0.44$

#25	#26	#27
7.04	7.05	6.93
7.34	7.11	7.05
7.35	7.01	7.17
7.11	7.06	7.02
7.06	7.07	7.08
$\bar{X} = 7.18$	$\bar{X} = 7.06$	$\bar{X} = 7.05$
$R = 0.31$	$R = 0.10$	$R = 0.15$

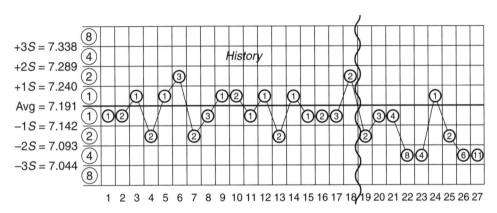

Completed Zone Chart.

BIBLIOGRAPHY

Jaehn, A. H. 1987. Zone Control Charts: A New Tool for Quality Control. *Tappi Journal* (February): 159–161.

APPENDIX A: TABLES

STANDARD NORMAL DISTRIBUTION

Z	x.x0	x.x1	x.x2	x.x3	x.x4	x.x5	x.x6	x.x7	x.x8	x.x9
4.0	0.00003	0.00003	0.00003	0.00003	0.00003	0.00003	0.00003	0.00002	0.00002	0.00002
3.9	0.00005	0.00005	0.00004	0.00004	0.00004	0.00004	0.00004	0.00004	0.00003	0.00003
3.8	0.00007	0.00007	0.00007	0.00006	0.00006	0.00006	0.00006	0.00005	0.00005	0.00005
3.7	0.00011	0.00010	0.00010	0.00010	0.00009	0.00009	0.00008	0.00008	0.00008	0.00008
3.6	0.00016	0.00015	0.00015	0.00014	0.00014	0.00013	0.00013	0.00012	0.00012	0.00011
3.5	0.00023	0.00022	0.00022	0.00021	0.00020	0.00019	0.00019	0.00018	0.00017	0.00017
3.4	0.00034	0.00032	0.00031	0.00030	0.00029	0.00028	0.00027	0.00026	0.00025	0.00024
3.3	0.00048	0.00047	0.00045	0.00043	0.00042	0.00040	0.00039	0.00038	0.00036	0.00035
3.2	0.00069	0.00066	0.00064	0.00062	0.00060	0.00058	0.00056	0.00054	0.00052	0.00050
3.1	0.00097	0.00094	0.00090	0.00087	0.00084	0.00082	0.00079	0.00076	0.00074	0.00071
3.0	0.00135	0.00131	0.00126	0.00122	0.00118	0.00114	0.00111	0.00107	0.00104	0.00100
2.9	0.00187	0.00181	0.00175	0.00170	0.00164	0.00159	0.00154	0.00149	0.00144	0.00140
2.8	0.00256	0.00248	0.00240	0.00233	0.00226	0.00219	0.00212	0.00205	0.00199	0.00193
2.7	0.00347	0.00336	0.00326	0.00317	0.00307	0.00298	0.00289	0.00280	0.00272	0.00264
2.6	0.00466	0.00453	0.00440	0.00427	0.00415	0.00402	0.00391	0.00379	0.00368	0.00357
2.5	0.00621	0.00604	0.00587	0.00570	0.00554	0.00539	0.00523	0.00509	0.00494	0.00480
2.4	0.00820	0.00798	0.00776	0.00755	0.00734	0.00714	0.00695	0.00676	0.00657	0.00639
2.3	0.01072	0.01044	0.01017	0.00990	0.00964	0.00939	0.00914	0.00889	0.00866	0.00842
2.2	0.01390	0.01355	0.01321	0.01287	0.01255	0.01223	0.01191	0.01160	0.01130	0.01101
2.1	0.01787	0.01743	0.01700	0.01659	0.01618	0.01578	0.01539	0.01500	0.01463	0.01426
2.0	0.02275	0.02222	0.02169	0.02118	0.02068	0.02018	0.01970	0.01923	0.01876	0.01831
1.9	0.02872	0.02807	0.02743	0.02680	0.02619	0.02559	0.02500	0.02442	0.02385	0.02330
1.8	0.03593	0.03515	0.03438	0.03363	0.03288	0.03216	0.03144	0.03074	0.03005	0.02938
1.7	0.04457	0.04363	0.04272	0.04182	0.04093	0.04006	0.03920	0.03836	0.03754	0.03673
1.6	0.05480	0.05370	0.05262	0.05155	0.05050	0.04947	0.04846	0.04746	0.04648	0.04551
1.5	0.06681	0.06552	0.06426	0.06301	0.06178	0.06057	0.05938	0.05821	0.05705	0.05592
1.4	0.08076	0.07927	0.07780	0.07636	0.07493	0.07353	0.07215	0.07078	0.06944	0.06811
1.3	0.09680	0.09510	0.09342	0.09176	0.09012	0.08851	0.08692	0.08534	0.08379	0.08226
1.2	0.11507	0.11314	0.11123	0.10935	0.10749	0.10565	0.10384	0.10204	0.10027	0.09853
1.1	0.13567	0.13350	0.13136	0.12924	0.12714	0.12507	0.12302	0.12100	0.11900	0.11702
1.0	0.15866	0.15625	0.15386	0.15151	0.14917	0.14686	0.14457	0.14231	0.14007	0.13786
0.9	0.18406	0.18141	0.17879	0.17619	0.17361	0.17106	0.16853	0.16602	0.16354	0.16109
0.8	0.21186	0.20897	0.20611	0.20327	0.20045	0.19766	0.19490	0.19215	0.18943	0.18673
0.7	0.24196	0.23885	0.23576	0.23270	0.22965	0.22663	0.22363	0.22065	0.21770	0.21476
0.6	0.27425	0.27093	0.26763	0.26435	0.26109	0.25785	0.25463	0.25143	0.24825	0.24510
0.5	0.30854	0.30503	0.30153	0.29806	0.29460	0.29116	0.28774	0.29434	0.28096	0.27760
0.4	0.34458	0.34090	0.33724	0.33360	0.32997	0.32636	0.32276	0.31918	0.31561	0.31207
0.3	0.38209	0.37828	0.37448	0.37070	0.36693	0.36317	0.35942	0.35569	0.35197	0.34827
0.2	0.42074	0.41683	0.41294	0.40905	0.40517	0.40129	0.39743	0.39358	0.38974	0.38591
0.1	0.46017	0.45621	0.45224	0.44828	0.44433	0.44038	0.43644	0.43251	0.42858	0.42466
0.0	0.50000	0.49601	0.49292	0.48803	0.48405	0.48006	0.47608	0.47210	0.46812	0.46414

POISSON UNITY VALUES, *Pa*

Pa	0.95	0.90	0.75	0.50	0.25	0.10	0.05
C							
0	0.051	0.105	0.288	0.693	1.386	2.303	2.996
1	0.355	0.532	0.961	1.678	2.693	3.890	4.744
2	0.818	1.102	1.727	2.674	3.920	5.322	6.296
3	1.366	1.745	2.523	3.672	5.109	6.681	7.754
4	1.970	2.433	3.369	4.671	6.274	7.994	9.154
5	2.613	3.152	4.219	5.670	7.423	9.275	10.513
6	3.286	3.895	5.083	6.670	8.558	10.532	11.842
7	3.981	4.656	5.956	7.669	9.684	11.771	13.148
8	4.695	5.432	6.838	8.669	10.802	12.995	14.434
9	5.426	6.221	7.726	9.669	11.914	14.206	15.705
10	6.169	7.021	8.620	10.668	13.020	15.407	16.962
11	6.924	7.829	9.519	11.668	14.121	16.598	18.208
12	7.690	8.646	10.422	12.668	15.217	17.782	19.442
13	8.464	9.470	11.329	13.668	16.310	18.958	20.668
14	9.246	10.300	12.239	14.668	17.400	20.128	21.886
15	10.035	11.135	13.152	15.668	18.486	21.292	23.098
16	10.831	11.976	14.068	16.668	19.570	22.452	24.302
17	11.663	12.822	14.986	17.668	20.652	23.606	25.500
18	12.442	13.672	15.907	18.668	21.731	24.756	26.692
19	13.254	14.525	16.830	19.668	22.808	25.902	27.879
20	14.072	15.383	17.755	20.668	23.883	27.045	29.062
21	14.894	16.244	18.682	21.668	24.956	28.184	30.241
22	15.719	17.108	19.610	22.668	26.028	29.320	31.416
23	16.548	17.975	20.540	23.668	27.098	30.453	32.586
24	17.382	18.844	21.471	24.668	28.167	31.584	33.752
25	18.218	19.717	22.404	25.667	29.234	32.711	34.916
26	19.058	20.592	23.338	26.667	30.300	33.836	36.077
27	19.900	21.469	24.237	27.667	31.365	34.959	37.234
28	20.746	22.348	25.209	28.667	32.428	36.080	38.389
29	21.594	23.229	26.147	29.667	33.491	37.198	39.541
30	22.444	24.113	27.086	30.667	34.552	38.315	40.690
31	23.298	24.998	28.025	31.667	35.613	39.430	41.838
32	24.152	25.885	28.966	32.667	36.672	40.543	42.982
33	25.010	26.774	29.907	33.667	37.731	41.654	44.125
34	25.870	27.664	30.849	34.667	38.788	42.764	45.226
35	26.731	28.556	31.792	35.667	39.845	43.872	46.404
36	27.594	29.450	32.736	36.667	40.901	44.978	47.540
37	28.460	30.345	33.681	37.667	41.957	46.082	48.676

POISSON UNITY VALUES, RQL/AQL

C	$\alpha = 0.05$ $\beta = 0.10$	$\alpha = 0.05$ $\beta = 0.05$	$\alpha = 0.05$ $\beta = 0.01$	np_1
0	44.89	58.404	89.781	0.052
1	10.946	13.349	18.781	0.355
2	6.509	7.699	10.280	0.818
3	4.890	5.675	7.352	1.366
4	4.057	4.646	5.890	1.970
5	3.549	4.023	5.017	2.613
6	3.206	3.604	4.435	3.286
7	2.957	3.303	4.019	3.981
8	2.768	3.074	3.707	4.695
9	2.618	2.895	3.462	5.426
10	2.479	2.750	3.265	6.169
11	2.379	2.630	3.104	6.924
12	2.312	2.528	2.968	7.690
13	2.240	2.442	2.852	8.464
14	2.177	2.367	2.752	9.246
15	2.122	2.302	2.665	10.035
16	2.073	2.244	2.588	10.831
17	2.029	2.192	2.520	11.633
18	1.990	2.145	2.458	12.442
19	1.954	2.103	2.403	13.254
20	1.922	2.065	2.352	14.072
21	1.892	2.030	2.307	14.894
22	1.865	1.999	2.265	15.719
23	1.840	1.969	2.226	16.548
24	1.817	1.942	2.191	17.382
25	1.795	1.917	2.158	18.218
26	1.775	1.893	2.127	19.058
27	1.757	1.871	2.098	19.900
28	1.739	1.850	2.071	20.746
29	1.723	1.831	2.046	21.594
30	1.707	1.813	2.023	22.444
31	1.692	1.796	2.001	23.298
32	1.679	1.780	1.980	24.152
33	1.665	1.764	1.960	25.010
34	1.653	1.750	1.941	25.870
35	1.641	1.736	1.923	26.731
36	1.630	1.723	1.906	27.594
37	1.619	1.710	1.890	28.460
38	1.609	1.698	1.875	29.327
39	1.599	1.687	1.860	30.196
40	1.590	1.676	1.846	31.066

t-DISTRIBUTION

Critical tail areas (α for one-tail tests, $\alpha/2$ for two-tail tests)

Degrees of freedom	0.10	0.05	0.025	0.010	0.005	0.0025
1	3.078	6.314	12.706	31.820	63.657	127.32
2	1.886	2.920	4.303	6.965	9.925	14.089
3	1.638	2.353	3.182	4.541	5.841	7.453
4	1.533	2.132	2.776	3.747	4.604	5.598
5	1.476	2.015	2.571	3.365	4.032	4.773
6	1.440	1.943	2.447	3.143	3.707	4.317
7	1.415	1.895	2.365	2.998	3.499	4.019
8	1.397	1.860	2.306	2.896	3.355	3.833
9	1.383	1.833	2.262	2.821	3.250	3.690
10	1.372	1.812	2.228	2.764	3.169	3.581
11	1.363	1.796	2.201	2.718	3.106	3.497
12	1.356	1.782	2.179	2.681	3.055	3.428
13	1.350	1.771	2.160	2.650	3.012	3.372
14	1.345	1.761	2.145	2.624	2.977	3.326
15	1.341	1.753	2.131	2.602	2.947	3.286
16	1.337	1.746	2.120	2.583	2.921	3.252
17	1.333	1.740	2.110	2.567	2.898	3.222
18	1.330	1.734	2.101	2.552	2.878	3.197
19	1.328	1.729	2.093	2.539	2.861	3.174
20	1.325	1.725	2.086	2.528	2.845	3.153
21	1.323	1.721	2.080	2.518	2.831	3.135
22	1.321	1.717	2.074	2.508	2.819	3.119
23	1.319	1.714	2.069	2.500	2.807	3.104
24	1.318	1.711	2.064	2.492	2.797	3.091
25	1.316	1.708	2.060	2.485	2.787	3.078
26	1.315	1.706	2.056	2.479	2.779	3.067
27	1.314	1.703	2.052	2.473	2.771	3.057
28	1.313	1.701	2.048	2.467	2.763	3.047
29	1.311	1.699	2.045	2.462	2.756	3.038
30	1.310	1.697	2.042	2.457	2.750	3.030
40	1.303	1.684	2.021	2.423	2.704	2.971
60	1.296	1.671	2.000	2.390	2.660	2.915
120	1.289	1.658	1.980	2.358	2.617	2.860
∞	1.282	1.645	1.960	2.326	2.576	2.807

CHI-SQUARE DISTRIBUTION

	Critical tail areas (α for one-tail tests, $\alpha/2$ for two-tail tests)							
Degrees of freedom	**0.995**	**0.975**	**0.950**	**0.90**	**0.100**	**0.050**	**0.025**	**0.005**
1	0.0^439	0.0^398	0.003	0.01	2.71	3.84	5.02	7.88
2	0.010	0.020	0.103	0.21	4.61	5.99	7.38	10.6
3	0.072	0.216	0.352	0.58	6.25	7.82	9.35	12.8
4	0.207	0.484	0.711	1.06	7.78	9.49	11.1	14.9
5	0.412	0.831	1.15	1.61	9.24	11.1	12.8	16.7
6	0.676	1.24	1.64	2.20	10.6	12.6	14.4	18.5
7	0.989	1.69	2.17	2.83	12.0	14.1	16.0	20.3
8	1.34	2.18	2.73	3.49	13.4	15.5	17.5	22.0
9	1.73	2.70	3.33	4.17	14.7	16.9	19.0	23.6
10	2.16	3.25	3.94	4.87	16.0	18.3	20.5	25.2
11	2.60	3.82	4.57	5.58	17.3	19.7	21.9	26.8
12	3.07	4.40	5.23	6.30	18.5	21.0	23.3	28.3
13	3.57	5.01	5.89	7.04	19.8	22.4	24.7	29.8
14	4.07	5.63	6.57	7.79	21.1	23.7	26.1	31.3
15	4.60	6.26	7.26	8.55	22.3	25.0	27.5	32.8
16	5.14	6.91	7.96	9.31	23.5	26.3	28.8	34.3
17	5.70	7.56	8.67	10.09	24.8	27.6	30.2	35.7
18	6.26	8.23	9.39	10.87	26.0	28.9	31.5	37.2
19	6.84	8.91	10.1	11.65	27.2	30.1	32.9	38.6
20	7.43	9.59	10.9	12.44	28.4	31.4	34.2	40.0
21	8.03	10.3	11.6	13.24	29.6	32.7	35.5	41.4
22	8.64	11.0	12.3	14.04	30.8	33.9	36.8	42.8
23	9.26	11.7	13.1	14.85	32.0	35.2	38.1	44.2
24	9.89	12.4	13.8	15.66	33.2	36.4	39.4	45.6
25	10.5	13.1	14.6	16.47	34.4	37.7	40.6	46.9
26	11.2	13.8	15.4	17.29	35.6	38.9	41.9	48.3
27	11.8	14.6	16.2	18.11	36.7	40.1	43.2	49.6
28	12.5	15.3	16.9	18.94	37.9	41.3	44.5	51.0
29	13.1	16.0	17.7	19.77	39.1	42.6	45.7	52.3
30	13.8	16.8	18.5	20.60	40.3	43.8	47.0	53.7
40	20.7	24.4	26.5	29.05	51.8	55.8	59.3	66.7
50	28.0	32.4	34.8	37.69	63.2	67.5	71.4	79.5
60	35.5	40.5	43.2	46.46	74.4	79.1	83.3	92.0
70	43.3	48.8	51.7	55.33	85.5	90.5	95.0	104.2
80	51.2	57.2	60.4	64.28	96.6	101.9	106.6	116.3
90	59.2	65.7	69.1	73.29	107.6	113.1	118.1	128.3
100	67.3	74.2	77.9	82.36	118.5	124.3	129.6	140.2

For number of degrees of freedom > 100, calculate the approximate normal deviation

$Z = \sqrt{2X^2} - \sqrt{2(df)-1}$

CONTROL CHART FACTORS I

Subgroup size	A_2	d_2	D_3	D_4	A_3	c_4	B_3	B_4
2	1.880	1.128	—	3.267	2.659	0.7979	—	3.267
3	1.023	1.693	—	2.574	1.954	0.8862	—	2.568
4	0.729	2.059	—	2.929	1.628	0.9213	—	2.266
5	0.577	2.326	—	2.114	1.427	0.9400	—	2.089
6	0.483	2.534	—	2.004	1.287	0.9515	0.030	1.970
7	0.419	2.704	0.076	1.924	1.182	0.9594	0.118	1.882
8	0.373	2.847	0.136	1.864	1.099	0.9650	0.185	1.816
9	0.337	2.970	0.184	1.816	1.032	0.9693	0.239	1.761
10	0.308	3.078	0.223	1.777	0.975	0.9727	0.284	1.716
11	NA	3.173	NA	NA	0.927	0.9754	0.321	1.679
12	NA	3.258	NA	NA	0.886	0.9776	0.354	1.646
13	NA	3.336	NA	NA	0.850	0.9794	0.382	1.618
14	NA	3.407	NA	NA	0.817	0.9810	0.406	1.594
15	NA	3.472	NA	NA	0.789	0.9823	0.428	1.572
16	NA	3.532	NA	NA	0.763	0.9835	0.448	1.552
17	NA	3.588	NA	NA	0.739	0.9845	0.466	1.534
18	NA	3.640	NA	NA	0.718	0.9854	0.482	1.518
19	NA	3.689	NA	NA	0.698	0.9862	0.497	1.503
20	NA	3.735	NA	NA	0.680	0.9869	0.510	1.490
21	NA	3.778	NA	NA	0.663	0.9876	0.523	1.477
22	NA	3.819	NA	NA	0.647	0.9882	0.534	1.466
23	NA	3.858	NA	NA	0.633	0.9887	0.545	1.455
24	NA	3.895	NA	NA	0.619	0.9892	0.555	1.445
25	NA	3.931	NA	NA	0.606	0.9896	0.565	1.438

Note: Factors A_2, D_3, and D_4 are not appropriate (NA) for these sample sizes. Use the average/standard deviation control chart where $n > 10$.

CONTROL CHART FACTORS II

Subgroup size	E_2	$\tilde{A}_2$
2	2.660	1.880
3	1.772	1.187
4	1.457	0.796
5	1.290	0.691
6	1.184	0.548
7	1.109	0.508
11	0.946	0.350
12	0.921	0.317
13	0.899	0.308
14	0.881	0.283
15	0.864	0.276
16	0.849	0.258
17	0.836	0.252
18	0.824	0.237
19	0.813	0.232
20	0.803	0.220
21	0.794	0.216
22	0.785	0.206
23	0.778	0.203
24	0.770	0.194
25	0.763	0.191

CRITICAL VALUES OF SMALLER RANK SUM FOR WILCOXON-MANN-WHITNEY TEST

n_1 (Smaller sample)

n_2	α	3	4	5	6	7	8	9	10	11	12	13	14	15
3	0.10	7												
	0.05	6												
	0.025	—												
4	0.10	7	13											
	0.05	6	11											
	0.025	—	10											
5	0.10	8	14	20										
	0.05	7	12	19										
	0.025	6	11	17										
6	0.10	9	15	22	30									
	0.05	8	13	20	28									
	0.025	7	12	18	26									
7	0.10	10	16	23	32	41								
	0.05	8	14	21	29	39								
	0.025	7	13	20	27	36								
8	0.10	11	17	25	34	44	55							
	0.05	9	15	23	31	41	51							
	0.025	8	14	21	29	38	49							
9	0.10	11	19	27	36	46	58	70						
	0.05	10	16	24	33	43	54	66						
	0.025	8	14	22	31	40	51	62						
10	0.10	12	20	28	38	49	60	73	87					
	0.05	10	17	26	35	45	56	69	82					
	0.025	9	15	23	32	42	53	65	78					
11	0.10	13	21	30	40	51	63	76	91	106				
	0.05	11	18	27	37	47	59	72	86	100				
	0.025	9	16	24	34	44	55	68	81	96				
12	0.10	14	22	32	42	54	66	80	94	110	127			
	0.05	11	19	28	38	49	62	75	89	104	120			
	0.025	10	17	26	35	46	58	71	84	99	115			
13	0.10	15	23	33	44	56	69	83	98	114	131	149		
	0.05	12	20	30	40	52	64	78	92	108	125	142		
	0.025	10	18	27	37	48	60	73	88	103	119	136		
14	0.10	16	25	35	46	59	72	86	102	118	136	154	174	
	0.05	13	21	31	42	54	67	81	96	112	129	147	166	
	0.025	11	19	28	38	50	62	76	91	106	123	141	160	
15	0.10	16	26	37	48	61	75	90	106	123	141	159	179	200
	0.05	13	22	33	44	56	69	84	99	116	133	152	171	192
	0.025	11	20	29	40	52	65	79	94	110	127	145	164	184

VALUES OF *i* AND *f* FOR CSP-1 PLANS

f	AOQL (%)															
	0.018	0.033	0.046	0.074	0.113	0.143	0.198	0.33	0.53	0.79	1.22	1.90	2.90	4.94	7.12	11.46
1/2	1540	840	600	375	245	194	140	84	53	36	23	15	10	6	5	3
1/3	2550	1390	1000	620	405	321	232	140	87	59	38	25	16	10	7	5
1/4	3340	1820	1310	810	530	420	303	182	113	76	49	32	21	13	9	6
1/5	3960	2160	1550	965	630	498	360	217	135	91	58	38	25	15	11	7
1/7	4950	2700	1940	1205	790	623	450	270	168	113	73	47	31	18	13	8
1/10	6050	3300	2370	1470	965	762	550	335	207	138	89	57	38	22	16	10
1/15	7390	4030	2890	1800	1180	930	672	410	255	170	108	70	46	27	19	12
1/25	9110	4970	3570	2215	1450	1147	828	500	315	210	134	86	57	33	23	14
1/50	11730	6400	4590	2855	1870	1477	1067	640	400	270	175	110	72	42	29	18
1/100	14320	7810	5600	3485	2305	1820	1302	790	500	330	215	135	89	52	36	22
1/200	17420	9500	6810	4235	2760	2178	1583	950	590	400	255	165	106	62	43	26
	0.010	0.015	0.025	0.040	0.065	0.10	0.15	0.25	0.40	0.65	1.0	1.5	2.5	4.0	6.5	10.0

AQL(%)

CUMULATIVE POISSON DISTRIBUTION

λ

x	0.1	0.2	0.3	0.4	0.5	0.6	0.7	0.8	0.9	1.0
0	0.905	0.819	0.741	0.670	0.607	0.549	0.497	0.449	0.407	0.368
1	0.995	0.983	0.963	0.938	0.910	0.878	0.844	0.809	0.773	0.736
2	1.000	0.999	0.996	0.992	0.986	0.977	0.966	0.953	0.937	0.920
3	1.000	1.000	1.000	0.999	0.998	0.997	0.994	0.991	0.987	0.981
4	1.000	1.000	1.000	1.000	1.000	1.000	0.999	0.999	0.998	0.996
5	1.000	1.000	1.000	1.000	1.000	1.000	1.000	1.000	1.000	0.999
6	1.000	1.000	1.000	1.000	1.000	1.000	1.000	1.000	1.000	1.000
7	1.000	1.000	1.000	1.000	1.000	1.000	1.000	1.000	1.000	1.000

λ

x	1.1	1.2	1.3	1.4	1.5	1.6	1.7	1.8	1.9	2.0
0	0.333	0.301	0.273	0.247	0.223	0.202	0.183	0.165	0.150	0.135
1	0.699	0.663	0.627	0.592	0.558	0.525	0.493	0.463	0.434	0.406
2	0.900	0.880	0.857	0.834	0.809	0.783	0.757	0.731	0.704	0.677
3	0.974	0.966	0.957	0.946	0.934	0.921	0.907	0.891	0.875	0.857
4	0.995	0.992	0.989	0.986	0.981	0.976	0.970	0.964	0.956	0.947
5	0.999	0.999	0.998	0.996	0.994	0.992	0.992	0.990	0.987	0.983
6	1.000	1.000	1.000	0.999	0.999	0.999	0.998	0.997	0.997	0.996
7	1.000	1.000	1.000	1.000	1.000	1.000	1.000	0.999	0.999	0.999
8	1.000	1.000	1.000	1.000	1.000	1.000	1.000	1.000	1.000	1.000
9	1.000	1.000	1.000	1.000	1.000	1.000	1.000	1.000	1.000	1.000

λ

x	2.1	2.2	2.3	2.4	2.5	2.6	2.7	2.8	2.9	3.0
0	0.123	0.111	0.100	0.091	0.082	0.074	0.067	0.061	0.055	0.050
1	0.380	0.355	0.331	0.308	0.287	0.267	0.249	0.231	0.215	0.199
2	0.650	0.623	0.596	0.570	0.544	0.518	0.494	0.470	0.446	0.423
3	0.839	0.819	0.799	0.779	0.758	0.736	0.714	0.692	0.670	0.647
4	0.938	0.928	0.916	0.904	0.891	0.877	0.863	0.848	0.812	0.815
5	0.980	0.971	0.970	0.964	0.958	0.951	0.943	0.935	0.926	0.916
6	0.994	0.993	0.991	0.988	0.986	0.983	0.979	0.956	0.971	0.967
7	0.999	0.998	0.997	0.997	0.996	0.995	0.993	0.992	0.990	0.988
8	1.000	1.000	0.999	0.999	0.999	0.999	0.998	0.998	0.997	0.996
9	1.000	1.000	1.000	1.000	1.000	1.000	1.000	0.999	0.999	0.999
10	1.000	1.000	1.000	1.000	1.000	1.000	1.000	1.000	1.000	1.000
11	1.000	1.000	1.000	1.000	1.000	1.000	1.000	1.000	1.000	1.000
12	1.000	1.000	1.000	1.000	1.000	1.000	1.000	1.000	1.000	1.000

CUMULATIVE POISSON DISTRIBUTION

λ

x	3.1	3.2	3.3	3.4	3.5	3.6	3.7	3.8	3.9	4.0
0	0.045	0.041	0.037	0.033	0.030	0.027	0.022	0.022	0.020	0.018
1	0.185	0.171	0.159	0.147	0.136	0.126	0.116	0.107	0.099	0.092
2	0.401	0.380	0.359	0.340	0.321	0.303	0.285	0.269	0.253	0.238
3	0.625	0.603	0.580	0.558	0.537	0.515	0.494	0.474	0.453	0.434
4	0.798	0.781	0.763	0.744	0.725	0.706	0.687	0.668	0.648	0.629
5	0.906	0.895	0.883	0.871	0.858	0.844	0.830	0.816	0.801	0.785
6	0.961	0.955	0.949	0.942	0.935	0.927	0.918	0.909	0.900	0.889
7	0.986	0.983	0.980	0.977	0.973	0.969	0.965	0.960	0.955	0.949
8	0.995	0.994	0.993	0.992	0.990	0.988	0.986	0.984	0.982	0.979
9	0.999	0.998	0.998	0.997	0.997	0.996	0.995	0.994	0.993	0.992
10	1.000	1.000	0.999	0.999	0.999	0.999	0.998	0.998	0.998	0.997
11	1.000	1.000	1.000	1.000	1.000	1.000	1.000	0.999	0.999	0.999
12	1.000	1.000	1.000	1.000	1.000	1.000	1.000	1.000	1.000	1.000
13	1.000	1.000	1.000	1.000	1.000	1.000	1.000	1.000	1.000	1.000
14	1.000	1.000	1.000	1.000	1.000	1.000	1.000	1.000	1.000	1.000

λ

x	4.1	4.2	4.3	4.4	4.5	4.6	4.7	4.8	4.9	5.0
0	0.017	0.015	0.014	0.012	0.011	0.010	0.009	0.008	0.007	0.007
1	0.085	0.078	0.072	0.066	0.061	0.056	0.052	0.048	0.044	0.040
2	0.224	0.210	0.197	0.185	0.174	0.163	0.152	0.143	0.133	0.125
3	0.414	0.395	0.377	0.360	0.342	0.326	0.310	0.294	0.279	0.265
4	0.609	0.590	0.570	0.551	0.532	0.513	0.495	0.476	0.458	0.441
5	0.769	0.753	0.737	0.720	0.703	0.686	0.668	0.651	0.634	0.616
6	0.879	0.868	0.856	0.844	0.831	0.818	0.805	0.791	0.777	0.762
7	0.943	0.936	0.929	0.921	0.913	0.905	0.896	0.887	0.877	0.867
8	0.976	0.972	0.968	0.964	0.960	0.955	0.950	0.944	0.938	0.932
9	0.991	0.989	0.987	0.985	0.983	0.981	0.978	0.975	0.972	0.968
10	0.997	0.996	0.995	0.994	0.993	0.992	0.991	0.990	0.988	0.986
11	0.999	0.999	0.998	0.998	0.998	0.997	0.997	0.996	0.995	0.995
12	1.000	1.000	1.000	0.999	0.999	0.999	0.999	0.999	0.998	0.998
13	1.000	1.000	1.000	1.000	1.000	1.000	1.000	1.000	0.999	0.999
14	1.000	1.000	1.000	1.000	1.000	1.000	1.000	1.000	1.000	1.000
15	1.000	1.000	1.000	1.000	1.000	1.000	1.000	1.000	1.000	1.000
16	1.000	1.000	1.000	1.000	1.000	1.000	1.000	1.000	1.000	1.000

F-TABLE

V_1 (corresponding to the larger sample variance—numerator)

α	V_2	2	3	4	5	6	7	8	9	10	15	∞
10	2	9.00	9.16	9.24	9.29	9.33	9.35	9.38	9.30	9.39	9.42	9.49
5		19.0	19.2	19.2	19.3	19.3	19.4	19.4	19.4	19.4	19.4	19.5
2.5		39.0	39.2	39.3	39.3	39.3	39.4	39.4	39.4	39.4	39.4	39.5
1		99.0	99.2	99.2	99.3	99.3	99.4	99.4	99.4	99.4	99.4	99.5
10	3	5.46	5.39	5.34	5.31	5.28	5.27	5.25	5.24	5.23	5.20	5.13
5		9.55	9.28	9.12	9.01	8.94	8.89	8.85	8.81	8.79	8.70	8.53
2.5		16.0	15.4	15.1	14.9	14.7	14.6	14.5	14.5	14.4	14.3	13.9
1		30.8	29.5	28.7	28.2	27.9	27.7	27.5	27.3	27.2	26.9	26.1
10	4	4.32	4.19	4.11	4.05	4.01	3.98	3.95	3.94	3.92	3.87	3.76
5		6.94	6.59	6.39	6.26	6.16	6.09	6.04	6.00	5.96	5.86	5.63
2.5		10.7	10.0	9.60	9.40	9.20	9.10	9.00	8.90	8.80	8.70	8.30
1		18.0	16.7	16.0	15.5	15.2	15.0	14.8	14.7	14.5	14.2	13.5
10	5	3.78	3.62	3.52	3.45	3.40	3.37	3.34	3.32	3.30	3.24	3.10
5		5.79	5.41	5.19	5.05	4.95	4.88	4.82	4.77	4.74	4.62	4.36
2.5		8.40	7.80	7.40	7.10	7.00	6.90	6.80	6.70	6.60	6.40	6.00
1		13.3	12.1	11.4	11.0	10.7	10.5	10.3	10.2	10.1	9.72	9.02
10	6	3.46	3.29	3.18	3.11	3.05	3.01	2.98	2.96	2.94	2.87	2.72
5		5.14	4.76	4.53	4.39	4.28	4.21	4.15	4.10	4.06	3.94	3.67
2.5		7.30	6.60	6.23	5.99	5.82	5.70	5.60	5.52	5.46	5.27	4.85
1		10.9	9.78	9.15	8.75	8.47	8.26	8.10	7.98	7.87	7.56	6.88
10	7	3.26	3.07	2.96	2.88	2.83	2.78	2.75	2.72	2.70	2.63	2.47
5		4.74	4.35	4.12	3.97	3.87	3.79	3.73	3.68	3.64	3.51	3.23
2.5		6.54	5.89	5.52	5.29	5.12	5.00	4.90	4.82	4.76	4.57	4.14
1		9.55	8.45	7.85	7.46	7.19	6.99	6.84	6.72	6.62	6.31	5.65
10	8	3.11	2.92	2.81	2.73	2.67	2.62	2.59	2.56	2.54	2.46	2.20
5		4.46	4.07	3.84	3.69	3.58	3.50	3.44	3.39	3.35	3.22	2.93
2.5		6.06	5.42	5.05	4.82	4.65	4.53	4.43	4.36	4.30	4.10	3.17
1		8.65	7.59	7.01	6.63	6.37	6.18	6.03	5.91	5.81	5.52	4.86
10	9	3.01	2.81	2.69	2.61	2.55	2.51	2.47	2.44	2.42	2.34	2.16
5		4.26	3.86	3.63	3.48	3.37	3.29	3.23	3.18	3.14	3.01	2.71
2.5		5.72	5.08	4.72	4.48	4.32	4.20	4.10	4.03	3.96	3.77	3.33
1		8.02	6.99	6.42	6.06	5.80	5.61	5.47	5.35	5.26	4.96	4.31
10	10	2.92	2.73	2.61	2.52	2.46	2.41	2.38	2.35	2.32	2.24	2.06
5		4.10	3.71	3.48	3.33	3.22	3.14	3.07	3.02	2.98	2.85	2.54
2.5		5.46	4.83	4.47	4.24	4.07	3.93	3.86	3.78	3.72	3.52	3.08
1		7.56	6.55	5.99	5.64	5.39	5.20	5.06	4.94	4.85	4.56	3.91
10	12	2.81	2.61	2.48	2.39	2.33	2.28	2.24	2.21	2.19	2.10	1.90
5		3.89	3.49	3.26	3.11	3.00	2.91	2.85	2.80	2.75	2.62	2.30
2.5		5.10	4.47	4.12	3.89	3.73	3.61	3.51	3.44	3.37	3.18	2.73
1		6.93	5.95	5.41	5.06	4.82	4.64	4.50	4.39	4.30	4.01	3.36
10	15	2.70	2.49	2.36	2.27	2.21	2.16	2.12	2.09	2.06	1.97	1.76
5		3.68	3.29	3.06	2.90	2.79	2.71	2.64	2.59	2.54	2.40	2.07
2.5		4.77	4.15	3.80	3.58	3.42	3.29	3.20	3.12	3.06	2.86	2.40
1		6.36	5.42	4.89	4.56	4.32	4.14	4.00	3.89	3.80	3.52	2.87
10	∞	2.30	2.08	1.94	1.85	1.77	1.72	1.67	1.63	1.60	1.49	1.00
5		3.00	2.60	2.37	2.21	2.10	2.01	1.94	1.88	1.83	1.67	1.00
2.5		3.69	3.12	2.79	2.57	2.41	2.29	2.14	2.11	2.05	1.83	1.00
1		4.61	3.78	3.32	3.02	2.80	2.64	2.51	2.41	2.32	2.04	1.00

Index